Ecology and Natural Resource Management

Ecology and Natural Resource Management

Edited by Paul Cooper

SYRAWOOD
PUBLISHING HOUSE
New York

Published by Syrawood Publishing House,
750 Third Avenue, 9th Floor,
New York, NY 10017, USA
www.syrawoodpublishinghouse.com

Ecology and Natural Resource Management
Edited by Paul Cooper

International Standard Book Number: 978-1-68286-575-0 (Hardback)

Cataloging-in-Publication Data

Ecology and natural resource management / edited by Paul Cooper.
p. cm.
Includes bibliographical references and index.
ISBN 978-1-68286-575-0
1. Ecology. 2. Natural resources--Management. 3. Conservation of natural resources. I. Cooper, Paul.
QH541 .E25 2018
577--dc23

Printed in China.

TABLE OF CONTENTS

Preface .... **VII**

Chapter 1 **Water Quality Assessment of *Aflaj* in the Mountains of Oman** **1**
Mohammed Saif Al-Kalbani, Martin F. Price, Mushtaque Ahmed, Asma Abahussain and Timothy O'Higgins

Chapter 2 **Increased Terrestrial Disturbances Modify Skeletal Properties and Mechanical Strength of Hard Corals** **17**
Shaaban A. Mwachireya, Tim R. McClanahan, Isabelle M. Cote and Brian E. Hartwick

Chapter 3 **Integrated use of Geological, Geophysical, Radiocarbon and Stable Isotopes Data for Tracing the Conduit Flow Paths in a Small Karstic Aquifer: Poocher Swamp Freshwater Lens, South Australia** **38**
Nara Somaratne and Simon Mann

Chapter 4 **Trajectory Modelling of Atmospheric Pollutants Associated with Aircraft Emissions for Air Quality Assessment** **53**
Marilia Mitidieri Fernandes de Oliveira, Nelson Francisco Favilla Ebecken,Jorge Luiz Fernandes de Oliveira, Marina Aires and José Maria de Castro Jr.

Chapter 5 **Forest Resource Management and Utilisation through a Gendered Lens in Namibia** **65**
Immaculate Mogotsi, Selma Lendelvo, Margaret Angula and Jesaya Nakanyala

Chapter 6 **Ecological Landscape Planning and Design Strategies for Mangrove Communities (Hara Forests) in South-Pars Special Economic Energy Zone, Asalouyeh- Iran** **77**
Mohammad R. Masnavi, Neda Amani and Ali Ahmadzadeh

Chapter 7 **Influence of Temperature and Humidity on the Physiological Indices of Stress in the Obudu Mountain Landscape Environment, Nigeria** **91**
Henry Ojobo, Bukar Usman Wakawa and Aminu Umar

Chapter 8 **Role of Environmental Pollutants in Skeletal Muscle Insulin Resistance and Mitochondrial Dysfunction** **101**
Lucia Chehade, Audrey Caron and Céline Aguer

Chapter 9 **Assessment of Water Resources and Analysis of Safe Yield and Reliability of Surface Water Reservoirs of Asmara Water Supply System** **120**
Kahsay N. Zeraebruk, Alfred O. Mayabi and John M. Gathenya

Chapter 10 **Heavy Precipitation Events in Marmara Region and connections with the North Atlantic and Arctic Oscillation Patterns** **135**
Sevinc A. Sirdas, Ali Behzat Diriker and Vinay Kumar

Chapter 11 **The Sustainable Livelihood Challenge of REDD+ Implementation in the Philippines** **152**
Rose Jane Peras, Juan Pulhin, Makoto Inoue, Abrar Jurar Mohammed, Kazuhiro Harada and Masatoshi Sasaoka

Chapter 12 **Evaluation and Prediction of Regional Water Resources Carrying Capacity: A Case Study of Shandong Province** **167**
Hongyan Liao, Yuabiao Zhang, Zhifeng Chen and Zexin Meng

Chapter 13 **Social and Economic Impacts of Community Managed Reforestation and Natural Regeneration of Forestry Development, the Case of Humbo District, Ethiopia** **180**
Asamere Wolde, Tadesse Amsalu and Molla Mekonnen Alemu

Chapter 14 **Characterization of Water Source Types and uses in Kirisia Forest Watershed, Samburu County, Kenya** **195**
John Warui Kiringe, Francis Mwaura and Fiesta Warinwa

Chapter 15 **Antifouling Effectiveness and Potential Toxicological Risk of an Elastomer-Based Coating against Zebra Mussels** **209**
Yves de Lafontaine and Julie Veillette

Permissions

List of Contributors

Index

PREFACE

This book has been a concerted effort by a group of academicians, researchers and scientists, who have contributed their research works for the realization of the book. This book has materialized in the wake of emerging advancements and innovations in this field. Therefore, the need of the hour was to compile all the required researches and disseminate the knowledge to a broad spectrum of people comprising of students, researchers and specialists of the field.

Ecology studies the interaction between organisms and their environment in a holistic manner. Some of the concerns discussed in the subject of ecology are biodiversity, population and distribution. Natural resource management is a significant aspect of ecology as it facilitates species and habitat conservation. The various studies that are constantly contributing towards advancing technologies and evolution of this field are examined in detail herein. The extensive content of this book provides the readers with a thorough understanding of the subject.

At the end of the preface, I would like to thank the authors for their brilliant chapters and the publisher for guiding us all-through the making of the book till its final stage. Also, I would like to thank my family for providing the support and encouragement throughout my academic career and research projects.

Editor

1

Water Quality Assessment of *Aflaj* in the Mountains of Oman

Mohammed Saif Al-Kalbani[1], Martin F. Price[1], Mushtaque Ahmed[2], Asma Abahussain[3] & Timothy O'Higgins[4]

[1] Centre for Mountain Studies, Perth College, University of the Highlands and Islands, Perth, UK

[2] Department of Soils, Water & Agricultural Engineering, College of Agricultural and Marine Sciences, Sultan Qaboos University, Sultanate of Oman

[3] Department of Natural Resources and Environment, College of Graduate Studies, Arabian Gulf University, Manama, Kingdom of Bahrain

[4]Scottish Association for Marine Sciences, University of the Highlands and Islands, Scottish Marine Institute, Argyll, UK

Correspondence: Mohammed Saif Al-Kalbani, Centre for Mountain Studies, Perth College, University of the Highlands and Islands, Crieff Road, Perth PH1 2NX, UK. E-mail: Kmohd2020@yahoo.com

Abstract

The research was conducted to assess the *aflaj* water quality in Al Jabal Al Akhdar, Oman. 9 *aflaj* were sampled during summer and winter seasons in 2012-2013 to evaluate for the physico-chemical characteristics of major quality parameters; and assess the suitability of *aflaj* for irrigation purposes. Samples collection, handling and processing followed the standard methods recommended by the *American Public Health Association* and analysed in quality assured laboratories using appropriate analytical methods and instrumental techniques. The quality parameters of the selected *aflaj* water indicated their suitability for irrigation as most of the quality parameters were within the permissible limits set by Omani regulations of wastewater reuse for irrigation. These selected water resources are excellent or good in quality for irrigation purposes based on the evaluation of different hazards parameters including the salinity-alkalinity hazards which indicate good to admissible water based on electrical conductivity and sodium adsorption classification; and water quality indices which reveal high or moderate classes, indicating the suitability of *aflaj* for irrigation of the majority of crops and soils. This study is a first comprehensive assessment towards providing indicators and classification indices on irrigation water quality of this fragile mountain ecosystem, which will be the basis for future planning decisions on agricultural demand management measures to protect these principal resources for agricultural production in Al Jabal Al Akhdar.

Keywords: *Aflaj*, Al Jabal Al Akhdar, hazards, irrigation water, mountains, Oman, water quality index

1. Introduction

Like other countries located in arid regions, the Sultanate of Oman suffers from rainfall scarcity and limited renewable water resources. Oman depends totally on groundwater and *aflaj* for domestic and irrigation purposes. *Aflaj* (singular *falaj*) are surface and/or underground channels fed by groundwater, springs, or streams, built to provide water to communities for domestic and/or agricultural use (Al-Marshudi, 2001; Zekri & Al-Marshudi, 2008). *Aflaj* have been constructed in Oman for thousands of years to tap concentrated lines of groundwater flow and guide them to the surface along a channel (often several kilometers long) at a lesser gradient than the water table (Al-Marshudi, 2007). On reaching ground level, the main channel splits into many smaller channels, which in turn divide to supply individual farms. *Aflaj* are managed by local people, with their own designated administrative structure, who are responsible for the overall organization of *falaj* affairs and water distribution for irrigation without government involvement in this organizational structure (Al-Marshudi, 2007; Zekri & Al-Marshudi, 2008). The *aflaj* system is mainly based on a time-share among water rights holders, but in some areas volume is used instead, especially during drought periods (Al-Ghafri et al., 2003; Zekri et al., 2014). Given the importance of *aflaj* as a unique Omani water resource, UNESCO has listed five of the *aflaj* in Oman on the World Heritage list (MRMWR, 2008a).

Most of the *aflaj* are located in the northern Oman Mountains; mountains cover 15% of the country total area. Al Jabal Al Akhdar (Green Mountain) is the largest structural domain, located in the central part of the northern Oman Mountains (Figure 1). It reaches heights between 1500 to 3000 m above sea level. Because of its altitude, temperatures are some 10 to 12°C lower than in the coastal plains (DGMAN, 2014). In general, the temperatures drop during winter to below 0°C and rise in summer to around 22°C. Rainfall is highly variable and irregular and is the main source of fresh water in the mountain, where the mean annual rainfall is about 300 mm (DGMAN, 2014; Al-Kalbani et al., 2014; Al-Kalbani et al., 2015a). Agriculture has been the principal traditional economic sector with around 70 % of the local inhabitants practice agriculture and animal husbandry (Al-Riyami, 2006; Al-Kalbani et al., 2015b). The mountain terraces produce a variety of perennial fruits, especially pomegranates as well as roses for producing rose water, as a unique business in the area. Tourism is a growing sector in the area where the number of tourists and tourism infrastructure has increased over the last few years (Al-Balushi et al., 2011; Ministry of Tourism, 2014; Al-Kalbani et al., 2015b).

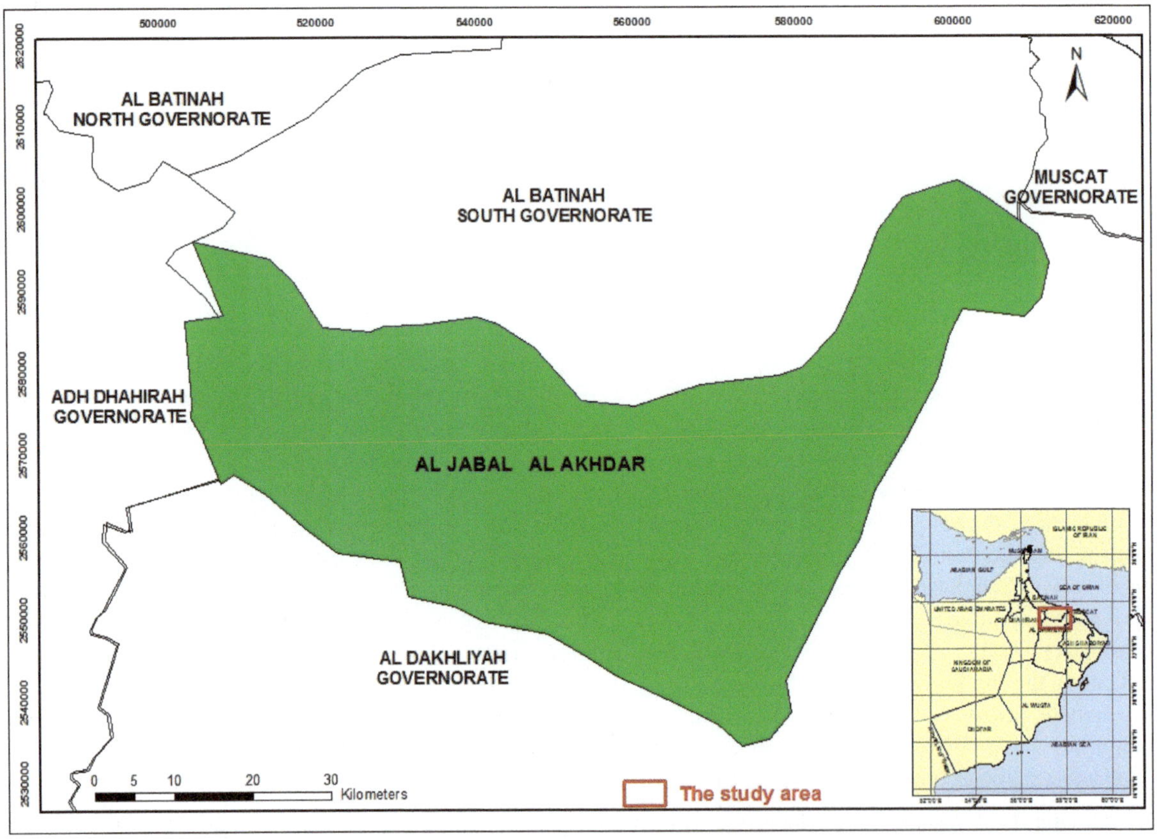

Figure 1. Location of the study area in the Sultanate of Oman map (*Source*: MECA, 2015)

Al Jabal Al Akhdar has experienced rapid socioeconomic development and urbanization in recent decades. These changes have influenced the water resources, which are the lifeline of its natural ecosystems and therefore human well-being. According to the National *Aflaj* Inventory conducted from March 1997 to June 1998, there were 72 *Aflaj* in Al Jabal Al Akhdar (MWR, 1999; MRMEWR, 2001). However, this number hasdecreased to 38 based on the recent study by Al-Kalbani et al. (2015b). Many studies have conducted on *aflaj* in Oman: their physical structure, method of construction and governance, irrigation scheduling, water right and market (e.g. Abdel Rahman & Omezzine, 1996; Norman et al., 1998; Al-Marshudi, 2007; Zekri et al., 2014). However, there is very little information assessing and classifying the water quality of *aflaj* and their suitability for irrigation and domestic purposes, especially in the mountains.

Irrigation water quality is an important tool in the assessment and sustainable management of water resources and agricultural production. The presence of excessive amounts of ions in irrigation water affects soil's physical and chemical properties, reduce soil productivity, create crop toxicity and eventually reduce yields (Kraiem et al., 2014; Nag, 2014; Varol & Davraz, 2015). Major water quality problems for irrigation are salinity, sodicity and alkalinity (Simsek & Gunduz, 2007; Sadashivaiah et al., 2008; Nazzal et al., 2014). An appropriate assessment of water for irrigation requires the determination of of physical, chemical and biological parameters that are greatly influenced by geological formations and anthropogenic activities (Sarath Prasanth et al., 2012; Al-Khashman & Jaradat, 2014; Al-Harbi et al., 2014) and directly related to the classifying of water quality (Saber et al., 2014; Aly, 2014; Aly et al., 2014). Several quality indices can be used to assess the suitability of water for irrigation; the most commonly used are salinity hazards, percent sodium (% Na), sodium adsorption ratio (SAR), residual sodium carbonate (RSC), soluble sodium percentage (SSP), residual sodium bicarbonate RSBC), permeability index (PI), potential salinity (PS), Kelley's index (KI) and magnesium hazard (MH) (Tatawat & Chandel, 2008; Sarath Prasanth et al., 2012; Al-Harbi et al., 2014; Nag, 2014). Water quality index (WQI) also provides a simple and concise method for assessing the water quality by integrating the water quality variables into one single number depending on several quality variables: salinity hazard, infiltration or permeability hazard, specific ion toxicity, trace element toxicity, and various miscellaneous effects to susceptible crops (Lou & Han, 2007; Simsek & Gunduz, 2007; Tatawat & Chandel, 2008; Mohammed Muthanna, 2011; Al-Bahrani et al., 2012; Adhikari et al., 2013; Aly et al., 2014).

This study interprets and classifies the hydro-chemical characteristics and WQI of *aflaj* water in Al Jabal Al Akhdar and evaluates their suitability for irrigation. Previous studies on water quality in this fragile mountainous region (e.g. Al-Haddabi, 2003; Ahmed et al., 2006; Al-Haddabi et al., 2009; Victor et al., 2009) focused only on the general physico-chemical characteristics of few *aflaj*. This paper presents a comprehensive assessment and classification of water quality of principal *aflaj* for agricultural activities in the area, using several quality parameters and classification indices.

2. Materials and Methods

2.1 Water Sampling and Analytical Methods

Nine *aflaj* were selected for water sampling (Figure 2; Table 1): these are the main *aflaj* that are most reliable for agriculture and active annually most of the time. The other *aflaj* were inactive or do not continuously flow during the whole year and are less reliable in terms of number of demand areas. At least two sampling points were identified along the channel of each *falaj*: the first at the source, and the second in a demand area (total area irrigated by the *falaj*. For longer *aflaj*, or those with many demand areas, a third sampling point was also used.

Table 1. Characteristics of the *aflaj* sampled in the study area (Data source: National *Aflaj* Inventory, March 1997-June 1998) (MWR, 1999)

Code	*Falaj* Name	Village	Type	Length (m)	*Falaj* mother well Coordinates (UTM)		Altitude (m)	No. of demand areas	Total demand area (m²)	Total annual water demand (m³/year)
					North	East				
F1	Masirat Al Jawamid	Masirat Al Jawamid	Ayni	153	2558682	554283	1420	1	21,147	41,583
F2	Masirat Ar Rawajih	Masirat Ar Rawajih	Ayni	1,979	2548880	570600	1212	4	39,208	76,802
F3	Al Azizi	Sayq	Ayni	428	2552062	564891	1942	3	118,690	234,740
F4	Qatam	Sayq	Ayni	331	2552199	564435	1914	4	2,984	5,889
F5	Al Awar	Al Ayn	Ayni	140	2551912	567800	1970	4	62,630	123,348
F6	Al Kabari	Ash Shirayjah	Ayni	1,077	2552208	568992	1941	1	272,564	539,879
F7	As Sawjrah	As Sawjrah	Ayni	22	2558641	568788	1863	2	12,541	24,723
F8	Wadi Bani Habib	Wadi Bani Habib	Ayni	833	2552346	562507	1935	9	6,512	12,839
F9	Al Khamirah Al Sufla	Sayq	Ghayli	128	2550653	565729	1864	4	9,457	18,611

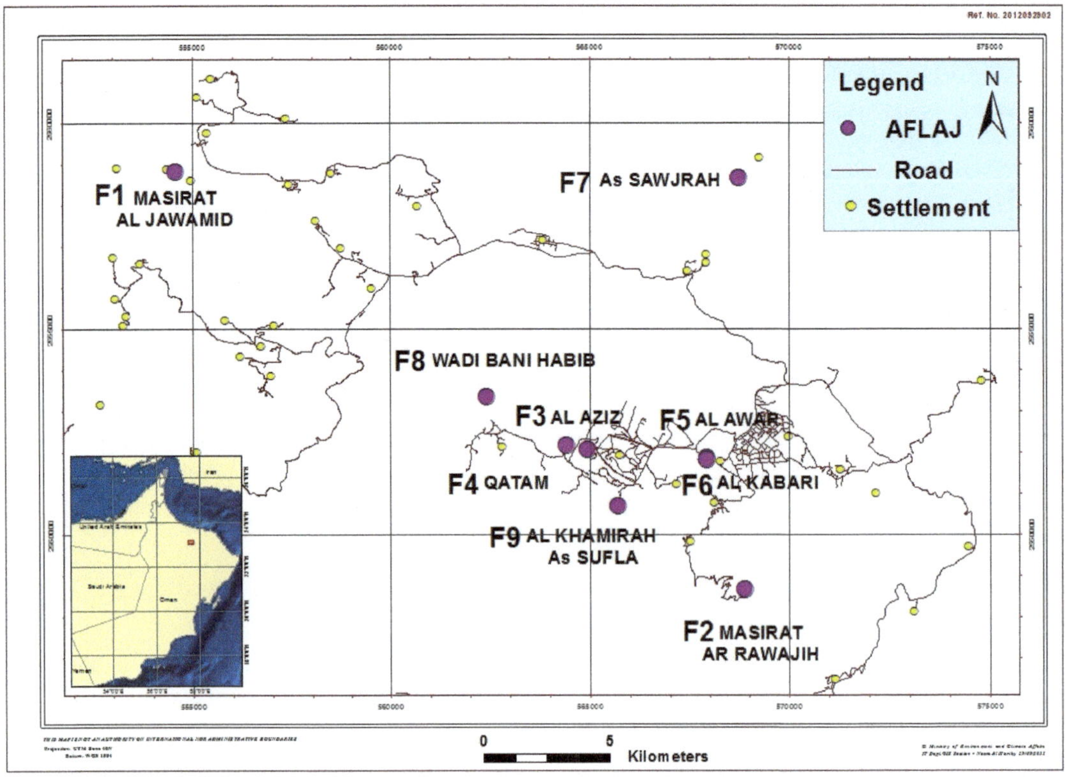

Figure 2. The location of sampled *aflaj* of the study area during summer and winter 2012-2013

Table 2. Determination of physiochemical parameters by different methods/instruments used

Parameters	Method/Instrument Used
Electrical Conductivity (EC)	Measured in the field using a battery-operated conductivity meter (SevenGo, Mettler-Toledo AG 8603 Schwerzenbach, Switzerland) and in the laboratory using the Orion Thermo 550A.
pH	Determined in the field using a pH meter (SevenGo, Mettler-Toledo GmbH, 8603 Schwerzenbach, Switzerland), and in the laboratory using a pH meter (Mettler Toledo)
Turbidity	Turbidity meter (Orion AQ 4500), Nephelometric Turbidity Units (NTU)
Total Dissolved Solids (TDS)	Gravimetric method
Alkalinity ($CaCO_3$, HCO_3 and CO_3)	Autotitration
Total Hardness (TH) (mg/l as $CaCO_3$)	Complexometric titration method using Ethylene Diamine Tetra Acetic Acid (EDTA)
Dissolved Oxygen (DO)	Measured in the field using Multi Probe System/data logger, YSI Incorporated 556 Instrument, Bramum Lane and in the laboratory using Mettler Toledo Seven Go Pro
Biochemical Oxygen Demand (BOD_5)	$BOD_5 = (D2 - D1) / P$
	D1: DO of diluted sample immediately after preparation, mg/l
	D2: DO of diluted sample after 5 days incubation at 20 °C, mg/l
	P: Decimal volumetric fraction of sample used
Sodium, Calcium, Magnesium, Potassium	Inductively Coupled Plasma (ICP-OES) (Perkin-Elmer Optima 3300 DV)
Fluoride, Chloride, Nitrate, Sulphate, Phosphate	Metrohm Professional Compact Ion Chromatography System 881 with Metrohm 858Professional Sample Processor
Heavy Metals	Inductively Coupled Plasma (ICP-OES) (Perkin-Elmer Optima 3300 DV)

The locations of sampled *aflaj*, mother well and points along their channels, were determined using GPS (etrex, Garmin). The sampling regime for all selected *aflaj* was 3 months in winter and 3 months in summer, taking into account that seasonal events such as rainfall and storms may influence sampling; and to obtain a reasonable range of data in each season. Sample collection, handling and processing followed the methods recommended by the *American Public Health Association* [APHA] (2005); water quality parameters were selected according to Chapmana and Kimstach (1996). Major physico-chemical and microbiological parameters were analysed in quality assured laboratories in Oman using the analytical methods and instrumental techniques shown in Table 2. The accuracy of the chemical analysis was verified by the calculation of ion-balance errors of 5% for all the sampled water resources. The respective values for all these parameters are compared with standard limits recommended by Omani standards for Un-bottled Drinking Water 8/2006 (MD, 2007) and the World Health Organization [WHO], 2011) for drinking water, and Omani regulations of Wastewater reuse for irrigation (MD, 1993). Statistical analysis was performed using descriptive statistics for all physico-chemical and microbiological parameters as well as correlation analyses using Pearson's coefficient (r) among the levels of parameters in water samples.

2.2 Hydrochemical Water Quality

Water quality indicators, including salinity hazards, percent sodium, sodium adsorption ratio, residual sodium carbonate, soluble sodium percentage, residual sodium bicarbonate, permeability index, potential salinity, Kelley's index, and magnesium hazard, were calculated for the water sampled in summer and winter from the selected *aflaj* of the study area using the equations in Table 3.

Table 3. Chemical composition indicators used for classifying irrigation water quality of the selected *aflaj* of the study (All ionic concentrations are in milliequivalents/l; meq/l)

Chemical Composition Indicator	**Equation**	**Reference**
%Na	$(Na^+ + K^+) / (Ca^{2+} + Mg^{2+} + Na^+ + K^+) * 100$	Wilcox, 1955
RSC (meq/l)	$(HCO_3^- + CO3^{2-}) - (Ca^{2+} + Mg^{2+})$	Richard, 1954
SSP (%)	(Soluble sodium concentration/total cations concentration) * 100	Todd, 2005
RSBC (meq/l)	$HCO_3^- - Ca^{2+}$	Richard, 1954
PI (%)	$Na^+ + (HCO_3^-)^{0.5} / (Ca^{2+} + Mg^{2+} + Na^+) * 100$	Doneen, 1964
PS (meq/l)	$Cl^- + 0.5\ SO_4^{2-}$	Richard, 1954
KI (Ratio)	$Na^+ / Ca^{2+} + Mg^{2+}$	Kelley, 1951
MH (%)	$Mg^{2+} / (Ca^{2+} + Mg^{2+}) * 100$	Szabolcs & Darab, 1964
Sodium Adsorption Ratio (SAR)	$\frac{[Na^+]}{\sqrt{[Ca^{2+}] + [Mg^{2+}]/2}}$	Todd, 2005

2.3 Irrigation Water Quality Index

WQI was used in this study to assess the quality of *aflaj* water and provide an overall indication for their suitability for irrigation purposes. The methodology requires that all five hazards (Appendix I) are simultaneously included in the analysis and combined to form a single WQI value, which is then assessed to determine the suitability of the irrigation water. The five hazards were grouped into five weighing coefficients, given the numbers 5, 4, 3, 2, 1, respectively, such that the most and the least important groups in irrigation water quality are given the highest (5) and lowest (1) points. For each hazard, several parameters were determined with different ranges and rating suitability. Three categories - high, medium and low - were given to the three rating suitability (3, 2, 1), respectively (Appendices I, II, III). After the total value of the index was computed, a suitability analysis was done based on the three different categories of WQI: low (< 22), medium (22-37) and high (> 37) (Simsek & Gunduz, 2007). The detailed calculation of all five hazard categories and the WQI are summarized in Appendix IV.

3. Results and Discussion

3.1 Physico-Chemical and Microbiological Parameters of Water Quality

The physico-chemical parameters of the selected *aflaj* of the study area are presented in Table 4, as mean, median, standard deviation, minimum and maximum over the summer and winter seasons of 2012-2013, and compared to the quality standards and guidelines for Omani Wastewater Reuse for irrigation and Discharge (Ministerial Decision, MD 145/1993) as there are no specific guidelines set for *aflaj* water.

Table 4. Physiochemical and microbiological characteristics of *aflaj* water quality for 22 sampling points from the study area during the two periods in 2012-2013

Variables	Mean	Median	Std. Deviation	Minimum	Maximum	Omani Standard	WHO Standard	MD 145/1993
EC (μS/cm)	564	538	159	294	896	160-1600	2000	2000
pH	8.11	8.18	0.26	7.41	8.51	6.5-9.0	6.5-9.5	6-9
Turbidity (NTU)	0.79	0.56	1.00	0.06	4.48	1 - < 5	NG	NG
TDS (mg/l)	345	329	100	172	562	120-1000	1000	1500
$CaCO_3$ (mg/l)	235	237	49	134	326	NG	NG	NG
HCO_3 (mg/l)	240	237	50	137	336	NG	NG	NG
Total Hardness (mg/l as $CaCO_3$)	272	265	58	158	376	≤ 200 - 500	500	NG
Sodium (mg/l)	19.68	16.11	11.50	7.48	48.88	≤ 200 - 400	200	200
Calcium (mg/l)	48.98	48.60	12.54	26.83	78.98	200	NG	NG
Magnesium (mg/l)	27.34	26.58	7.66	13.44	46.03	150	NG	150
Potassium (mg/l)	3.79	4.04	3.07	0.51	10.08	NG	NG	NG
Fluoride (mg/l)	0.12	0.12	0.04	0.04	0.18	1.5	1.5	1
Chloride (mg/l)	26.29	19.71	18.02	12.08	76.68	≤ 250 - 600	250	650
Nitrate (NO_3) (mg/l)	6.54	2.98	9.30	0.34	35.03	50	50	50
Sulphate (SO_4) (mg/l)	26.59	20.91	15.93	10.89	74.16	≤ 250 – 400	400	400
Phosphate (PO_4) (mg/l)	0.27	0.08	0.81	0.03	3.80	NG	NG	NG
Dissolved Oxygen (mg/l)	5.85	5.96	0.82	4.29	7.30	NG	NG	NG
BOD5 (mg/l)	1.91	1.77	0.76	0.73	3.84	NG	NG	15
Coliforms (MPN/100 ml)	133.68	155.93	82.12	3.75	> 200.50	10	10	200
E-Coli (MPN/100 ml)	7.30	2.25	10.83	0.00	37.50	0	0	NG

Omani Standard: Un-bottled Drinking Water Standard (No. 8/2006), WHO Standard: World Health Organization Drinking Water Standard, MD: Ministerial Decision (145/1993): Regulation for wastewater discharge and reuse standards, NG: No guideline is recommended.

According to the Omani regulations of wastewater reuse for irrigation water, the Electrical Conductivity (EC) of irrigation water has a maximum limit of 2000 μS/cm. The results of EC measured in all sampled *aflaj* water ranged from 341 to 793 μS/cm (mean 528.18 μS/cm) in summer, and from 246 to 999 μS/cm (mean 599.91 μS/cm) in winter. None of the *aflaj* water samples exceed the maximum limit of 2000 μS/cm specified in the Omani regulations of wastewater reuse for irrigation water. During summer, the pH values ranged from 7.37 to 8.41 and during winter they ranged from 7.44 to 8.61. These pH values are within the limits of the recommended Omani wastewater maximum quality regulations for irrigation water (pH 6-9). The measured turbidity (TR) in all selected *aflaj* during summer ranged from 0.060 to 6.67 NTU (mean 1 NTU) and from 0.050 to 2.29 NTU (mean 0.57 NTU) in winter.

Total dissolved solids (TDS) in the selected *aflaj* water ranged from 183.20 to 475.42 mg/l (mean 300.44 mg/l) in summer and 160.16 to 649.35 mg/l (mean 389.95 mg/l) in winter. These levels are well below the permissible level of TDS of 1500 mg/l in the Omani regulations of wastewater reuse for irrigation water. The total hardness (TH) measurements in the *aflaj* samples ranged from 157.43 to 348.19 mg/l; and from 159.49 to 402.84 mg/l with means of 256.86 mg/l; and 286.17 mg/l during summer and winter, respectively. Adopting Sawyer (2003) classification criteria, these water resources of the entire study area are hard to very hard as the total hardness (mg/l as $CaCO_3$) is in the range of 150-300 and more than 300 mg/l. The hardness of the water is due to the presence of alkaline earths such as calcium and magnesium, and anions such as carbonate, bicarbonate, chloride and sulphate.

According to Omani regulation of wastewater reuse, the results of cations and anions concentrations in the selected *aflaj* water studied showed no values exceeding the permissible limits and all quality values are within these standards.The mean concentration of cation (in mg/l) in the selected *aflaj* water sampled during summer and winter was in order $Ca^{2+}>Mg^{2+}>Na^{+}>K^{+}$. The mean concentration of anions (in mg/l) in the *aflaj* water sampled during summer and winter was in order $HCO_3^{-}>SO_4^{2-}>Cl^{-}>NO_3^{-}>CO_3^{2-}>PO_4^{3-}>F^{-}$. None of the cation and anion concentrations in the samples exceeded the permissible limits according to the Omani regulation of wastewater reuse. Monitoring programs of phosphates particularly for cyanobacteria blooms are very important since some of the *aflaj* waters in the area are used for washing cloths by detergents containing phosphates which can be a source of pollution and contaminate these water resources and promote algal blooms growth.

Although the amount of dissolved oxygen (DO) often gives a good indication of water quality, the Omani regulations of wastewater reuse do not recommend guidelines regarding the acceptability of low levels. Generally, concentrations in unpolluted waters are usually close to, but less than 10 mg/l; concentrations below 5 mg/l may adversely affect the functioning and survival of biological communities (Chapman & Kimstach, 1996). All samples had DO concentrations less than 10 mg/l, but some samples were below 5 mg/l, taking into account changes in field water temperatures. Measurements of DO in the *aflaj* water samples during summer and winter had mean values of 7.93 and 3.77 mg/l; they ranged from 6.16 to 9.89 mg/l; and from 2.41 to 4.70 mg/l, respectively.

Omani regulations of wastewater reuse recommend 15 mg/l as a guideline values for biochemical oxygen demand (BOD_5). However, unpolluted waters typically have BOD_5 values of 2 mg/lor less whereas those receiving wastewater may have values up to 10 mg/lor more (Chapman & Kimstach, 1996). In the selected *aflaj* water samples, BOD5 concentrations were generally very low with means of 1.71 and 2.12 mg/l during summer and winter. All recorded values were below the range of BOD5 from 15 to 20 mg/l recommended in the Omani regulations of wastewater reuse for irrigation.

Most heavy metals in *aflaj* water sampled during summer and winter were below the detection levels of the ICP instrument. These include cadmium, chromium, cobalt, copper, vanadium, lithium, selenium, titanium and beryllium. In some *aflaj* water samples, concentrations of manganese, molybdenum and arsenic were just above the limit specified in the Omani standards of wastewater reuse for irrigation water. Other heavy metals were found in the *aflaj* water samples but within the safe limits for irrigation water. The chemical weathering and soil leaching are the primary natural sources that contribute in the presence of trace metals in water (Fetter, 2001; Şen, 2014).

Although all the physico-chemical parameters of *aflaj* water are within the permissible limits set by Omani standard for Un-bottled Drinking Water 8/2006 (MD, 2007) and the World Health Organization (WHO, 2011) for drinking water, most of the *aflaj* water resources studied were contaminated with coliform and *E. coli* bacteria. Of the 22 *aflaj* sampling points, 12 showed more than 200.5 total numbers of coliform bacteria in summer, and 9 in winter. Most of the *aflaj* water samples showed the presence of *E. coli* bacteria. These results indicate that *aflaj* waters are unacceptable and hazardous for drinking according to Omani standard for Un-bottled Drinking Water 8/2006 (MD, 2007) and the World Health Organization (WHO, 2011). The Omani and WHO standards allow the Most Probable Number (MPN) of 10/100 ml. In both guidelines, total *E. coli* should be 0 MPN/100 ml of a sample.

3.2 Hydro-Chemical Water Quality

Table 5 shows the mean, median, standard deviation, minimum and maximum of chemical composition indicators including percent sodium (%Na), residual sodium carbonate (RSC), soluble sodium percentage (SSP), residual sodium bicarbonate (RSBC), permeability index (PI), Kelley's index (KI), and magnesium hazard (MH) in the *aflaj* water samples in summer and winter. The selected *aflaj* of the study area have excellent or good quality; none of the water samples exceeded the limits and all were under the satisfactory category values, indicating their suitability for irrigation for the most crops and soils, based on % Na, SSP, RSC, RSBC, PI, PS and MH and their irrigation water classification criteria (Table 6).

Table 5. Descriptive statistics of chemical composition indicators for 22 sampling points of the selected *aflaj* water of the study area sampled during summer and winter 2012-2013

Chemical Composition Indicators		Mean	Median	Std. Deviation	Minimum	Maximum
%Na	Summer	11.43	10.48	4.15	6.29	23.71
	Winter	19.82	18.58	6.56	9.36	30.48
RSC (meq/l)	Summer	-1.21	-0.98	0.59	-2.72	-0.63
	Winter	-0.68	-0.62	0.73	-2.83	0.75
SSP (%)	Summer	10.60	9.53	3.76	5.85	21.63
	Winter	10.60	9.53	3.76	5.85	21.63
RSBC (meq/l)	Summer	0.90	0.94	0.54	-0.12	1.80
	Winter	1.74	1.87	0.69	0.39	3.22
PI (%)	Summer	45.77	46.81	4.02	36.30	54.39
	Winter	54.02	53.29	5.38	46.54	64.54
PS (meq/l)	Summer	0.90	0.74	0.50	0.43	2.36
	Winter	1.13	0.81	0.84	0.47	3.50
KI (Ratio)	Summer	0.12	0.11	0.05	0.06	0.28
	Winter	0.22	0.20	0.09	0.10	0.39
MH (%)	Summer	42.21	41.21	7.49	28.22	59.19
	Winter	54.00	54.47	6.42	40.49	63.74

Table 6. Classification of irrigation water based on different hazards

Parameter	Range	Water class	*Aflaj* Samples
%Na	< 20	Excellent	F1, F2, F3, F4, F5, F6, F7, F8
(After Wilcox, 1955)	20-40	Good	F9
	40-60	Permissible	
	60-80	Doubtful	
	> 80	Unsuitable	
RSC (meq/l) (After Richard, 1954)	< 1.25	Good	F1, F2, F3, F4, F5, F6, F7, F8, F9
	1.25-2.50	Doubtful	
	> 2.5	Unsuitable	
SSP (%)	0-20	Excellent	F1, F2, F3, F4, F5, F6, F7, F8
(After Todd, 2005)	20-40	Good	F9
	40-60	Permissible	
	60-80	Doubtful	
	80-100	Unsuitable	
KI (Ratio) (After Kelley, 1944)	<1	Suitable	F1, F2, F3, F4, F5, F6, F7, F8, F9
	>1	Unsuitable	
MH (%)	<50	Suitable	F1, F2, F3, F4, F5, F6, F7
(After Szabolcs & Darab, 1964)	>50	Unsuitable	F8, F9
PI (%) (After Doneen, 1964)	50-75	Suitable	F1, F2, F3, F4, F5, F6, F7, F8, F9
	25	Unsuitable	

The assessment of irrigation water quality based on the combination of salinity hazard using Electrical Conductivity (EC) and alkalinity hazard using Sodium Adsorption Ratio (SAR) is another classification for the suitability of water for irrigation. As presented in section 4.1, all the *aflaj* water have mean EC values below 700 µS/cm indicating good water quality for irrigation (Tatawat & Chandel, 2008). SAR in *aflaj* water samples ranged from 0.61 to 4.70; and from 0.53 to 2.73 with average values of 1.21 and 1.35 during summer and winter, respectively. Using the combined results of EC measurements and the SAR values, the analytical data plot on the EC-SAR diagram illustrates that most of the summer and winter water samples from the *aflaj* (Figure 3) fall in the field of C2-S1 (good: medium salinity/low sodium hazards) suitability based on the EC and SAR classification of irrigation water (Appendix V). These categories of water quality can be used for irrigation of most crops and majority of soils (Richard, 1954; Simsek & Gunduz, 2007).

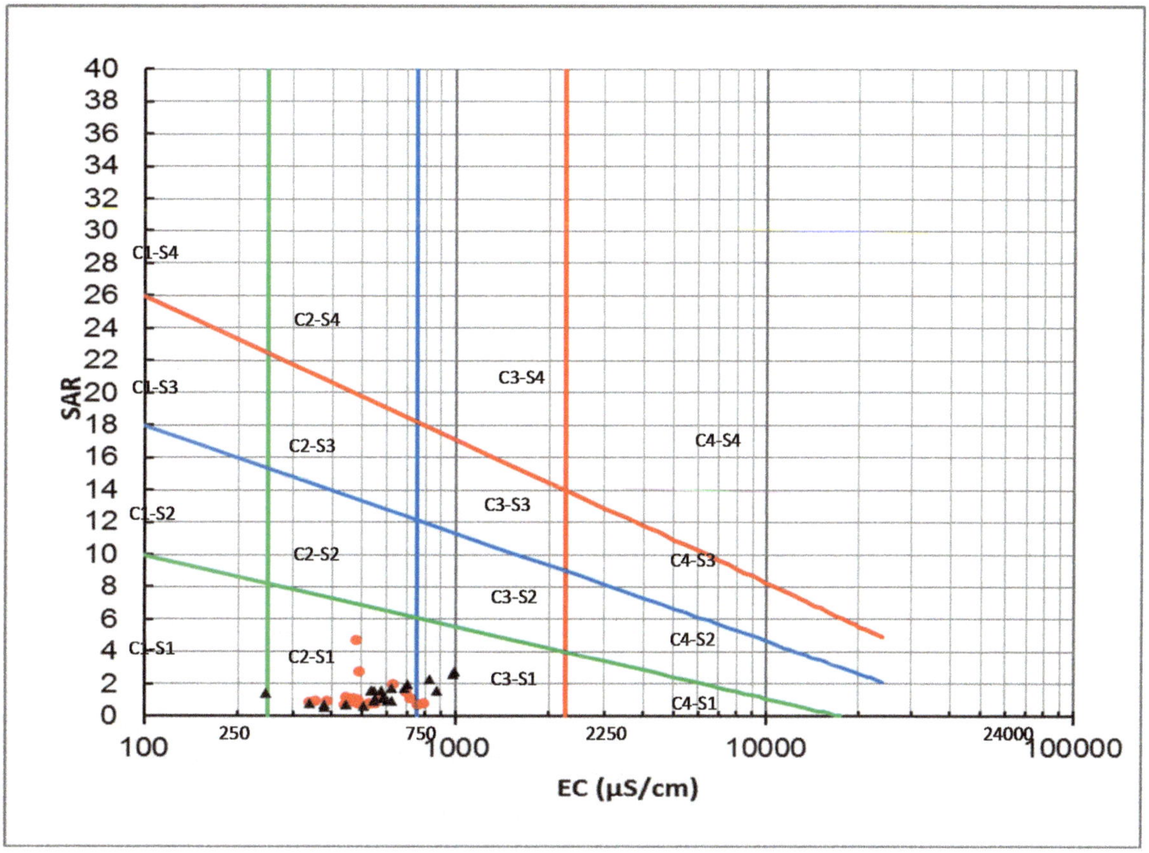

Figure 3. EC and SAR classification of *aflaj* water sampled during summer (in red circle) and winter (in black triangle) in 2012-2013

3.3 Correlation of Water Quality Parameters

Correlation analyses using Pearson's coefficient (r) among the levels of parameters in water samples indicates the existence of an association, and thus, a single parameter can act remarkably as a reliable indicator of the presence of a number of parameters (El Maghraby et al., 2013; Mohammed et al., 2014; Varol & Davraz, 2015). In the present study, statistical analysis has shown that some of the parameters correlate significantly with one another. The terms strongly, moderately and weakly correlations in this study refer to $r > 0.7$, $r = 0.5\text{-}0.7$, and $r < 0.5$, respectively.

The correlation matrix among water quality parameters of the selected *aflaj* sampling points (Table 7) showed the highest positive correlation between EC and TDS ($r = 0.978$), highly statistically significant at $p < 0.01$. Other positive strongly correlations, highly statistically significant at $p < 0.01$ include: between TDS with TH, Na^+, Mg^{2+}, Cl^-, NO_3^- and SO_4^{2-}; between EC with TH, Na^+, Mg^{2+}, K^+, Cl^-, HCO_3^-, NO_3^- and SO_4^{2-}; between HCO_3^- with TH and Ca^{2+}; between TH with Na^+, Mg^{2+} and Cl^-; between Na^+ with Mg^{2+}, Cl^-, NO_3^- and SO_4^{2-}; and between Mg^{2+} with Cl^-, NO_3^- and SO_4^{2-}. Moderately positive correlations ($p < 0.01$) were found between TDS with K^+ and HCO_3^-;

between pH with CO_3^{2-}; between HCO_3^- with Mg^{2+}; between TH with NO_3^- and SO_4^{2-} and between Na^+ with K^+. Other positively weakly correlated and statistically significant ($p < 0.01$) relationships were found between EC with Ca^{2+} and K^+; between HCO_3^- with Na^+, Cl^- and NO_3^-; between TR with Ca^{2+} and K^+; between Ca^{2+} with NO_3^-, and between Mg^{2+} with K^+.

Table 7. Correlation matrix among quality parameters for *aflaj* water of the study area sampled during summer and winter 2012-2013

	TR	TDS	EC	pH	HCO_3	CO_3	TH	Na	Ca	Mg	K	F	Cl	NO_3	SO_4
TR															
TDS	-.197														
EC	-.187	.978**													
pH	.222	-.351*	-.376*												
HCO_3	-.243	.632**	.727**	-.667**											
CO_3	.139	-.086	-.096	.681**	-.341*										
TH	-.263	.897**	.905**	-.536**	.826**	-.228									
Na	-.124	.861**	.827**	-.093	.398**	.123	.765**								
Ca	.057	.288	.436**	-.398**	.702**	-.247	.431**	.082							
Mg	-.290	.861**	.871**	-.177	.639**	.109	.853**	.813**	.272						
K	-.207	.578**	.493**	-.196	.251	-.063	.464**	.601**	-.307*	.440**					
F	-.133	.083	.092	.132	.175	.058	.167	.059	.179	.211	-.040				
Cl	-.057	.829**	.836**	-.008	.390**	.136	.705**	.909**	.296	.807**	.335*	.082			
NO_3	-.110	.725**	.764**	-.182	.432**	.038	.656**	.745**	.431**	.701**	.189	.037	.835**		
SO_4	-.031	.735**	.756**	.113	.305*	.201	.603**	.856**	.296	.744**	.316*	.107	.951**	.853**	

*significantly correlated at 0.05 level, **significantly correlated at 0.01 level.

These correlation relationships between parameters in *aflaj* water samples clearly identify the main elements that are normally found in the studied water resources. The strongest significant relationship ($r > 0.80$) between Mg^{2+} and Cl^-, and the moderately significant relationships between total hardness with Ca^{2+} ($r > 0.50$) and the strongest with Mg^{2+} ($r > 0.80$) in *aflaj* water samples, indicate that the hardness of the water was permanent in nature. The statistically significant and strong correlation ($r > 0.90$) between Cl^- and Na^+ confirms their common origin: the dissolution of the halite resulting from the action of water on salts. The concentrations of SO_4^{2-} are tightly correlated to the presence of Na^+ and Mg^{2+}, which is explained by the dissolution of evaporate minerals.

3.4 Irrigation Water Quality Index

Based on the assessment criteria of EC, infiltration hazard, sodium as SAR, chloride, boron, pH, the concentrations of bicarbonate, nitrate-nitrogen and trace elements, values of the WQI of the *aflaj* water samples in the study area were not significantly different between summer and winter ($F = 1.181$, $P = 0.283 > 0.05$). The WQI of the selected *aflaj* ranged from 35.20 to 40.31 (mean 39.57, median 39.86, standard deviation 1.05) during summer, and from 38.53 to 40.67 (mean 39.84, median 39.92, standard deviation 0.53) during winter.

These results show that the water of the selected *aflaj* was of high or medium suitability for irrigation, falling within the 3 or 2 rating categories of irrigation water classification criteria. In the selected 22 *aflaj* water sampling points during summer, 21 were classified as high in quality and only one as medium. The selected *aflaj* channel points sampled during the winter were all classified as highly suitable for irrigation based on WQI classification criteria (Figure 4).

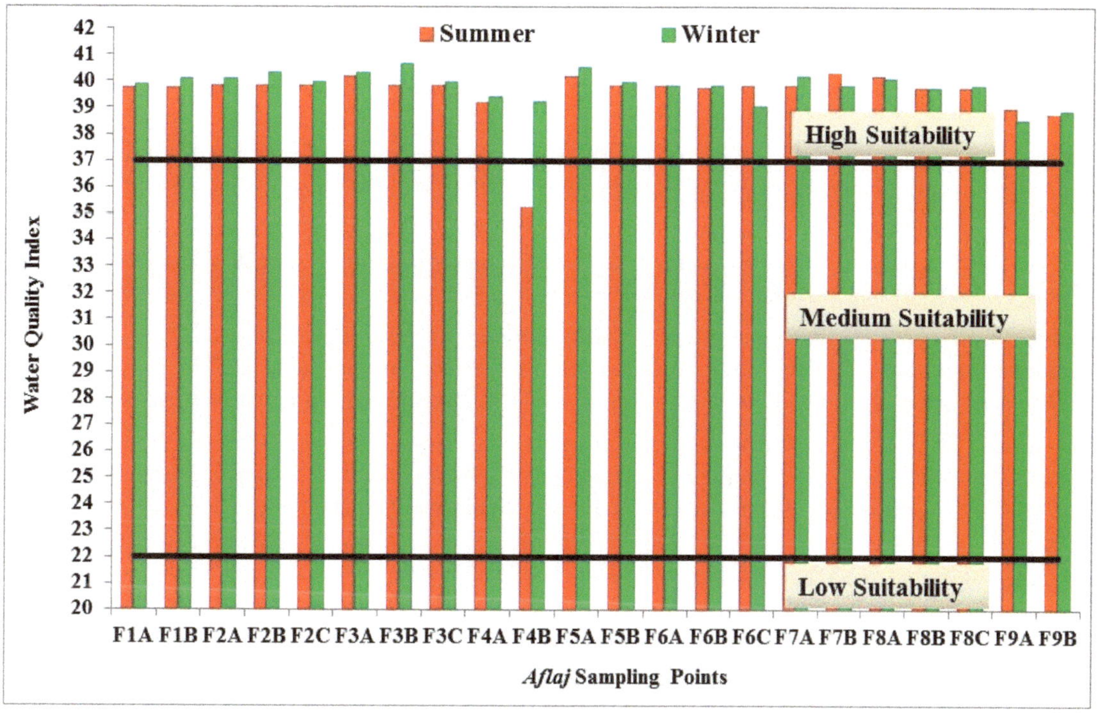

Figure 4. WQI of the selected *aflaj* water sampling points during summer and winter 2012-2013

4. Conclusions and Recommendations

Water quality assessment of the selected *aflaj* in Al Jabal Al Akhdar area indicated that quality parameters are within the permissible limits set by Omani regulations of wastewater reuse for irrigation. However, most of the *aflaj* are contaminated with *E. coli* bacteria; indicating unacceptable for drinking as per the guidelines of Omani and WHO standards. Overall, the selected *aflaj* are excellent or good in quality for irrigation purposes based on the evaluation of different hazards parameters including percent sodium, residual sodium carbonate, soluble sodium percentage, residual sodium bicarbonate, permeability index, Kelley's index, and magnesium hazard; indicating their suitability for irrigation for the majority of crops and soils. The salinity-alkalinity hazards assessment showed that the *aflaj* water are C2-S1 (Good) based on EC and SAR classification; such slightly high salinity and low sodium water can be used for irrigation on almost all types of soil with little danger of exchangeable sodium. All computed water quality indices showed that most of the *aflaj* have high suitability for irrigation and only one has moderate suitability, and no serious problems with respect to irrigation quality. To keep all *aflaj* of the study area under good water quality for domestic and irrigation purposes, the study recommends further corrective demand management measures, such as water conservation, reuse of treated wastewater effluents, reusing greywater, redesigning septic tanks and protecting *aflaj* mother well and their channels. Water quality monitoring programmes should be also carried out on a regular basis to ensure the suitability of water for domestic and agricultural uses.

Acknowledgments

The authors would gratefully acknowledge the Ministry of Regional Municipalities and Water Resources (MRMWR), and *aflaj* committee of the study area for their logistic support and permission to sample *aflaj* water. Our grateful thanks also go to Sultan Qaboos University and MRMWR, Muscat, Oman for their assistance and facilities provided to analyse the water samples in their laboratories. Finally, we would like to send our greatest appreciations to the Ministry of Environment and Climate Affaires, Muscat, Oman for funding this research.

References

Abdel Rahman, H. A.,&Omezzine, A. (1996). *Aflaj* water resources management: tradable water rights to improve irrigation productivity in Oman. *Water International, 21*(2), 70-75.

Adhikari, K., Chakraborty, B., & Gangopadhyay, A. (2013). Assessment of irrigation potential of ground water using water quality index tool. *Asian Journal of Water, Environment and Pollution, 10*(3), 11-21.

Ahmed, M., Victor, R., Al Haddabi, M., & Al-Handali, J. (2006). Water Quality Assessment in Al Jabal Al Akhdar Region of Oman for Sustainable Water Resources Management. Paper presented at the IAESTED International Conference on Environmentally Sound Technology in Water Resources Management, ESTW. Gabarone, Botswana, 11-13 September 2006.

Al-Bahrani, H. S., Razzaq, K. A. A., & Saleh, S. A. H. (2012). Remote sensing of water quality index for irrigation usability of the Euphrates River. *WIT Transactions on Ecology and the Environment, 164,* 55-66.

Al-Balushi, A. S., Al Mukhtar, B., Al-Hayis, A. J., Orfan, M., Hamza, J., Al-Busaidi, Y. S., & Al-Busaidi N. H. (2011). Socioeconomic Study of Tourism Development in Al Jabal Al Akhdar. Muscat, Oman: Sultan Qaboos University and Ministry of Tourism.

Al-Ghafri, A. S., Inoue, T., & Nagasawa, T. (2003). Irrigation scheduling of *aflaj* of Oman: Methods and Modernization. In Z. Adeel (Ed.), *Sustainable Management of Marginal Drylands - Application of Indigenous Knowledge for Coastal Drylands* (pp. 147-166). United Nations University, Tokyo, Japan.

Al-Haddabi, L. H. S. (2003). *Environmental Assessment of an Area Associated with a Diesel Contaminated Stream in Jebel AL-Akhdar, Sultanate of Oman* (Unpublished master's thesis). Department of Biology, College of Science, Sultan Qabios University, Muscat, Sultanate of Oman.

Al-Haddabi, L. H. S., Victor, R., & Pillay, A. (2009). Assessment of Water Quality in a Mountain Aquatic Resource in Al-Jabal Al-Akhdar (Oman) after Ten Years of Diesel Contamination. In R. Victor & M. D. Robinson (Eds.), *Proceedings of the International Conference on Mountains of the World: Ecology, Conservation and Sustainable Development*, Held on 10-14 February 2008 (pp. 121-126). Center for Environmental Studies and Research, Sultan Qaboos University, Muscat, Oman.

Al-Harbi, M., Al-Ruwaih, F. M., & Alsulaili, A. (2014). Statistical and analytical evaluation of groundwater quality in Al-Rawdhatain field. *Environmental Progress and Sustainable Energy, 33*(3), 895-904.

Al-Kalbani, M. S., Price, M. F, Abahussain, A., Ahmed, M., & O'Higgins, T. (2014). Vulnerability assessment of environmental and climate change impacts on water resources in Al Jabal Al Akhdar, Sultanate of Oman. *Water (Switzerland), 6(10)*, 3118-3135.

Al-Kalbani, M. S., John, C., Martin, F. P. (2015a). Recent Trends in Temperature and Precipitation in Al Jabal Al Akhdar, Sultanate of Oman, and the Implications for Future Climate Change. *J Earth Sci Clim Change, 6*, 295. http://dx.doi.org/10.4172/2157-7617.1000295.

Al-Kalbani, M. S., Price, M. F., O'Higgins, T., Ahmed, M. & Abahussain, A. (2015b). Integrated environmental assessment to explore water resources management in Al Jabal Al Akhdar, Sultanate of Oman. *Regional Environmental Change*, First online: 11 September 2015.

Al-Khashman, O. A., & Jaradat, A. Q. (2014). Assessment of groundwater quality and its suitability for drinking and agricultural uses in arid environment. *Stochastic Environmental Research and Risk Assessment, 28*(3), 743-753.

Al-Marshudi, A. S. (2001). Traditional irrigated agriculture in Oman: Operation and management of the *aflaj* system. *Water International, 26*(2), 259-264.

Al-Marshudi, A. S. (2007). The falaj irrigation system and water allocation markets in Northern Oman. *Agricultural Water Management, 91*(1-3), 71-77.

Al-Marshudi, A. S. (2008). Economic instruments for water management in the Sultanate of Oman. *Water International, 33*(3), 361-368.

Al-Riyami, Y. N. S. (2006). Investment opportunities in Agriculture Sector in Al Jabal Al Akhdar. In Oman chamber of commerce and industry (Ed.), Working paper presented in the Symposium on Economic Development of Al Jabal Al Akhdar, Nizwa, Oman, 11th September 2006. (In Arabic).

Aly, A. A. (2014). Hydrochemical characteristics of Egypt western desert oases groundwater. *Arabian Journal of Geosciences*. http://dx.doi.org/10.1007/s12517-014-1680-8.

Aly, A. A., Al-Omran, A. M., Alharby, M. M. (2014). The water quality index and hydrochemical characterization of groundwater resources in Hafar Albatin, Saudi Arabia. *Arabian Journal of Geosciences.*

APHA (American Public Health Association). (2005). S*tandard methods for the examination of water and wastewater* (21st ed.). American Public Health Association, American Water Works Association and Water Environment Federation Publication, Washington D.C., USA.

Chapman, D., & Kimstach, V. (1996). The selection of water quality variables. In D. Chapman (Ed.), *Water Quality Assessments* (pp. 59-126). E & FN Spon, London, UK.

DGMAN (Director General of Meteorology and Air Navigation). (2014). Climatic Data: Rainfall and Temperatures Data Series. [unpublished]. Public Authority of Civil Aviation, Muscat, Sultanate of Oman.

Doneen, L. D. (1964). *Notes on water quality in agriculture*. Published as a Water Science and Engineering Paper 4001. Department of Water Science and Engineering, University of California, Davis, USA.

El Maghraby, M. M. S., El Nasr, A. K. O. A., & Hamouda, M. S. A. (2013). Quality assessment of groundwater at south Al Madinah Al Munawarah area, Saudi Arabia. *Environmental Earth Sciences, 70*(4), 1525-1538.

Fetter, C. W. (2001). *Applied hydrogeology* (4th ed.). Upper Saddle River, N.J.: Prentice Hall.

Kelley, W. P. (1951). *Alkali, Soils, their Formation, Properties and Reclamation.* Reinhold Pub, New York, USA.

Kelley, W. P. (1944). Permissible composition and concentration of irrigation water, *Proc Am Soc Civ Eng, 66,* 607–613.

Kraiem, Z., Zouari, K., Chkir, N., & Agoune, A. (2014). Geochemical characteristics of arid shallow aquifers in Chott Djerid, south-western Tunisia. *Journal of Hydro-Environment Research, 8(4)*, 460-473.

Lou, J., & Han, J. (2007). Assessing water quality of drinking water distribution system in the South Taiwan. *Environmental monitoring and assessment, 134*(1-3), 343-354.

MD (Ministerial Decision). (2007). *Omani Standard 8/2006 for Un-Bottled Drinking Water*. Issued by Ministerial Decision (MD 2/2007) dated on 15 January 2007, based on the International Guidelines for Drinking Water Quality vol. 1. Recommendations - World Health Organization, 2004. Published by Directorate General for Specifications and Measurements, Ministry of Commerce and Industry, Muscat, Sultanate of Oman.

MD (Ministerial Decision). (1993). Regulations for Wastewater Re-use and Discharge. Ministry of Regional Municipalities and Environment, Muscat, Sultanate of Oman.: Issued by Ministerial Decision 145/93 dated on 13 June 1993.

MECA (Ministry of Environment and Climate Affaires). (2015). Unpublished data and maps. Ministry of Environment and Climate Affaires, Muscat, Sultanate of Oman.

Ministry of Tourism. (2014). Number of Tourists and Tourism Projects in Al Jabal Al Akhdar [Unpublished Data]. Retrieved April 14, 2014, from Department of Statistics and Information, Ministry of Tourism, Muscat, Oman.

Mohammed, A. A. J., Rahman, I. A., & Lim, L. H. (2014). Groundwater quality assessment in the urban-west region of Zanzibar Island. *Environmental monitoring and assessment, 186*(10), 6287-6300.

Mohammed, Muthanna, N. (2011). Quality assessment of Tigris River by using Water Quality Index for irrigation purpose. *European Journal of Scientific Research, 57*(1), 15-28.

MRMEWR (Ministry of Regional Municipalities, Environment and Water Resources). (2001). National *Aflaj* Inventory, summary report. MRMEWR, Muscat, Sultanate of Oman.

MRMWR (Ministry of Regional Municipalities and Water Resources). (2008). *Aflaj* Oman in the World Heritage List. MRMWR, Muscat, Sultanate of Oman.

MWR (Ministry of Water Resources). (1999). *Aflaj* Inventory Project: *Aflaj* System Reports & Plots, Wilayat Nizwa. Ministry of Water Resources, Muscat, Sultanate of Oman.

Nag, S. K. (2014). Evaluation of Hydrochemical Parameters and Quality Assessment of the Groundwater in Gangajalghati Block, Bankura District, West Bengal, India. *Arabian Journal for Science and Engineering, 39*(7), 5715-5727.

Nazzal, Y., Ahmed, I., Al-Arifi, N. S. N., Ghrefat, H., Zaidi, F. K., El-Waheidi, M. M., Batayneh, A., Zumlot, T. (2014). A pragmatic approach to study the groundwater quality suitability for domestic and agricultural usage, Saq aquifer, northwest of Saudi Arabia. *Environmental monitoring and assessment, 186*(8), 4655-4667.

Norman, W. R., Shayya, W. H, Al-Ghafri, A. S., McCann, I. R. (1998). *Aflaj* irrigation and on-farm water management in northern Oman. *Irrigation and Drainage Systems, 12*(1), 35-48.

Richard, L. A. (1954). Diagnosis and Improvement of Saline Alkali Soils (pp. 98-99). IBH Publishing Co. Ltd., New Delhi, India.

Saber, M., Abdelshafy, M., Faragallah, M. E. A., Abd-Alla, M. H. (2014). Hydrochemical and bacteriological analyses of groundwater and its suitability for drinking and agricultural uses at Manfalut District, Assuit, Egypt. *Arabian Journal of Geosciences, 7*, 4593-4613. http://dx.doi.org/10.1007/s12517-013-1103-2.

Sadashivaiah, C., Ramakrishnaiah, C. R., & Ranganna, G. (2008). Hydrochemical analysis and evaluation of groundwater quality in Tumkur Taluk, Karnataka State, India. *International Journal of Environmental Research and Public Health, 5*(3), 158-164.

Sarath Prasanth, S. V., Magesh, N. S, Jitheshlal, K. V., Chandrasekar, N., & Gangadhar, K. (2012). Evaluation of groundwater quality and its suitability for drinking and agricultural use in the coastal stretch of Alappuzha District, Kerala, India. *Applied Water Science, 2*(3), 165-175.

Sawyer, C. N. (2003). *Chemistry for environmental engineering and science* (International ed.). McGraw-Hill, Boston; London.

Şen, Z. (2014). *Practical and applied hydrogeology* (1st ed.). Elsevier, Amsterdam; New York.

Simsek, C., & Gunduz, O. (2007). IWQ Index: A GIS-integrated technique to assess irrigation water quality. *Environmental monitoring and assessment, 128*(1-3), 277-300.

Szabolcs, I., & Darab, C. (1964). The influence of irrigation water of high sodium carbonate content of soils. Proc. *Int. Congress Trans, 2*, 803-812.

Tatawat, R. K., & Chandel, C. P. S. (2008). A hydrochemical profile for assessing the groundwater quality of Jaipur City. *Environmental monitoring and assessment, 143*(1-3), 337-343.

Todd, D. K. (2005). Groundwater hydrology, 3rd edn. Wiley, Hoboken, New York; Great Britain, p 656.

Varol, S., & Davraz A. (2015). Evaluation of the groundwater quality with WQI (Water Quality Index) and multivariate analysis: a case study of the Tefenni plain (Burdur/Turkey). *Environmental Earth Sciences, 73*(4), 1725-1744.

Victor, R., Ahmed, M., Al Haddabi, M., & Jashoul, M. (2009). Water Quality Assessments and Some Aspects of Water Use Efficiency in Al Jabal AlAkhdar. In R. Victor & M. D. Robinson (Eds.), *Proceedings of the International Conference on Mountains of the World: Ecology, Conservation and Sustainable Development*, Held 10-14 February 2008 (pp. 165-170). Center for Environmental Studies and Research, Sultan Qaboos University, Muscat, Oman.

WHO (World Health Organization). (2011). *Guidelines for drinking-water quality* (4th ed.). World Health Organization, Geneva, Switzerland.

Wilcox, L. V. (1955). *Classification and Use of Irrigation Water.* US Department of Agriculture, Washington, D.C., 969,19.

Zekri, S., & Al-Marshudi, A. S. (2008). A millenarian water rights system and water markets in Oman. *Water International, 33*(3), 350-360.

Zekri, S., Powers, D., & Al-Ghafri, A. S. (2014). Century old water markets in Oman. In K. W. Easter & Q. Huang (Eds.), *Water Markets for the 21st Century - What Have We learned?* Global Issues in Water Policy 11, Chapter 8 (pp. 149-162). Springer Science Dordrecht Heidelberg, New York and London.

Appendices

Appendix I: Classification for irrigation WQI parameters (Simsek & Gunduz, 2007)

Hazard	Weight	Indicator	Rating		
			3	2	1
Salinity hazard	5	Electrical conductivity (µS/cm)	EC < 700	700≤EC≤3,000	EC > 3,000
Infiltration and permeability hazard	4	See Table 5			
Specific ion toxicity	3	Sodium adsorption ratio	SAR < 3.0	3.0≤SAR≤9.0	SAR > 9.0
		Boron (mg/l)	B < 0.7	0.7≤B≤3.0	B > 3.0
		Chloride (mg/l)	Cl < 140	140≤Cl≤350	Cl > 350
Trace element toxicity	2	See Table 6			
Miscellaneous effects to sensitive crops	1	Nitrate Nitrogen (mg/l)	NO_3-N < 5.0	5.0≤ NO_3-N≤30.0	NO_3-N > 30.0
		Bicarbonate (mg/l)	HCO_3< 90	90≤HCO_3≤500	HCO_3> 500
		pH	7.0≤pH≤8.0	6.5≤pH<7.0 and 8.0<pH≤8.5	pH <6.5 or pH > 8.5

Appendix II: Classification for infiltration and permeability hazard (Simsek & Gunduz, 2007)

	Sodium Adsorption Ratio (SAR)					Rating
	< 3	3-6	6-12	12-20	> 20	
Electrical	> 700	>1,200	>1,900	>2,900	>5,000	3
Conductivity	700-200	1,200-300	1,900-500	2,900-1,300	5,000-2,900	2
(μS/cm)	<200	<300	<500	<1,300	<2,900	1

Appendix III: Classification for trace element toxicity (Simsek & Gunduz, 2007)

Parameter (mg/l)	**Rating**		
	3	**2**	**1**
Aluminum (Al)	$Al < 5.0$	$5.0 \leq Al \leq 20.0$	$Al > 20.0$
Arsenic (As)	$As < 0.1$	$0.1 \leq As \leq 2.0$	$As > 2.0$
Beryllium (Be)	$Be < 0.1$	$0.1 \leq Be \leq 0.5$	$Be > 0.5$
Cadmium (Cd)	$Cd < 0.01$	$0.01 \leq Cd \leq 0.05$	$Cd > 0.05$
Chromium (Cr)	$Cr < 0.1$	$0.1 \leq Cr \leq 1.0$	$Cr > 1.0$
Cobalt (Co)	$Co < 0.05$	$0.05 \leq Co \leq 5.0$	$Co > 5.0$
Copper (Cu)	$Cu < 0.2$	$0.2 \leq Cu \leq 5.0$	$Cu > 5.0$
Fluoride (F)	$F < 1.0$	$1.0 \leq F \leq 15.0$	$F > 15.0$
Iron (Fe)	$Fe < 5.0$	$5.0 \leq Fe \leq 20.0$	$Fe > 20.0$
Lead (Pb)	$Pb < 5.0$	$5.0 \leq Pb \leq 10.0$	$Pb > 10.0$
Lithium (Li)	$Li < 2.5$	$2.5 \leq Li \leq 5.0$	$Li > 5.0$
Manganese (Mn)	$Mn < 0.2$	$0.2 \leq Mn \leq 10.0$	$Mn > 10.0$
Molybdenum (Mo)	$Mo < 0.01$	$0.01 \leq Mo \leq 0.05$	$Mo > 0.05$
Nickel (Ni)	$Ni < 0.2$	$0.2 \leq Ni \leq 2.0$	$Ni > 2.0$
Selenium (Se)	$Se < 0.01$	$0.01 \leq Se \leq 0.02$	$Se > 0.02$
Vanadium (V)	$V < 0.1$	$0.1 \leq V \leq 1.0$	$V > 1.0$
Zinc (Zn)	$Zn < 2.0$	$2.0 \leq Zn \leq 10.0$	$Zn > 10.0$

Appendix IV: Summary of the five hazard categories, weighing coefficient and formula used for the calculation of each parameter group and irrigation WQI (Simsek & Gunduz, 2007)

Category	**weighing coefficient (w)**	**Formula used**	**Description**
Salinity (EC)	5	$G_1 = w_1 r_1$	w: weight value of this hazard r: rating value (Appendix I)
Infiltration & permeability hazard (EC-SAR)	4	$G_2 = w_2 r_2$	w: weight value of this hazard r: rating value (Appendix II)
Specific ion toxicity (SAR, Cl, B)	3	$G_3 = w_3/3 \sum (r_j)$	J: incremental index w: weight value (Appendix I) r: rating value of each parameter
Trace element toxicity (elements in Appendix III)	2	$G_4 = w_4/N \sum (r_k)$	k: incremental index N: total number of trace elements w: weight value of this group r: rating value of each parameter (Appendix III)
Miscellaneous effects to sensitive crops (NO_3-N, HCO_3, pH)	1	$G_5 = w_5/3 \sum (r_m)$	m: incremental index w: weight value of this group r: rating value of each parameter (Appendix I)
Irrigation Water Quality Index		$WQI = \sum (G_i)$	i: incremental index G: hazard category

Appendix V: Classification of irrigation water based on EC and SAR parameters (After Richard, 1954)

Water Class	Suitability	Water Class	Suitability
C1 – S1	Excellent	C3 – S1	Admissible
C1 – S2	Good	C3 – S2	Marginal
C1 – S3	Admissible	C3 – S3	Marginal
C1 – S4	Poor	C3 – S4	Poor
C2 – S1	Good	C4 – S1	Poor
C2 – S2	Good	C4 – S2	Poor
C2 – S3	Marginal	C4 – S3	Very Poor
C2 – S4	Poor	C4 – S4	Very Poor

2

Increased Terrestrial Disturbances Modify Skeletal Properties and Mechanical Strength of Hard Corals

Shaaban A. Mwachireya[1], Tim R. McClanahan[2], Isabelle M. Cote[3] & Brian E. Hartwick[3]

[1] Kenya Marine & Fisheries Research Institute, PO BOX 81651, Mombasa 80100, Kenya

[2] Wildlife Conservation Society, Marine Programs, Bronx, NY. 10460, USA

[3] Department of Biological Sciences, Simon Fraser University, 8888 University Drive, Burnaby, BC, V5A 1S6 Canada

Correspondence: Shaaban A. Mwachireya, Kenya Marine & Fisheries Research Institute, PO BOX 81651, Mombasa 80100, Kenya. Email: mwachireya@yahoo.com

Abstract

Skeleton properties determine coral survival by influencing the range of hydraulic conditions colonies can withstand, selection of suitable habitat, ability to compete for space and light, repair damage and the overall fitness and ecological success of scleractinian corals. Skeletal properties of 16 coral species comprising 3 growth forms collected from Kenyan coral reef lagoons were investigated and found to vary considerably not only between species but between reefs as well, with corals exposed to both sediment and nutrients showing consistent lower skeleton density and strength but high porosity compared to those from sediment-unaffected reefs. Further, high skeletal density and strength but low porosity values were measured in branching relative to other growth forms. The present findings also suggest that the negative effects of nutrients on skeleton properties may be counteracted by high hydrodynamic energy, resulting in stronger skeletons in high hydrodynamic energy-nutrient-polluted reef habitats relative to pristine reefs. These findings have important ecological and management implications with regard to the existence, persistence, productivity and protective value of reefs, damage risks, maintenance and conservation of biological diversity with respect to future global climate change events. Consequently, appropriate watershed, reef and fisheries management options that address the impacts of local anthropogenic stresses (sediments, nutrients, overexploitation) would be expected to alleviate the effects of these disturbances and have the potential to minimize future large-scale coral reef damage resulting from increased and frequent global climate change events, such as increased ocean acidification (due to elevated atmospheric CO_2) and sea surface temperature.

Keyword: coral skeleton, density, growth form, porosity, sediment, strength, anthropogenic disturbances

1. Introduction

1.1 Impacts of Disturbances on Coral Reef

Coral reefs are maintained by the close balance between constructive (by carbonate-producing corals and coralline algae) and destructive (carbonate-removal by physicochemical and biological factors) processes (McKibben, 2012; McClanahan & Abunge, 2014). However, disturbances that impair carbonate production (calcification) or facilitate carbonate destruction (erosion) can potentially push coral reefs into 'phase shifts' and states of degradation (Tribollet & Golubic, 2011; Doropoulos, Ward, Diaz-Pullido, Hoegh-Guldberg, & Mumby, 2012). There is growing concern that rapid changes in global and local environmental factors are increasing the severity of reef damage (Hoegh- Guldberg et al., 2007; Selig, Casey, & Bruno, 2012; McClanahan, Starger, & Baker, 2014) therefore threatening the ecology of reefs and the livelihood of coastal inhabitants (McKibben, 2012), particularly in tropical and sub-tropical shallow nearshore reefs adjacent to highly modified catchment areas (De'ath, Fabricius, Sweatman, & Puotine, 2012; Maina et al., 2013; Bartley et al., 2014; Ramos-Sharrón, Terres-Pulliza, & Hernandez-Delgado, 2015). Predicted increases in SST and atmospheric carbon dioxide (ocean acidification) and water quality changes (Cooper, De'ath, Fabricius, & Lough, 2008; Doropoulos et al., 2012) will likely combine to cause significant negative effects on calcification, carbonate production and reef framework integrity (Kleypas et al., 1999; Hoegh-Guldberg et al., 2007; McClanahan, Atewebarhan, & Omukoto,

2008; Godinot, Tribollet, Grover & Ferrier-Pagés, 2012) therefore undermining individual coral fitness and skeleton characteristics (Pisapia, Anderson, & Pratchett, 2014). Compromised coral skeleton integrity could adversely impact the mechanical properties of corals with severe consequences on resistance and resilience of coral reefs (Madin, Hughes & Connolly, 2012; Reyes-Nivia, Diaz-Pullido, Kline, Hoegh-Guldberg & Dove, 2013), ultimately affecting reef productivity and services.

1.2 Why Coral Skeletal Properties?

Skeletal material deposition and properties have important ecological and economic implications in coral fitness and the subsequent formation and maintenance of coral reef structures (Chamberlain, 1978; Schuhmacher & Plewka, 1981a: Cordero, 2013) as well as the socio-economics of coastal inhabitants (McClanahan and Cinner, 2012). Skeletal properties also play a significant role in the ability of corals to repair damage, withstand impacts of bioeroding organisms (Pisapia et al., 2014; Chen, Li, & Yu, 2013; Cameron & Edmunds, 2014) and the maintenance and structuring of coral assemblages. Apart from preventing assemblage dominance by small subsets of species, skeleton characteristics may also be of significance in (i) determining boulder movements and damage during high hydrodynamic energy events (Spiske et al., 2008), (ii) explaining the differential mortality of corals (Roughael & Inglis, 1998) and (iii) understanding maintenance mechanisms of biological diversity, patchiness and changes in community composition in reefs after physical disturbance events (Yost et al., 2013). Mechanical properties of coral skeletons are also important determinants in the survival of corals by limiting colony or branch size, range of hydraulic conditions colonies can withstand, influencing competition for space and light, selection of suitable habitat as well as affecting growth form characteristics (Allemand, E. Tambutté, Ziccola, & S. Tambutté, 2011; Yost et al. Chindapol et al., 2013; 2014). Previous studies have shed some light on the consequences of skeleton property characteristics and morphological modifications (Chamberlain, 1978; Schuhmacher & Plewka 1981a; Madin et al., 2012; Reyes-Nivia et al., 2013), importance of skeletal strength through geological time (Schumacher & Plewka, 1981b; Tomiak et al., 2013; Robinson et al., 2014) as well as changes in colony and reef framework strength (Caroselli et al., 2011; Chen et al., 2013). However, few studies have been undertaken to elucidate the influence of water quality parameters on coral skeleton modifications, especially the effects of increased sediment and nutrient concentrations on coral skeleton mechanical properties.

1.3 Terrestrial Run-Off and Coral Skeleton Properties

Anthropogenically enhanced delivery of sediments and nutrients represents a well-recognized threat to nearshore coral reef communities globally (Mellela, Perry, & Haley, 2004; Ramos-Sharrón et al., 2015). Increased levels of terrestrial inputs have not only been reported to cause modified coral growth and poor health conditions, but have also been shown to be trapped and incorporated into coral skeletons (Edinger et al., 2000; Mellela, Lewis, & Croke, 2013; Shirai et al., 2014; Risk, 2014) through various mechanisms (Corrège, 2006; Prouty et al., 2013; Risk, 2014). Although incorporated materials have been used to infer a variety of information relating to past environmental changes (Fleitmann et al., 2007; Lewis et al., 2012; Inoue et al., 2014), their effects on coral skeleton structure, level of coral architectural modification and overall skeleton mechanical properties have scantly been investigated. Moreover, studies undertaken so far have reported conflicting results with Buddemeier, Maragos, & Knudson, 1974; Barnes & Devereaux, 1988; Motai et al., 2012 implicating skeletal property changes to the packing of $CaCO_3$ crystals (micro-architecture) while others have linked skeleton changes to the arrangements of corallite septa, thecae and dissepiments (Dodge et al. 1993; Bucher, Harriot & Roberts, 1998; Cordero, 2013) or meso-architecture. Further, Buddemeir and Kinze (1975) failed to ascertain whether observed variations in *Porites lobata* skeleton properties resulted from micro- or meso-architectural changes. More case studies are thus needed to increase our understanding on the impacts of terrestrial-derived disturbances on coral skeleton properties. This is important in the identification, formulation and implementation of appropriate measures for the management and conservation of shallow nearshore reefs impacted by terrestrially derived disturbance, measures that will potentially minimize and/or mitigate the impacts of future climate change events on coral reef communities.

Corals reefs in the northern end of Kenya's fringing reef are located in an ecotone between the seasonal influence of river discharge and Somali current/upwelling in the north, and low river discharge/runoff in the south. The Malindi fringing reef has, for many years, experienced the influx of sediment and nutrient input from the Sabaki River (McClanahan & Obura, 1997; Fleitmann et al, 2007) but only a few studies of the effects of this disturbance have been undertaken (Giesen & van der Kerkhof, 1984; van der Kerkhof and Giesen, 1984; van Katwijk et al., 1993; Obura, 1995; McClanahan & Obura, 1997). River Sabaki discharge has increased from 28 $m^3 s^{-1}$ in the 1950s through 49 $m^3 s^{-1}$ during the period 1957-1970 to about 73 $m^3 s^{-1}$ in the period 2001-2004 (Kitheka, Nthenge & Obiero, 2003) with potential severe consequences on nearby coral reef communities. The synergy between this and future global climate change is potentially alarming in that it could damage reefs, impede reef recovery, interfere with natural regeneration and accelerate the destruction of reefs. Consequently, a study of the effect posed by river discharge in the Malindi coral reef system was undertaken and compared to other unfished reefs along the Kenyan coast in order to evaluate the response of coral skeleton properties and mechanical strength to the influence of land-derived disturbances.

1.4 Hypotheses and Research Design

The presence of reefs under different forms of management (protected or unprotected) as well as reefs under the influence of various forms of terrestrially derived disturbances (river discharge, nutrients, sediments) along the Kenya coast offers a natural experimental design and unique opportunity to study the effects of these disturbances. The study was designed to investigate the impact of terrestrial input on mechanical properties and strength of hard corals in Kenyan protected reef lagoons in order to test the following three (3) main hypotheses

1) Reefs impacted by terrestrial run-off will show reduced coral skeleton matrix volume and bulk density
2) Terrestrial run-off impacted reefs will be dominated by corals with porous and weaker skeletons
3) Terrestrial run-off will modify the mechanical properties and strength of coral growth forms.

2. Materials and Methods

2.1 Study Sites

Four study reefs, Mombasa, Watamu, Shimoni and Malindi (Figure 1) were chosen for this investigation based on their accessibility, type of exploitation, terrestrial influence and study history. All reefs are fully protected, effectively excluding fishing and shell collection for over 15 years. Malindi reef experiences a pulse of land-derived sediments and nutrients during the September-November short rain period (NE monsoon season) via the Sabaki River. Watamu reef forms part of the Watamu Marine Park and Reserve including the highly ramified 360 km^2 groundwater-water sustained Mida mangrove creek. Mombasa MNP is a long stretch of lagoonal fringing reef ~ 6 km long and 1 km from the shore, and experiences occasional water exchange with the ocean via a depression/channel across and through the reef and two creeks (Tudor and Mtwapa) on either side. Shimoni reef lies within the Kisite/Mpunguti Marine Park situated on the southernmost part of the Kenyan coast ~ 8 km offshore from Shimoni town and on the seaward side of Wasini Island.

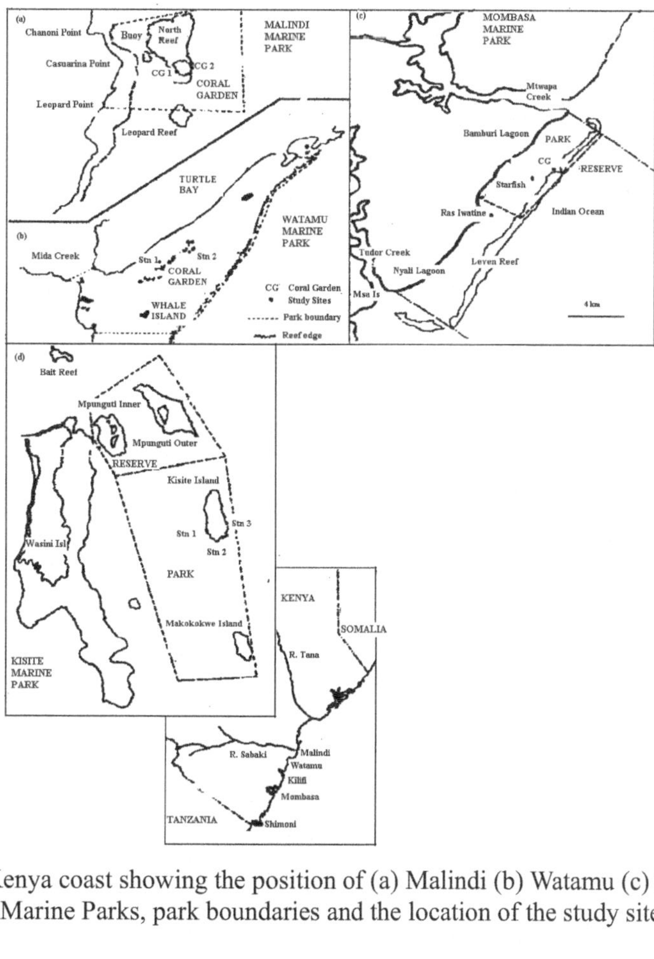

Figure 1. Map of the Kenya coast showing the position of (a) Malindi (b) Watamu (c) Mombasa and (d) Kisite Marine Parks, park boundaries and the location of the study sites

2.2 Oceanographic Studies

Environmental sampling was conducted twice per month. Triplicate temperature and salinity measurements were taken using an automatic probe and current velocity with a modified current drogue. Chlorophyll *a*, phosphate and nitrate concentrations were determined spectrophotometrically after filtration of triplicate 1L water samples through 0.45 μm pore size glass filters. Nitrates were first reduced to nitrites by passing water samples through a column containing copper coated cadmium fillings followed by diazotization with sulfanilamide and coupling with N (1-naphthyl)-ethylenediamine. For phosphate analysis, samples were reacted with a composite reagent containing molybdic acid, ascorbic acid and trivalent antimony. The absorbances of the resultant colored complexes were then measured with a spectrophotometer. Analytical grade potassium nitrate and potassium dihydrogen phosphates were used as standards for nitrates and phosphates analysis, respectively, standard curves were then generated for each nutrient species and the concentrations calculated using the slope and y-intercept of these graphs.

Sedimentation (deposition) rates were measured using replicate sediment traps retrieved from the sites every two (2) weeks, sediment then dried, weighed and sedimentation rates calculated. Total suspended sediment (TSS) concentration was determined by filtering triplicate 1-litre seawater samples through pre-weighed 0.45μm pore size glass filters, filters oven dried overnight at 60 ˚C and weighed after drying for gravimetric determination of TSS. Additionally, trapped and bottom sediments (collected using sediments corers) were analyzed for acid-insoluble residue (acid insolubles) and all sediment categories thereafter analyzed for organic matter. Briefly, organic matter was determined as loss in weight after sample combustion in a furnace at 500 ˚C for 4 hrs and acid-insoluble fraction was determined by digesting 5-10g sediment samples with dilute (5 %) hydrochloric acid (HCl) and weighing the residue after drying.

2.3 Coral Skeleton Characteristics

2.3.1 Bulk Density

Two concepts of coral skeletal density have been described in the literature; bulk-density or the variation in thickness and spacing of the principal skeleton components i.e. meso-architecture and microdensity or the arrangement and packing of calcium carbonate crystals (Barnes & Devereux, 1988; Bucher et al., 1998). These can be measured using methods that either preserve (direct volume measurements, X-ray and CT densitometry) or destroy (density of powdered coral skeleton) meso-architecture of skeleton (Buddemeier et al. 1974; Bucher et al., 1998; Barnes & Devereux, 1988; Cordero, 2013). Water displacement methods also provide microdensity estimates (Barnes & Devereux, 1988; Bucher et al., 1998; Cordero, 2013) when the voids in the skeleton are not allowed to fill with water. Coral density was obtained from coral pieces collected and transported immersed in sea water to avoid air entering the skeleton pores. At the laboratory sea water was gradually replaced with distilled water over a period of two (2) weeks, thereafter corals were soaked in bleach and acetone to remove organic matter. Corals were then washed in distilled water to remove bleach and acetone, care being taken to ensure complete immersion of samples throughout the procedure. Thereafter, saturated coral weight was determined by weighing samples in distilled water using a thin (~ 0.4-0.5 mm diameter) copper wire and a spring balance and sample volume then calculated by employing Archimedean techniques. Corals pieces were later dried in an oven at 100 ˚C to constant weight. After measurement of the dry weight of coral pieces, a water-proof coating was applied by quickly dipping coral samples in molten wax (at 105-110 ˚C) so that a thin wax layer prevented the ingress of water into the coral sample. Waxed corals were then weighed in air and then in water (buoyant weight) and the bulk density (dry weight/enclosed volume) calculated as

$$\text{wt in air} \div [(\text{waxed wt in air} - \text{waxed wt in water}) \times \text{water density}]. \quad (1)$$

2.3.2 Matrix Volume, Microdensity and Porosity

From the above data, skeletal matrix volume and volume of skeletal voids were calculated according to Bucher et al (1998) for the determination of porosity and microdensity. Bulk and microdensity were then treated as measures of meso- and micro-architecture respectively (Barnes & Devereux, 1988; Bucher et al., 1998; Cordero, 2013).

2.3.3 Skeletal Strength

Strength of coral skeletons was estimated using the porosity values obtained from Archimedes Principle and wax techniques according to the formula of Ryshkewitch (1953) for crystalline ceramic materials.

$$\sigma_P = \sigma_0 e^{-nP} \qquad (2)$$

where σ_0 represents the strength of a solid material; P is porosity; n is a constant between 4 (maximum strength) and seven (minimum strength) and σ_P is strength at porosity = P. The compressive strength of nonporous engineering/construction limestone (45-140 MPa) was used as σ_0 due to its similar composition and crystal size with coral skeleton, the lower constant being used since the fine grained variety of limestone gives the highest value in compressive strength and was therefore deemed closer to that of nonporous coral skeletons (Chamberlain, 1978). Compressive strength of carbonate materials is typically an order of magnitude greater than tensile strength and both scale well and similarly with density (Chamberlain, 1978), thus the present skeleton strength estimates can still be useful in comparing the vulnerability of coral skeletons despite being derived from the compressive strength formula. The relationship between skeletal parameters was then explored through regression and correlation analysis with the aim of defining their relationship with skeletal strength and sedimentation.

2.3.4 Statistical and Data Analysis

Data was analyzed using JMP 7 and coral skeleton properties were compared between reefs using one-way ANOVA, a 2-way ANOVA was used to investigate the interaction between reefs and growth forms and post-hoc multiple comparison tests. Transformation of data was performed after tests for homogeneity and homoscedasticity using Bartlett's test. Tukey's HSD test was used to find differences between means after detecting significant differences through ANOVA tests. Relationships between coral strength growth parameters were assessed with correlations analysis using average values for whole reefs. Relationships between the three strength parameters (strength, bulk density and microdensity) were investigated for all the data from the four reefs together as well as for each individual reef.

3. Results

3.1 Oceanographic Studies

No significant differences in temperature, salinity and phosphates were detected between reefs but chlorophyll *a*, current velocity and nitrates were all found to be significantly high in Malindi but low in Shimoni compared to the rest of the study reefs. Higher TSS and sedimentation (trapped sediments) rates were measured in Malindi and Watamu relative to the other study reefs however no differences in TSS organics and trapped sediment (deposition) insoluble residue were detected between reefs (Table 1). Trapped sediment organic content, bottom sediment insoluble residue and organic matter content were all found to be higher in Malindi but low in Shimoni compared to the rest of the study reefs. Chlorophyll *a* also differed significantly between reefs and was found to be higher in Watamu and Malindi but lower in Mombasa and Shimoni.

Table 1. Environmental parameters measured for each reef from the study. All values are averages (and standard deviations) from the NE and SE monsoon periods

Parameter	Malindi	Watamu	Mombasa	Shimoni	F	p
SST (°C)	28.01±1.42	27.58 ±1.53	27.83 ±1.13	27.71±1.26	1.18	ns
Salinity (‰)	34.18±0.67	34.68±0.69	33.98±0.82	34.48±0.66	1.13	ns
Current (m s^{-1})	0.26±0.03	0.17±0.04	0.17±0.04	0.15±0.04	1.86	*
Chlorophyll *a* (μg l^{-1})	0.62±0.22	0.69±0.24	0.33±0.25	0.41±0.25	4.13	*
Nitrates (mg l^{-1})	1.46±0.48	1.19±0.28	0.89±0.39	0.80±0.37	2.35	*
Phosphates (mg l^{-1})	0.97±0.08	0.83±0.08	0.76±0.07	0.69±0.11	0.44	ns
TSS (g l^{-1})	28.45±4.75	28.10±4.33	23.88±1.13	21.48±0.55	7.00	*
TSS Organics (%)	5.01±2.88	4.31±2.19	4.72±1.90	4.17±1.05	1.76	ns
Trapped (g m^{-2} day^{-1})	10.08±1.78	7.05±1.45	2.29±0.68	3.12±0.40	11.5	*
Trapped Insolubles (%)	9.66±0.81	7.07±0.91	6.43±1.48	8.22±1.11	7.52	ns
Traped Organics (%)	7.57±0.99	4.14±0.76	5.65±2.29	3.91±1.75	7.52	*
Bottom Insolubles (%)	8.32±0.45	6.28±0.54	5.69±0.50	3.47±0.48	4.74	*
Bottom organics (%)	3.52±2.08	2.31±0.59	2.64±1.55	2.59±1.14	4.74	*

significant levels indicated by * < 0.05, ns = non significant.

3.2 Coral Skeleton Characteristics

3.2.1 Bulk Density

Bulk density was observed to be low in Malindi corals compared to all other reefs, Tables 2&3 and Figure 2 (Watamu >Mombasa > Shimoni > Malindi) and was found to be high in branching forms compared to both massive and encrusting forms (branching > massive > encrusting, Figure 3). Malindi reef exhibited the lowest bulk density values for all the growth forms except for massive corals (Figure 3a). Species comparison revealed that *Alveopora fenestrata* (0.9±0.1 g cm^{-3}) and *Echinopora gemmacea* (1.7±0.1 g cm^{-3}) had the lowest and highest bulk densities, respectively, and both values were measured in the Malindi reef impacted by river discharge (Table 4).

Table 2. Analysis of variance for the influence of reef, species and interactions on skeletal bulk density, porosity, microdensity and strength of selected corals from Malindi, Watamu, Mombasa and Shimoni protected reef lagoons

Parameter	Factor	df	SS	F	p	Tukey's HSD
Bulk density	Reef	3	0.36	6.77	*	SHM > MLD/WTM < MSA
	Species	14	6.41	25.7	*	*P.eydouxi, Acropora sp*>all,
	R x S	42	2.68	3.57	*	see Table 4
Microdensity	Reef (R)	3	1.86	17.26	*	MSA/WTM <MLD/SHM
	Species(S)	14	1.49	2.99	*	*P.rus*>all, *A.humilis*<all
	R x S	42	6.34	4.27	*	see Table 5
Matrix Volume	Reef (S)	3	2.95	3.27	**	SHM >MLD, WTM > MSA
	Species (R)	14	38.71	9.20	*	*P. eydouxi, Acropora* sp >all
	R x S	42	54.47	4.31	*	see Table 6
Porosity	Reef (R)	3	2783.3	20.53	*	MSA/WTM < MLD, SHM
	Species (S)	14	11232.6	17.76	*	*P. daedelea* > all
	R x S	42	7386.2	63.89	*	see Table 7
Strength	Reef	3	8822.9	12.81	*	MSA>WTM< SHM>MLD
	Species	15	2606.4	18.93	*	*A. humilis* > all
	R x S	45	7882.6	3.82	*	see Table 8

*denote significance at 0.01) with Tukey's HSD multiple.

Table 3. Analysis of variance for the influence of reef, growth form and interactions on bulk density, porosity and microdensity of selected corals from Malindi, Watamu, Mombasa and Shimoni protected reef lagoons

Parameter	Factor	df	SS	F	p	Tukey's HSD
Bulk density	Reef	3	0.84	3.7	*	WTM>MSA/SHM>MLD
	Growth form	2	2.58	17.1	**	b>e/m
	R x G	6	0.67	1.48	0.19	ns
Matrix Volume	Reef	3	4.82	2.5	0.06	ns
	Growth form	2	20.0	15.6	**	e>m/b
	R x G	6	10.0	2.6	*	See Figure 3b
Porosity	Reef (R)	3	3702	10.0	**	MLD>WTM>SHM>MSA
	Growth form	2	7633	31.1	**	m>e>b
	R x G	6	169.4	0.23	0.96	ns
Microdensity	Reef	3	33.8	14.2	**	SHM/MLD>WTM/MSA
	Growth form	2	82.2	51.9	**	e/m>b
	R x G	6	11.6	2.43	*	see Figure 3c
Strength	Reef	3	2309	8.4	**	MSA>WTM>SHM>MLD
	Growth form	2	5291	28.9	**	b>e>m
	R x G	6	791.5	1.45	0.19	ns

* and **denote significance at 0.05 and 0.001, respectively, ns=not significant) with Tukey's HSD multiple.

Table 4. Bulk density data (g cm^{-3}, measured without exposure to air) for selected coral species from Kenyan protected reefs and a summary of statistical analysis

Species	Malindi	Watamu	Mombasa	Shimoni	F	p
Acropora sp	1.29±0.13	1.68±0.19	1.26±0.14	1.61±0.17	18.2	*
Acropora humilis	1.26±0.12	1.41±0.07	1.49±0.06	1.36±0.04	4.22	*
Acropora robusta	1.24±0.19	1.29±0.18	1.29±0.06	1.64±0.07	5.38	*
Porites lutea	1.26±0.27	1.27±0.07	1.13±0.02	1 .04±0.04	11.2	*
Porites rus	1.03±0.07	1.29±0.04	1.29±0.28	1.10±0.13	0.68	ns
Porites cylindrica	1.06±0.23		1.09±0.05	0.99±0.12	0.36	ns
Pocillopora eydouxi	1.37±0.11	1.57±0.03	1.50±0.22	1.63±0.18	1.53	ns
Pocillopora damicornis	1.32±0.20	1.42±0.12	1.34±0.09	1.53±0.15	1.35	ns
Montipora digitata	1.00±0.07		1.43±0.33	1.05±0.04	8.89	*
Favia sp	0.99±0.12	1.15±0.17	1.13±0.19	0.93±0.04	2.23	ns
Favites sp	1.21±0.07	1.56±0.04	1.34±0.04	1.26±0.09	5.79	*
Pavona clavus	1.27±0.26	1.49±0.05	1.07±0.06	1.27±0.04	5.17	*
Pavona decussatta	1.09±0.05		1.53±0.08	1.13±0.13	19.6	*
Alveopora fenestrata	0.76±0.09	2.14±1.04	0.92±0.18	0.85±0.13	86.4	*
Galaxea fascicularis	0.91±0.06	1.18±0.20	1.46±0.10	1.02±0.21	4.38	*
Echinopora gemmacea	1.72±0.10	1.38±0.12	1.41±0.04	1.55±0.16	5.47	*

Values are means (with standard deviations) and significance levels are indicated by: * < 0.05, ns = not significant.

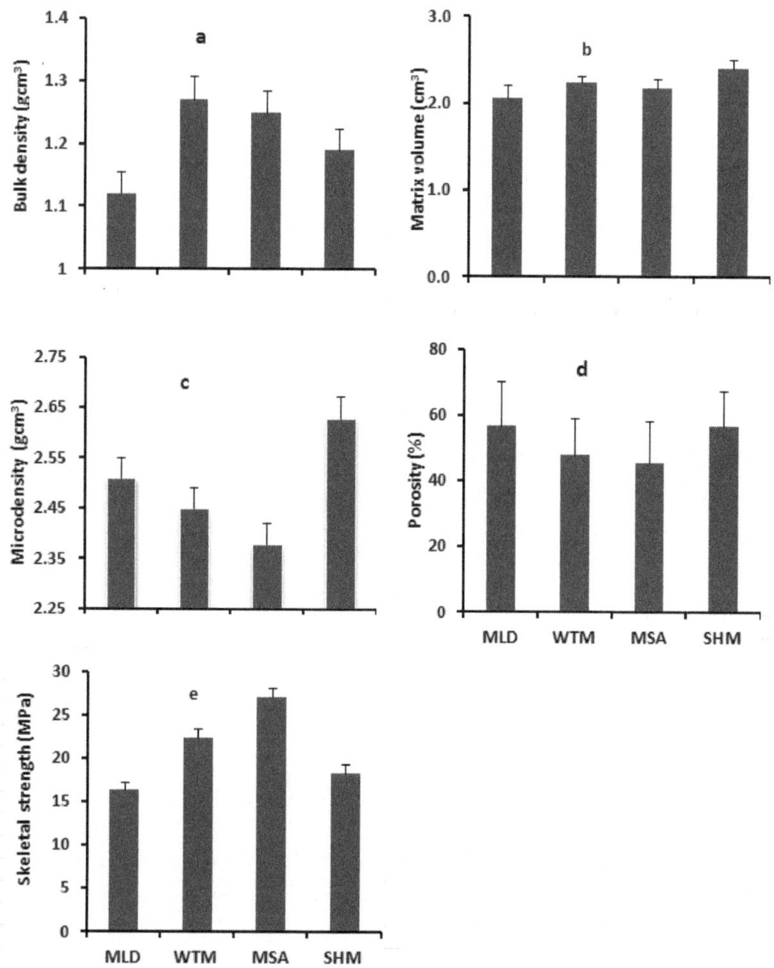

Figure 2. Coral skeleton property means from Malindi, Watamu, Mombasa and Shimoni coral reefs (a) bulk density (b) matrix volume (c) microdensity (d) porosity and (e) skeletal strength

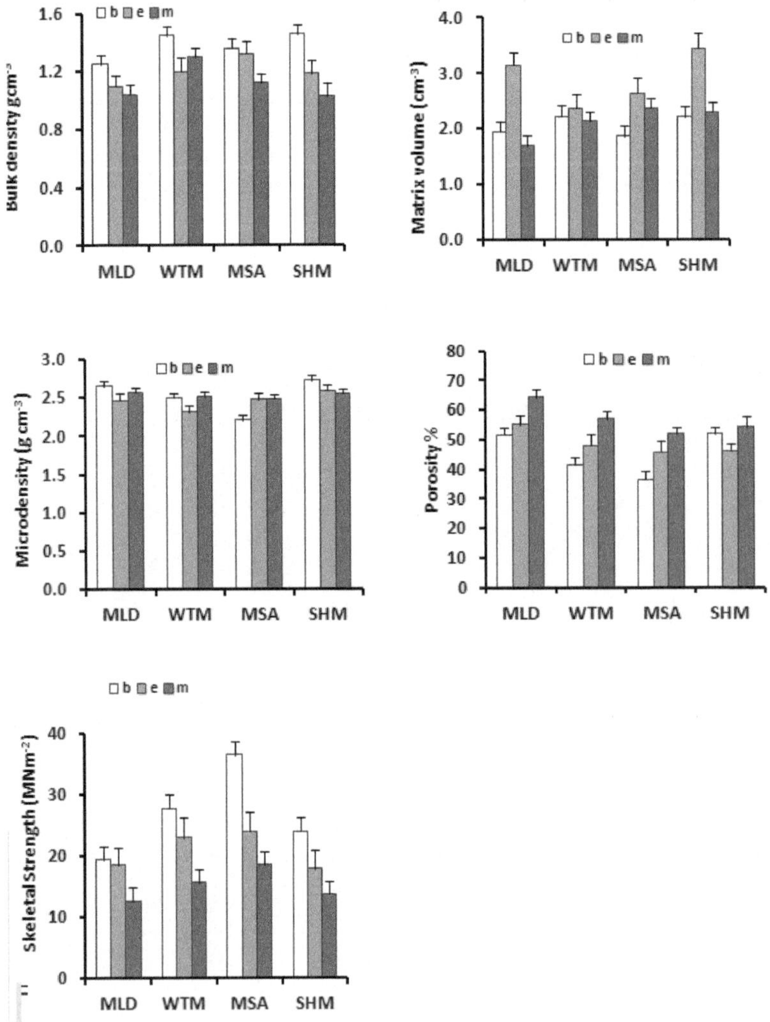

Figure 3. Skeletal characteristics of coral growth forms (b, branching, m, massive and e, encrusting) from Kenyan protected reef lagoons (a) bulk density, (b) matrix volume, (c) microdensity, (d) porosity and (e) skeletal strength

3.2.2 Matrix Volume, Microdensity and Porosity

Low matrix volume was measured in Malindi relative to other reefs (Figure 2b, Table 2) but was high in encrusting compared to other growth forms (Table 3). The highest and lowest matrix volume values were both measured in Malindi from *E. gemmacea* (5.6±0.4) and *Favia sp* (1.00±0.25), respectively (Table 5). High matrix volume was measured in Shimoni for all species except *A. robusta and E. gemmacea.* Like bulk density, microdensity also showed significant differences between reef and species and between growth forms ($p < 0.001$, Table 2) with the highest and lowest microdensity being recorded for *Pocillopora eydouxi* (Shimoni) and *M. digitata* (Malindi), respectively (Table 6). Low microdensity was measured in Mombasa (2.4±0.1 g cm^{-3}) relative to all other reefs (Mombasa<Watamu<Malindi<Shimon, Figure 2c). In fact Mombasa reef exhibited low microdensity values for all coral species except *P. rus, M. digitata, Favia sp* and *Pavona clavus*. Within reefs, the general trend was that massive and the encrusting forms showed lower microdensity compared to branching types (Figure 3c) except in the Watamu reefs.

Coral skeletons in Malindi reef showed higher porosity compared to all other reefs (Malindi>Shimoni>Watamu>Mombasa, Figure 2d). Individually, all species studied had higher porosity in Malindi compared to all other studied reefs except *Acropora humilis M. digitata, Pavona decusatta* and *Pavona clavus* (Table 7). Species-reef comparisons revealed that the most porous species was *Galaxea fuscicularis* in Watamu (74.5±10.5%) and the least porous was *Acropora humilis* (22.8± 1.4%) in Mombasa. Massive corals were found to be the most porous (58.1± 1.2 %) relative to other growth forms. Highest and lowest porosity were measured in massive (Malindi, 64.5± 2.6%) and branching (Mombasa, 36.5 ± 2.6%) corals, respectively (Figure 3d).

3.2.3 Skeleton Strength

Significant differences in coral skeletal strength between reefs and species as well as reef-species interactions were detected. Stronger coral skeletons were observed in Mombasa (except *Pavona clavus*) compared to all other reefs, with corals from the Malindi reef impacted by river discharge generally showing weaker skeletons (Mombasa>Watamu>Shimoni>Malindi, Figure 2e). Higher coral strength values were also measured in branching corals relative to massive and encrusting forms, (Figure 3e, Table 8). Reef-species interaction showed that *Montipora digitata* (Mombasa) had the strongest skeleton whereas the lowest strength was measured in *G. fascicularis* from Malindi.

Table 5. Matrix volume (cm^3) for selected coral species from Kenyan protected reefs and a summary of statistical analysis. Values are means (with standard deviations)

Species	Malindi	Watamu	Mombasa	Shimoni	F	p
Acropora sp	2.24±0.09	2.17±0.29	1.91±0.72	2.67±0.63	0.65	ns
Acropora humilis	1.83±0.14	1.83±0.14	2.08±0.30	2.17±0.52	0.77	ns
Acropora robusta	2.67±0.12	1.83±0.14	1.92±0.38	1.92±0.38	1.07	ns
Porites lutea	1.67±1.23	2.08±0.29	1.92±0.14	2.00±0.66	0.19	ns
Porites rus	1.42±0.14	1.92±0.72	1.08±0.14	1.33±0.38	2.07	ns
Pories cylindrica	1.25±0.50		1.92±0.52	1.58±0.38	1.50	ns
Pocillopora eydouxi	1.75±0.25	3.17±0.14	2.42±0.80	2.17±0.52	4.25	*
Pocillopora damicornis	1.67±0.58	2.42±0.29	1.75±0.25	3.00±0.25	8.67	*
Montipora digitata	1.50±0.25		1.13±0.18	2.08±0.58	4.53	*
Favia sp	1.00±0.25	2.58±0.52	2.92±0.76	2.75±0.25	9.61	*
Favites sp	2.67±1.53	2.00±0.50	1.50±0.25	3.25±0.25	2.41	ns
Pavona clavus	2.65±1.35	2.42±0.76	2.00±0.25	2.50±0.43	0.56	ns
Pavona decussata	1.83±0.52		2.33±0.14	2.25±0.90	0.58	ns
Alveopora fenestrata	1.42±0.14	1.83±1.51	2.83±0.76	1.75±0.50	5.78	*
Galaxea fascicularis	1.25±0.66	2.00±0.66	3.00±0.25	1.83±0.14	6.63	*
Echinopora gemmacea	5.58±0.38	2.33±0.14	2.08±0.80	4.25±1.64	9.43	*

Significance levels are indicated measured at $p = 0.05$, ns = not significant.

Table 6. Microdensity ($g\ cm^{-3}$) for selected coral species from Kenyan protected reefs and a summary of statistical analysis. Values are means (with standard deviations)

Species	Malindi	Watamu	Mombasa	Shimoni	F	p
Acropora sp	2.66±0.31	2.63±0.13	2.06±0.10	2.58±0.15	1.25	ns
Acropora humilis	2.42±0.26	2.68±0.09	2.28±0.38	2.65±0.09	1.95	ns
Acropora robusta	2.62±0.21	2.81±0.16	2.13±0.15	2.75±0.08	26.1	*
Porites lutea	2.79±0.01	2.55±0.12	2.33±0.49	2.65±0.18	1.14	ns
Porites rus	2.73±0.08	2.12±0.10	2.85±0.13	2.87±0.13	30.5	*
Porites cylindrica	2.70±0.45		2.41±0.10	2.81±0.19	1.56	ns
Pocillopora eydouxi	2.42±0.14	2.45±0.06	2.40±0.04	2.90±0.21	10.7	*
Pocillopora damicornis	2.67±0.47	2.31±0.12	2.26±0.21	2.56±0.20	1.46	ns
Montipora digitata	2.03±0.06		2.45±0.07	2.59±0.25	9.93	*
Favia sp	2.74±0.07	2.49±0.35	2.55±0.11	2.31±0.08	2.54	ns
Favites sp	2.60±0.17	2.59±0.09	2.38±0.20	2.64±0.07	1.97	ns
Pavona clavus	2.33±0.22	2.51±0.32	2.67±0.26	2.49±0.08	0.99	ns
Pavona decussata	2.54±0.23		2.57±0.11	2.67±0.21	1.59	ns
Alveopora fenestrata	2.56±0.39	2.41±0.39	1.63±0.15	2.32±0.29	5.45	*
Galaxea fascicularis	2.58±0.21	2.51±0.06	2.54±0.11	2.77±0.07	2.51	ns
Echinopora gemmacea	2.57±0.11	2.57±0.22	2.37±0.39	2.61±0.01	0.59	ns

Significance levels are indicated measured at $p = 0.05$, ns = not significant.

Table 7. Porosity (%) for selected coral species from Kenyan protected reefs and a summary of statistical analysis. Values are means (with standard deviations)

Species	Malindi	Watamu	Mombasa	Shimoni	F	p
Acropora sp	50.5±9.5	35.9±4.5	39.0±7.2	37.1±10.0	2.01	ns
Acropora humilis	47.3±7.8	47.2±4.8	22.8±10.4	48.7±2.2	9.35	*
Acropora robusta	52.1±9.1	53.7±3.9	26.1±6.7	45.4±9.1	8.49	*
Porites lutea	68.2±23.4	45.9±2.8	46.5±7.0	60.5±3.6	5.76	*
Porites rus	67.3±1.01	38.5±4.6	54.6±9.6	61.7±2.9	15.1	*
Porites cylindrica	63.9±12.7		54.5±3.9	64.5±6.8	1.28	ns
Pocillopora eydouxi	43.2±3.3	35.6±1.1	36.7±8.1	43.7±9.0	1.71	ns
Pocillopora damicornis	48.9±15.2	38.3±6.4	39.9±8.9	40.0±5.0	0.75	ns
Montipora digitata	52.8±6.2		22.1±2.9	59.5±4.7	43.8	*
Favia sp	65.9±3.5	53.9±3.3	56.1±5.7	59.8±2.9	4.15	*
Favites sp	68.9±16.8	61.1±11.2	39.9.5±3.2	52.4±2.8	4.42	*
Pavona clavus	45.4±11.0	43.9±5.9	63.3±3.9	49.0±1.7	5.35	*
Pavona decussata	56.6±4.5		40.4±3.1	61.8±6.31	5.88	*
Alveopora fenestrata	66.1±2.3	72.1±10.1	45.7±5.8	62.7±9.2	8.77	*
Galaxea fascicularis	74.5±10.5	52.7±9.2	48.5±4.5	63.2±8.5	5.65	*
Echinopora gemmacea	33.0±2.9	46.2±1.5	39.9±8.9	40.5±6.0	2.77	ns

Significance levels are indicated by: * < 0.05, ns = not significant.

Table 8. Coral skeleton strength (MPa) for selected species from Kenyan protected reefs and a summary of statistical analysis. Values are means (with standard deviations)

Species	Malindi	Watamu	Mombasa	Shimoni	F	p
Acropora sp	19.4±6.5	33.6±5.7	30.2±8.9	33.4±13.5	1.58	ns
A. humilis	21.9±7.2	21.5±4.3	59.1±24.2	20.0±1.8	6.59	*
A. robusta	18.1±6.6	16.5±2.6	50.4±13.0	23.8±8.8	10.04	*
Porites lutea	11.5±7.4	19.1±2.1	20.4±9.0	12.5±1.7	1.70	ns
Porites rus	9.5±0.4	30.4±5.8	16.6±6.5	11.9±1.4	13.5	*
Porites cylindrica	11.8±6.2		15.9±2.6	10.8±2.9	1.22	*
Pocillopora. eydouxi	25.0±3.3	33.7±1.4	33.3±10.2	25.6±9.9	1.26	ns
Pocillopora. damicornis	22.3±13.3	30.9±8.2	29.5±9.4	28.6±5.7	0.21	ns
Montipora digitata	17.3±4.4		62.6±0.4	13.1±2.6	245.3	*
Favia sp	10.2±1.4	16.3±2.3	15.1±3.2	12.9±1.5	4.59	*
Favites sp	15.5±2.9	18.3±2.4	24.7±3.2	17.3±1.9	8.62	*
Pavona clavus	24.4±5.6	24.6±5.8	12.7±0.9	19.7±1.4	4.21	*
Pavona decussatta	14.7±2.9		27.9±3.5	12.1±2.9	24.1	*
Alveopora fenestrata	9.9±0.9	8.3±3.5	23.0±5.5	11.9±4.6	8.44	*
Galaxea fascicularis	7.5±2.9	17.8±7.1	20.3±3.7	11.7±4.3	4.49	*
Echinopora. gemmacea	37.5±4.2	22.1±1.4	29.5±7.5	28.3±1.9	4.25	*

Significance levels are at $p < 0.05$. Values obtained from the relationship between porosity and strength of carbonate materials.

3.2.4 Coral Skeleton Properties and Environmental Factors

All skeletal parameters in the sediment-impacted reef were adversely affected by the presence of terrestrial input (sediments and nutrients) and there was a fairly strong inverse relationship between bulk density and porosity (Figure 4) but strong direct proportionality between skeleton strength with bulk density and matrix volume (Figure 5). Coral microdensity was however poorly correlated with coral matrix volume, porosity and skeletal strength (Table 9) suggesting that variations in coral strength found in this study were mostly and to a large extent mainly influenced by the variation in bulk density and matrix volume rather than by the variations in microdensity of coral skeleton. From these observations we can infer that the arrangement of fundamental coral skeleton elements were the main cause of density variations unlike other studies where skeletal chemistry was suggested as the main reason. Our results show that elevated sediment levels alone or in combination with nutrients negatively impact mechanical properties of coral skeleton and that water motion may play an important role in mitigating these negative impacts. Overall, our study reveals that disturbances of terrestrial origin not only cause declines in calcification rates, but can also cause morphological and mechanical changes on coral skeletons to more porous and potentially weaker and fragile forms, consequently undermining the integrity and biodiversity of coral reefs.

Table 9. Correlation analysis between coral skeleton parameters from four Kenyan protected lagoons. Relationships were fitted to the model $y=a+bx$ with sites arranged in order of decreasing sediment influence

Reef	Code	n	Variables		a	b	r^2	r	p
			x	y					
Malindi	MLD	57	Density	Porosity	1.99	-0.02	0.64	0.80	*
			Density	Strength	0.71	0.03	0.73	0.86	*
			Microdensity	Strength	3.45	0.05	0.19	0.43	*
			Matrix vol	Microdensity	1.71	0.13	0.14	0.12	ns
			Microdensity	Porosity	3.96	6.70	0.25	0.50	*
Watamu	WTM	48	Density	Porosity	2.45	-0.02	0.82	0.91	*
			Density	Strength	0.80	0.02	0.64	0.79	*
			Microdensity	Strength	2.41	0.003	0.01	0.11	ns
			Matrix vol	Microdensity	0.41	0.69	0.20	0.04	ns
			Microdensity	Porosity	2.29	0.003	0.001	0.10	ns
Mombasa	MSA	48	Density	Porosity	1.95	-0.02	0.75	0.87	*
			Density	Strength	0.86	0.02	0.73	0.86	*
			Microdensity	Strength	2.73	0.02	0.39	0.63	*
			Matrix vol	Microdensity	3.30	-0.45	0.03	0.18	ns
			Microdensity	Porosity	1.4	0.2	0.44	0.67	*
Shimoni	SHM	48	Density	Porosity	2.62	-0.06	0.89	0.95	*
			Density	Strength	0.66	0.03	0.47	0.69	*
			Microdensity	Strength	2.59	0.001	0.001	0.03	ns
			Matrix vol	Microdensity	24.3	- 0.71	0.02	0.14	ns
			Microdensity	Porosity	2.64	-0.002	0.02	0.04	ns

n = number of individuals, r^2, Pearson's coefficient of determination, r, Pearson's correlation coefficient, ns not significant, *, $p < 0.001$).

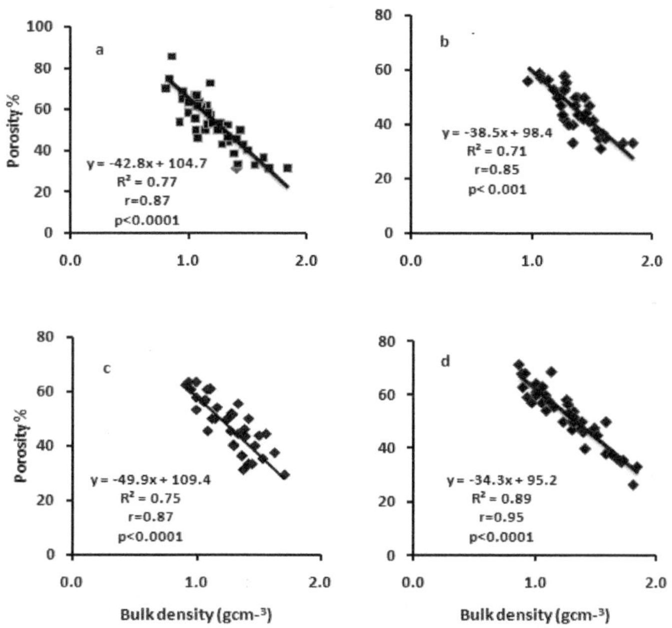

Figure 4. Relationship between coral skeleton porosity and bulk density for selected coral species from (a) Malindi, (b) Watamu, (c) Mombasa and (d) Shimoni protected reefs. Included are regression and correlation coefficients and significance tests

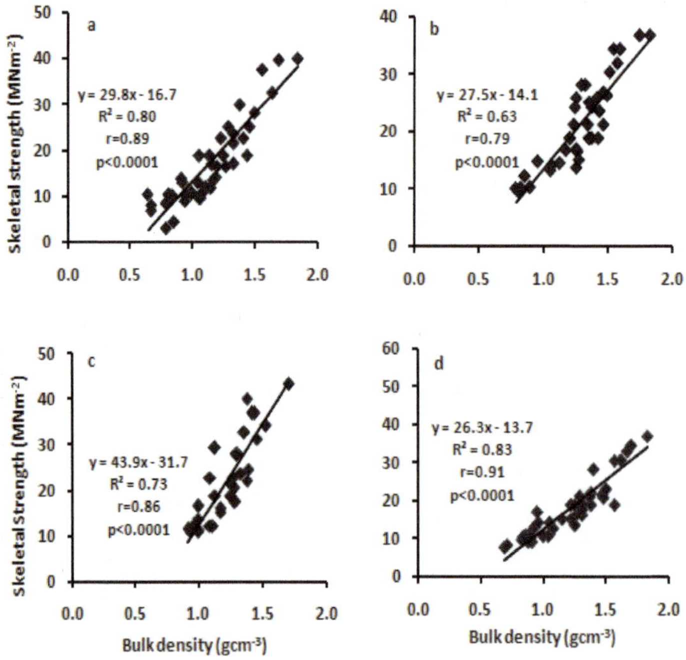

Figure 5. Relationship between strength and the bulk density of coral skeleton from (a) Malindi, (b) Watamu, (c) Mombasa and (d) Shimoni protected reefs, included regression and correlation coefficients and significance tests

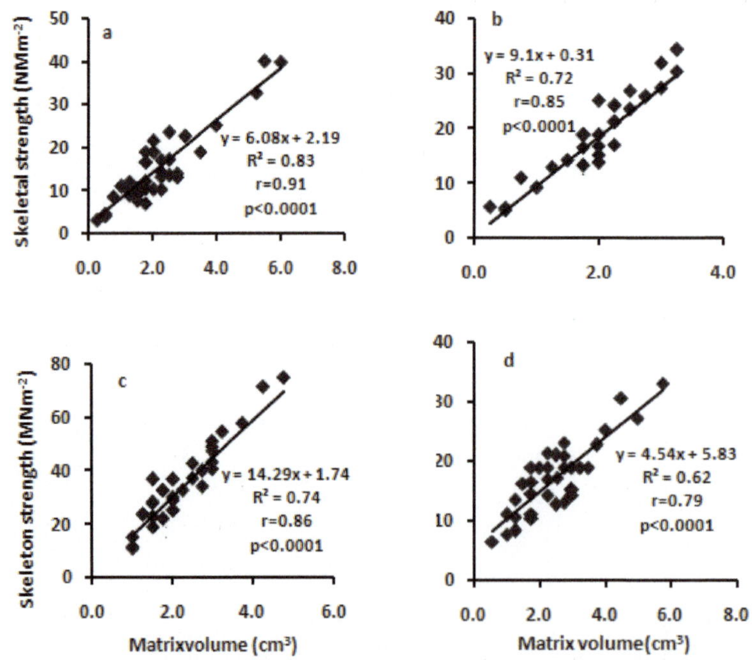

Figure 6. Relationship between strength and the matrix volume of coral skeleton from (a) Malindi, (b) Watamu, (c) Mombasa and (d) Shimoni protected reefs, included regression and correlation coefficients and significance tests

4. Discussion

The future of corals reefs is threatened by global climate change events associated with increased ocean temperature, reduced seawater *p*H (ocean acidification) and local environmental degradation (Fantazzini et al., 2015; de Carlo, Gaetani, Holcomb & Cohen, 2015; Collard et al., 2016). Predicting the future existence and management of coral reefs therefore requires knowledge of the role played by (manageable) local environmental factors on coral calcification and the integrity of carbonate skeletons and reefs. In the present study significant variations were detected in coral skeletal properties as a consequence of increased terrestrial input. Skeleton property variations found in this study are similar to those reported previously by other workers (Roche, Adel, Johnson & Perry, 2011; Cordero, 2013; Silbinger, Guadayol, Thomas, & Donahue, 2014; Tambutté et al., 2015; De Carlo et al., 2015). However, Walsh, Brading, Suggett and Smith. (2012) found no differences in coral skeleton characteristics from high and low turbid reefs and attributed this to branch orientation towards light (microhabitat characteristics).

Perhaps the most striking result from the present study is the consistently observed low bulk density, skeletal strength and high porosity of coral skeletons in the reef exposed to river discharge, similar to other reports on corals growing in disturbed sites (Carricart-Ganivet & Merino, 2001; Ammar, Mohammed & Mahmoud, 2005; Dunn, Sammarco & LaFleur Jr, 2012; Fantazzini et al., 2013). This may be attributed to compromised calcification process leading to the deposition of skeleton material of lower density and concomitant increase in linear growth ("stretching modulation" or Janus effect) as an adaptation to reduced resources, light and space (Carricart-Ganivet & Merino, 2001; Amman et al., 2005; Brahmi, Kopp, Domart-Coulon, Stolarski & Meibom, 2012; Cordero, 2013). Alternatively, this may also reflect growth-induced effects of higher levels of nutrients (Edinger et al., 2000). Several mechanisms have been proposed to explain the pathways involved in the detrimental effects of terrestrially derived disturbances and river discharge on coral skeletons (Edinger et al., 2000; Roche et al., 2011; Dunn et al., 2012; Cordero, 2013). Under optimum conditions of clear and oligotrophic waters, coral skeleton density approaches that of pure aragonite (2.94 gcm^{-3}) but skeletal material alterations e.g. from unstable aragonite (ρ = 2.95 gcm^{-3}) to stable calcite (ρ = 2.60-2.80 gcm^{-3}) or dolomitization (ρ = 2.90 gcm^{-3}) due to changing environmental variables has been shown to highly influence $CaCO_3$ depositional cycles (Caroselli et al., 2011; Roche et al., 2011; Browne, Tay, Low, Larson, & Todd, 2015), skeletal architecture, structure and composition (Chamberlain 1978; Belda, Cuff & Yellowlees, 1993; Spiske, Böröcz, & Bahlburg 2008; Godinot et al., 2011; Dunn et al., 2012), overall calcification and carbonate production, (Lewis et al., 2012; Cordero, 2013). The resultant consequences of this on reef diversity and productivity as well as the livelihood of adjacent coastal communities have already been highlighted (McClanahan & Cinner, 2012).

The present results revealed three major environmental factors mediating changes in coral skeleton properties, namely; hydrodynamic energy, nutrient and sediment/turbidity levels. Malindi and Watamu reefs exhibited almost similar nutrient and TSS characteristics, however, terrestrial influence was greater in Malindi relative to all other reefs (high organic and acid insolubles content), possibly explaining the significantly compromised skeleton properties observed in Malindi corals. The combined effect of nutrient and sediment (Malindi reef) caused greater (negative) impact on coral mechanical properties relative to reefs impacted by either nutrients or high hydrodynamic energy alone. Stronger (less porous, denser) coral skeletons measured in Mombasa and Watamu relative to the Shimoni 'pristine' reefs likely resulted from the high hydrodynamic energy and low turbidity/sediment levels measured at these sites (Samuel and Monismith, 2013; Baldock et al., 2014, Lowe & Falter, 2015) compared to Shimoni reef despite the proximity to mangrove creeks and potential nutrient impacts. This is similar to the results of Sawall et al. (2011) who found that eutrophication was not always a major influence on coral performance although Dunn et al. (2012) measured reduced coral density under increased nutrient levels in a controlled laboratory experiment. Our results suggest that sediment-nutrient interaction is a major synergistic factor in influencing coral skeletal properties even in the presence of relatively high hydrodynamic energy although the impacts of nutrients levels seem to be counteracted by high hydrodynamics (Mombasa and Watamu data). However, Stambler, Popper, Dubinsky and Stimson (1991) observed no changes in the strength of carbonate skeletons in a water motion gradient probably because coral skeleton mechanical integrity is determined mainly by the strength of reef substrate to which corals are attached (J. Madin, Dell, E. Madin, & Nash, 2013).

High nutrient and sediment levels have both been reported to negatively impact calcification (Edinger et al., 2000; Dunn et al., 2012; Godinot, Ferrier-Pages, Montagna & Grover, 2011; Godinot et al., 2012) via their action as crystallization inhibitors ("crystal poison") in the $CaCO_3$ crystal production and arrangement process (Cuif, Dauphin, Nehrke, Nouet & Perez-Huerta, 2012; Venn et al. 2013; Mass et al., 2014). By distorting the orderly array of $CaCO_3$ crystal packing (Venn et al. 2013; Mass et al., 2014), nutrients have been reported to modify the number and thickness of individual coral skeletal element as well as the arrangement, structure and chemistry of deposited calcium carbonate crystals (Belda et al., 1993; Mass, Drake, Peters, Jiang, & Falkowski, 2014) resulting in the production of less dense, more porous and structurally weaker calcium carbonate/calcium phosphate skeletons (Edinger et al., 2000; Godinot et al., 2011; Dunn et al., 2012; Madin et al., 2013) similar to present observation in Malindi reefs. Such skeleton pattern modifications have also been associated with increased soft tissue growth, zooxanthellae population and coral-zooxanthellae symbiotic activities (Edinger et al., 2000; Dunn et al., 2012; Tanaka, Inoue, Nakamura, Suzuki, & Sakai, 2014), CO_2 utilization and production of OH− ions (Bednarz, Neumann, Niggl, & Wild, 2012). Increased photosynthetic activities not only reduce available energy for calcification but also disrupt the normal and stable intra-coelenteron *p*H.

Accordingly, increased soft tissues growth must be accommodated by a similar increase in skeleton growth. If this happens without a concomitant increase in calcification rate, porous skeletons are thus produced to the detriment of skeleton density and strength (Belda et al., 1993; Dunn et al., 2012). Further, increased OH− ion concentration in coral's internal environment must be upregulated or counteracted by the generation of additional H+ ions from increased calcification (Laurent, Tambutte S, Tambutte E, Allemand, & Venn, 2013; Venn et al., 2013) causing acidification of the coral internal environment and subsequent reduction in calcification and skeleton density and strength. This is of particular concern considering future raises in ocean acidification (Doropoulos et al., 2012; Tambutté et al., 2015), frequency and intensity of high energy hydrodynamic events (Spiske et al., 2008) and increases in bioeroders populations (De Carlo et al., 2015). However, the creation of complex 3-D internal habitats by bioeroders has been shown to be important in coral host-symbiot dynamics and ultimately on the survival of the coral holobiont (Yost et al., 2013). Although the specific nutrient species involved in the inhibition of $CaCO_3$ precipitation have not yet been identified, phosphate and polyphosphate groups have been suggested as the likely candidates (Belda et al., 1993; Osinga et al., 2011; Hoac, Kiffer-Moreira, Millan & McKee, 2013) due to their structural similarity with aragonite crystals. Depressed skeleton mechanical integrity observed in Malindi reefs are similar to those observed elsewhere in polluted waters (Edinger et al., 2000; Dunn et al., 2012; Ezzat, Towle, Irisson, Langdon & Ferrier-Pages, 2016) and may be similarly associated with (a) decreased light energy for photosynthesis and calcification, (b) metabolic drain on corals (photosynthesis versus respiration) causing reduced available energy for calcification and (c) heterotrophy on increased particulate matter enhancing skeletal and tissue growth without a corresponding increase in calcification rate (Edinger et al., 2000; Carricart-Ganivet and Merino, 2001; Fantazzini et al., 2015). These scenarios may possibly explain the observed differences between Malindi and Watamu reefs. Malindi reef sediments consistently exhibited high proportions of organic matter and acid insolubles compared to the rest of the studied reefs, a positive indicator of increased terrestrial influence as well as a potential heterotrophic energy source.

Despite the high hydrodynamic energy measured in Malindi reef, coral skeletons here were found to be relatively more porous contrary to previous results (Madin et al., 2013; Baldock et al., 2014). We hypothesize that the hydrodynamic energy levels in Malindi were not high enough to elicit reduction in skeleton porosity through thickening or counteract the (negative) impact imposed by high nutrient and sediment levels, hence the low skeleton porosity. On the other hand, improved skeleton properties measured in branching relative to other growth forms may be an adaptation to the comparatively greater bending forces experienced by branching corals (Chamberlain, 1978; Walsh et al., 2012) coupled with their ability to increase strength by filling skeletal voids with organic cement as they age (Roche et al., 2011), unlike massive corals that have their strength provided for mainly by their geometry. Further, Walsh et al. (2012) found massive corals to have greater variability within skeleton variables (skeletal density, porosity and microdensity) and also deposit aragonite of significantly lower density than that in the skeletons of branching corals. The effects of these three environmental factors on skeleton mechanical strength may be appreciated by comparing reef pairs; hydrodynamics (Mombasa and Shimoni), nutrients (Watamu and Mombasa) and sediment (Malindi and Watamu). These comparisons suggest that the coral skeleton modifications observed in sediment-exposed Malindi reef are to a greater extent the result of high sediment levels due to Sabaki River discharge other than any of the other factor measured in the present study. Our study revealed minimal influence of microdensity on skeletal properties compared to bulk density suggesting that variations in skeletal parameters observed were probably due to changes at the meso-architectural level. This is similar to the results of other workers (Dodge et al., 1993; Bucher et al., 1998; Cordero, 2013) implying that terrestrially-derived stressors affect the arrangements of coral colony fundamental units (corallites including septa, thecae and dissepiments).

Whereas inherent and environmental factors may induce species-specific and habitat variations in skeletal properties respectively, care should be exercised when comparing results from different studies due to a number of reasons. Sampling and measurement techniques employed, and the size and position of specimen on coral colony may affect estimated skeletal property values (Roche et al., 2011; Dunn et al, 2012). For example, although larger coral specimens usually produce low measurement errors they are also likely to give low density/high porosity values due to the presence of more borings. Differences between findings may also emanate from the specimen's relation to individual calices, seasonal banding or relative position on colony (Barnes & Devereaux, 1988; Cordero, 2013) as well as age difference in coral samples (Caroselli et al., 2011; Roche et al., 2011). Moreover, compared to other techniques, the wax method used presently has several sources of limitation: (1) trapped air in skeleton voids during submersion tends to cause underestimation of porosity, (2) controlling viscosity of wax and size of wax coating on samples, (3) incomplete hydrated enclosed voids introduce errors in porosity calculations and (4) possible removal of acetone soluble skeletal structural compounds (Barnes & Devereux, 1988; Bucher et al., 1998).

Predicted increases in ocean pH (acidification) associated with increased atmospheric CO_2 levels and other climate change events (such as high SST and sea level) will likely cause weaker and more vulnerable coral skeletons (Kleypas et al., 1999: Doropolous et al., 2012; Tambutté et al., 2015; Collard et al., 2016) in future, immensely interfering with reef ecology and the provision of goods and services. Although declines in density and strength may sometimes be beneficial at the species level (Chamberlain, 1978; Schuhmacher 1984; Walsh et al., 2012), however, at the community level this may cause nearshore reefs, where local stresses are already a great concern, to be of reduced biodiversity, productivity and protective value (De Carlo et al., 2015; Ezzat et al., 2016). It is also worth noting that the relationship between porosity and strength is exponential (Roche et al., 2011; Dunn et al. 2012) hence coral skeletons in the low porosity-high strength end of the spectrum would be more vulnerable to synergistic effects of boring organisms and hydrodynamic forces (Chamberlain 1978; Maina et al., 2013; Bartley et al., 2014; Ramos-Sharrón et al., 2015) posing great concern to reef areas already weakened by anthropogenic perturbations.

This study results indicate that near shore reefs are under severe threat from local environmental factors and are likely to suffer most in future from climate change events such as ocean acidification, increased SST and high energy hydrodynamic events. However, it should be noted that not all reefs are at risk of degradation from disturbances, since exposure and site-specific conditions also determine susceptibility. Poorly flushed and shallow reefs close to river mouths, with disturbance history and low abundances of herbivores are the most vulnerable (Wooldridge et al., 2011). Disturbance vulnerability may also depend on the species composition of a reef as a result of the species-specific response of skeleton properties to environmental change. These species-specific responses have been attributed to different clades of *Symbiodinium sp* or other potential factors including the presence or absence of known or unknown enzymes e.g., Ca^{2+} -ATPase or carbonic anhydrase (Lesser, Weis, Patterson, & Jokiel, 1994, Goiran Al-Moghrabi, Allemand, & Jaubert, 1996, McClanahan et al., 2014). Previous studies have shown that reefs dominated by corals with weaker skeletons will suffer the severest losses in the event of hydrodynamic disturbances, elevated SST, CO_2 or increased bioeroder populations (Spiske et al., 2008; Reyes-Nivia et al., 2013; Madin et al., 2013;Wiedenmann et al.2013; D'Angelo & Wiedenmann, 2013; Perry et al., 2014; Perry & Harborne, 2016). Understanding the dynamics and interactions of environmental disturbances, their impact mechanisms and the tolerance/adaptability as well as resilience among coral species is therefore key in deriving focused and effective management plans for the future of healthy and productive reefs. For instance, the average strength, composition and abundance of species may be useful in ranking reefs according to their relative susceptibility and overall risk of damage (Craik, Kenchington, & Kelleher, 1990). Exposure, type and stress frequency data and the proportion of colonies impacted, combined with information on injury regeneration and fragment survival would in turn aid in evaluating the recovery process of reef assemblages. Identification of the susceptibility of corals to local stresses via proxies such as skeletal characteristics will therefore enable coral managers and conservationist to direct attention toward areas most at need and those with potential to become refugia in the future (Walsh et al 2012; Wiedenmann et al., 2013) providing shelter and maintenance of genetic diversity. This is especially important since it is projected that only minor benefits will accrue from current carbon emission reduction strategies to coral reef health, and will be noticeable only in the long term, probably 50-60 years after drastically reducing emissions (Donner, Skirving, Little, Oppenheimer, Hoegh-Guldberg, 2005; Ortiz et al., 2013; 2014). Consequently, appropriate watershed management strategies combined with reef and fisheries management options would be expected to alleviate the effects of terrestrially derived disturbances and have the potential to minimize predicted large-scale coral reef degradations arising from intense and frequent global climate change events.

Acknowledgement

This research was kindly and generously supported by grants from the British Ecological Society (BES), the World Wildlife Fund (WWF), the International Development Research Centre (IDRC) Doctoral Research Award and the Simon Fraser University (SFU) President's Award to the first author in support of his PhD program. Permission for research in the marine parks was provided by the Kenya Wildlife Service (KWS). We are also grateful to the logistic support provided by KWS rangers and laboratory and field assistance from the staff of Kenya Marine and Fisheries Research Institute (KMFRI). Excellent fieldwork was provided by Masudi J. Zamu and Anthony M. Nzioka. All authors participated in the birth of the concept, project and experimental design and development of manuscript. Data collection was done by the first author.

References

Allemand, D., Tambutté, É., Zoccola, D., & Tambutté, S. (2011). Coral calcification, cells to reefs. In *Coral reefs: an ecosystem in transition* (pp. 119-150). Springer, Netherlands. http//dx.doi.org/10.1007/978-94-007-0114-4_9

Ammar, M. S. A., Mohammed, T. A., & Mahmoud, M. A. (2005). Skeletal density (strength) of some corals in an actively flooding and a non-flooding site, South Marsa Alam, Red Sea, Egypt. *Journal of Egyptian-German Society of Zoology, 46D*, 125-139

Baldock, T. E., Karampour H., Sleep, R., Vyltla A., Albermani F., Golshani, A., Callaghan, P., ... Mumby, P. J. (2014). Resilience of branching and massive corals to wave loading under sea level rise–A coupled computational fluid dynamics-structural analysis. *Marine Pollution Bulletin, 86*(1), 91-101. http//dx.doi.org/10.1016/j.marpolbul.2014.07.038

Barnes, D. J., & Devereux, M. J. (1988). Variation in skeletal architecture associated with density banding in hard coral *Porites*. *Journal of Experimental Marine Biology and Ecology, 121*, 37-54 http//dx.doi.org/10.1016/0022-0981(88)90022-6

Bartley, R., Bainbridge, Z. T., Lewis, S. E., Kroon, F. J., Wilkinson, S. N., … Silburn, D. M. (2014). Relating sediment impacts on coral reefs to watershed sources, processes and management: A review. *Science of the Total Environment, 468*, 1138-1153. http://dx.doi.org/10.1016/j.scitotenv.2013.09.030

Bednarz, V. N., Naumann, M. S., Niggl, W., & Wild, C. (2012). Inorganic nutrient availability affects organic matter fluxes and metabolic activity in the soft coral genus *Xenia*. *Journal of Experimental Marine Biology and Ecology, 215*(20), 3672-3679. http//dx.doi.org/10.1242/jeb.072884

Belda, C. A., Cuff, C., & Yellowlees, D. (1993). Modifications of shell formation in the giant clam *Tridacna gigas* at elevated nutrient levels in sea water. *Marine Biology, 117*, 251-257. http//dx.doi.org/10.1007/BF00345670

Brahmi, C., Kopp, C., Domart-Coulon, I., Stolarsk, I. J., & Meibom A. (2012). Skeletal growth dynamics linked to trace-element composition in the scleractinian coral *Pocillopora damicornis*. *Geochimica et Cosmochimica Acta, 99*, 146-158. http://dx.doi.org/10.1016/j.gca.2012.09.031

Browne N. K., Tay, J. K., Low, J., Larson, O., & Todd, P. A. (2015). Fluctuations in coral health of four common inshore reef corals in response to seasonal and anthropogenic changes in water quality. *Marine Environment Research, 105*, 39-52. http://dx.doi.org/10.1016/j.marenvres.2015.02.002

Bucher, D. J., Harriot, V. J., & Roberts, L. G. (1998). Skeletal microdensity, porosity and bulk density of acroporid corals. *Journal of Experimental Marine Biology Ecolology, 228*, 117-136. http://dx.doi.org/10.1016/S0022-0981(98)00020-3

Buddemeir, R. W., & Kinze III, R. A. (1975). The chronometric reliability of contemporary corals. In G. D. Rosenberg & S. K. Runcorn (Eds.) *Growth rhythms and history of Earth's rotation* (pp. 135-147). John Wiley and Sons, London Uk.

Buddemeir, R. W., Maragos, J. E., & Knutson, D. W. (1974). Radiographic studies of reef coral exoskeletons: rates and patterns of coral growth. *Journal of Experimental Marine Biology Ecology, 14*, 179-20 http://dx.doi.org/10.1016/0022-0981(74)90024-0

Cameron, C. M., & Edmunds, P. J. (2014). Effects of simulated fish predation on small colonies of massive *Porites spp*. and *Pocillopora meandrina*. *Marine Ecology Progress Series, 508*, 139-148. http://dx.doi.org/10.3354/meps10862

Carilli, J., Donner, S. D., & Hartmann, A. C. (2012). Historical temperature variability affects coral response to heat stress. *Plos One, 7*(3), e34418. http://dx.doi.org/10.1371/journal.pone.0034418

Caroselli, E., Prada, F., Pasquini, L., Marzano, F. N., Zaccanti, F., Falini, G., …. Goffredo S. (2011). Environmental implications of skeletal microdensity and porosity variations in two scleractinian corals. *Zoology, 114*, 255-264. http://dx.doi.org/10.1016/j.zool.2011.04.003

Carricart-Gavinet, J. P., & Merino, M. (2001). Growth response of the reef building coral *Montastrea annularis* along a gradient of continental influence in Southern Gulf of Mexico. *Bulletin of Marine Science, 66*(1), 133-146.

Chamberlain, J. A. Jr. (1978). Mechanical properties of coral skeleton: compressive strength and adaptive resistance. *Paleobiology, 4*, 419-435. http://dx.doi.org/10.1017/S0094837300006163

Chen, T., Li, S., & Yu, K. (2013). Macrobioerosion in Porites corals in subtropical northern South China Sea: a limiting factor for high-latitude reef framework development. *Coral Reefs, 32*(1), 101-108. http://dx.doi.org/10.1007/s00338-012-0946-4

Chindapol, N., Kaandorp, J. A., Cronemberger, C., Mass, T., & Genin, A. (2013). Modelling growth and form of the scleractinian coral *Pocillopora verrucosa* and the influence of hydrodynamics. *PLoS Computational Biology, 9*(1), e1002849. http://dx.doi.org/10.1371/journal.pcbi.1002849

Collard, M., Rastrick, S. P., Calosi, P., Demolder, Y., Dille, J., Findlay H. S., ... Dehairs F. (2016). The impact of ocean acidification and warming on the skeletal mechanical properties of the sea urchin Paracentrotus lividus from laboratory and field observations. *ICES Journal of Marine Science: Journal du Conseil, 73*(3), 727-738. http://dx.doi.org/10.1093/icesjms/fsv018

Cooper, T. F., De'ath, G., Fabricius, K. E., & Lough, J. M. (2008). Declining coral calcification in massive *Porites* in two nearshore regions of the northern Great Barrier Reef. *Global Change Biology, 14*(3), 529-538. http://dx.doi.org/10.1111/j.1365-2486.2007.01520.x

Cordero, J. P. (2013). Assessing the impacts of ocean acidification, global warming and terrestrial runoff on the cross-shelf variability of coral calcification in the central Great Barrier Reef. PhD Thesis Australian National University, 175p

Corrège, T. (2006). Monitoring of terrestrial input by massive corals. *Journal of Geochemical Exploration, 88*, 380-383. http://dx.doi.org/10.1016/j.gexplo.2005.08.080

Craik, W. R., Kenchington, R., & Kelleher, G. (1990). Coral-reef management. In Z. Dubinsky (Ed.), *Ecosystems of the world, coral reefs*. Elsevier Science Publishers, NY p453-467.

Cuif, J. P., Dauphin, Y., Nehrke, G., Nouet, J., & Perez-Huerta, A. (2012). Layered growth and crystallization in calcareous biominerals: Impact of structural and chemical evidence on two major concepts in invertebrate biomineralization studies. *Minerals, 2*(1), 11-39. http://dx.doi.org/10.3390/min2010011

D'Angelo, C., & Wiedenmann, J. (2014). Impacts of nutrient enrichment on coral reefs: new perspectives and implications for coastal management and reef survival. *Current Opinion in Environmental Sustainability, 7*, 82-93. http://dx.doi.org/10.1016/j.cosust.2013.11.029

De'ath, G., Fabricius, K. E., Sweatman, H., & Puotinen, M. (2012). The 27–year decline of coral cover on the Great Barrier Reef and its causes. *Proceedings of the National Academy of Science, 109*(44), 17995-17999. http://dx.doi.org/10.1073/pnas.1208909109

DeCarlo, T. M., Cohen, A. L., Barkley, H. C., Cobban, Q., Young, C., Shamberger, K. E., ... Golbuu, Y. (2015). Coral macrobioerosion is accelerated by ocean acidification and nutrients. *Geology, 43*, 7-10., http://dx.doi.org/10.1130/G36147.1

Dodge, R. E., Szmant, A. M., Garcia, R., Swart, P. K., Forester, A., & Leder, J. J. (1993). Skeletal structure basis of density banding in the reef coral *Montastrea annularis*. *Proc 7th International Coral Reef Symposium, 1*, 186-195.

Donner, S. D., Skirving, W. J., Little, C. M., Oppenheimer, M., & Hoegh-Guldberg, O. V. E. (2005). Global assessment of coral bleaching and required rates of adaptation under climate change. *Global Change Biology, 11*(12), 2251-2265.

Doropoulos, C., Ward, S., Diaz-Pulido, G., Hoegh-Guldberg, O., & Mumby P. J. (2012). Ocean acidification reduces coral recruitment by disrupting intimate larval-algal settlement interactions. *Ecology Letters, 15*(4), 338-346. http://dx.doi.org/10.1111/j.1461-0248.2012.01743.x

Dunn, J. G., Sammarco, P. W., & LaFleur Jr, G. (2012). Effects of phosphate on growth and skeletal density in the scleractinian coral *Acropora muricata*: A controlled experimental approach. *Journal of Experimental Marine Biology and Ecology, 411*, 34–44. http://dx.doi.org/10.1016/j.jembe.2011.10.013

Edinger, E. N., Limmon, G. V., Jompa, J., Widjatmokos, W., Heikoop, J. M., & Risk, M. J. (2000). Normal coral growth rates on dying reefs: Are coral growth rates good indicators of reef health? *Marine Pollution Bulletin, 40*(5), 404-425. http://dx.doi.org/10.1016/S0025-326X(99)00237-4

Erler, D. V., Wang, X. T., Sigman, D. M., Scheffers, S. R., Martínez-García, A., & Haug, G. H. (2016). Nitrogen isotopic composition of organic matter from a 168 year-old coral skeleton: Implications for coastal nutrient cycling in the Great Barrier Reef Lagoon. *Earth and Planetary Science Letters, 434*, 161-170. http://dx.doi.org/10.1016/j.epsl.2015.11.023

Ezzat, L., Towle, E., Irisson, J. O., Langdon, C., & Ferrier-Pagès, C. (2016). The relationship between heterotrophic feeding and inorganic nutrient availability in the scleractinian coral *T. reniformis* under a short-term temperature increase. *Limnology and Oceanography, 61*(1), 89-102. http://dx.doi.org/10.1002/lno.10200

Fantazzini, P., Mengoli, S., Evangelisti, S., Pasquini, L., Mariani, M., Brizi, L., ... Levy, O. (2013). Time-Domain NMR study of Mediterranean scleractinian corals reveals skeletal-porosity sensitivity to environmental changes. *Environmental Science Technology, 47*, 12679-12686.

Fantazzini, P., Mengoli, S., Pasquin,i L., Bortolotti, V., Brizi, L., Mariani M, ... Prada, F (2015) Gains and losses of coral skeletal porosity changes with ocean acidification acclimation. *Nature Communications, 6*. http://dx.doi.org/10.1038/ncomms8785

Fleitmann, D., Dunbar, R. D., McCulloch, M., Mudelsee, M., Vuilee, M., McClanahan, T. R., ... Eggins S (2007). East African soil erosion recorded in a 300 year old coral colony from Kenya. *Geophysical Research Letters, 34*, 1-5. http://dx.doi.org/10.1029/2006GL028525

Giesen, W., & van de Kerkhof, K. (1984). The impact of river discharges on a Kenyan coral reef ecosystem-the physical processes. Part II: Effects on the Malindi-Watamu coastal environment. Report No. 194. Laboratory of Aquatic Ecology, Catholic University of Nijmegen, The Netherlands

Godinot, C., Ferrier-Pagès, C., Montagna, P., & Grover, R. (2011). Tissue and skeletal changes in the scleractinian coral *Stylophora pistillata* Esper 1797 under phosphate enrichment. *Journal Experimental Marine Biology and Ecology, 409*, 200-207. http://dx.doi.org/10.1016/j.jembe.2011.08.022

Godinot, C., Tribolle,t A., Grover, R., & Ferrier-Pagès C. (2012). Bioerosion by euendolith decreases in phosphate-enriched skeletons of living corals. *Biogeoscience Discussions, 9*, 2424-2444. http://dx.doi.org/10.5194/BGD-9-2425-2012

Goiran, C., Al-Moghrabi, S., Allemand, D., & Jaubert, J. (1996). Inorganic carbon uptake for photosynthesis by the symbiotic coral/dinoflagellate association I. Photosynthetic performances of symbionts and dependence on sea water bicarbonate. *Journal of Experimental Marine Biology Ecology, 199*(2), 227-248. http://dx.doi.org/10.1016/0022-0981(95)00201-4

Hoac, B., Kiffer-Moreira, T., Millán, J. L., & McKee, M. D., (2013) Polyphosphates inhibit extracellular matrix mineralization in MC3T3-E1 osteoblast cultures. *Bone, 53*(2), 478-486. http://dx.doi.org/10.1016/j.bone. 2013.01.020

Hoegh-Guldberg, O., Mumby, P. J., Hooten, A. J., Steneck, R. S., Greenfield, P., Gomez, E., ... Caldeira, K. (2007). Coral reefs under rapid climatic change and ocean acidification. *Science, 318*, 1737-1742. http://dx.doi.org/10.1126/science.1152509

Inoue, M., Ishikawa, D., Miyaji, T., Yamazaki, A., Suzuki, A., Yamano, H., ... Watanabe, T. (2014). Evaluation of Mn and Fe in coral skeletons (Porites spp.) as proxies for sediment loading and reconstruction of 50 yrs of land use on Ishigaki Island, Japan. *Coral Reefs, 33*(2), 363-373. http://dx.doi.org/10.1007/s00338-014-1128-3

Kitheka, J. U., Nthenge, P., & Obiero, M. (2003) Sabaki estuary sediment transport dynamics and resultant sediment fluxes in the monsoon-driven Malindi Bay, Kenya. LOICZ-START AFRICAT Project Report-II-Sabaki p 1-57.

Kleypas, J. A., Buddemeier R. W., Archer, D., Gattuso, J. P., Langdon, C., & Opdyke B. N. (1999). Geochemical Consequences of Increased Atmospheric Carbon Dioxide on Coral Reefs. *Science, 284*(5411), 118-120. http://dx.doi.org/10.1126/science.284.5411.118

Laurent, J., Tambutté, S., Tambutté, É., Allemand, D., & Venn, A. (2013). The influence of photosynthesis on host intracellular pH in scleractinian corals. *Journal of Experimental Biology, 216*(8), 1398-1404. doi: 10.1242/jeb.082081

Lesser, M. P., Weis, V. M., Patterson, M. R., & Jokiel, P. L. (1994). Effects of morphology and water motion on carbon delivery and productivity in the reef coral, *Pocillopora damicornis* (Linnaeus): Diffusion barriers, inorganic carbon limitation, and biochemical plasticity. *Journal of Experimental Marine Biology and Ecology, 178*, 153-179. http://dx.doi.org/10.1016/0022-0981(94)90034-5

Lewis, S. E., Brodie, J. E., McCulloch, M. T., Mallela, J., Jupiter, S. D., Williams, H. S., ... Matson, E. G. (2012). An assessment of an environmental gradient using coral geochemical records, Whitsunday Islands,

Great Barrier Reef, Australia. *Marine Pollution Bulletin, 65*(4), 306-319. http://dx.doi.org/10.1016/j.marpolbul.2011.09.030

Lowe, R. J., & Falter, J. L. (2015.) Oceanic Forcing of Coral Reefs. *Annual Review of Marine Sciences, 7*, 43-66. http://dx.doi.org/10.1146/annurev-marine-010814-015834

Madin, J. S., Dell, A. I., Madin, E. M. P., & Nash M. C. (2013). Spatial variation in mechanical properties of coral reef substrate and implications for coral colony integrity. *Coral Reefs, 32*(1), 173-179. http://dx.doi.org/10.1007/s00338-012-0958-0

Madin, J. S., Hughes, T. P., & Connolly, S. R. (2012). Calcification, storm damage and population resilience of tabular corals under climate change. *PLoS One, 7*(10), e46637. http://dx.doi.org/10.1007/s00338-012-0958-0

Maina, J., de Moel, H., Zinke, J., Madin, J., McClanahan, T. R., & Vermaat, J. E. (2013). Human deforestation outweighs future climate change impacts of sedimentation on coral reefs. *Nature Communications, 4*, http://dx.doi.org/10.1038/ncomms2986

Mallela, J., Lewis, S. E., & Croke, B. (2013). Coral skeletons provide historical evidence of phosphorus runoff on the Great Barrier Reef. *PloS one, 8*(9), e75663. http://dx.doi.org/10.1371/journal.pone.0075663

Mallela, J., Perry, C. T., & Haley, M. P. (2004). Reef morphology and community structure along a fluvial gradient, Rio Bueno, Jamaica. *Caribbean Journal of Science, 40*(3), 299-311.

Mass, T., Drake, J. L., Peters, E. C., Jiang, W., & Falkowski, P. G. (2014). Immunolocalization of skeletal matrix proteins in tissue and mineral of the coral *Stylophora pistillata. Proceedings of the National Academy of Science, 111*(35), 12728-12733. http://dx.doi.org/10.1073/pnas.1408621111

McClanahan, T. R. & Obura, O. (1995). Status of Kenyan coral reefs. *Coastal Management, 23*, 57-76.

McClanahan, T. R., & Abunge, C. A. (2014). Catch rates and income are associated with fisheries management restrictions and not an environmental disturbances, in a heavily exploited tropical fishery. *Marine Ecology Progress Series, 513*, 201-210. http://dx.doi.org/10.3354/meps10925

McClanahan, T. R., & Cinner, J. E. (2012) Adapting to a changing environment:Confronting the consequences of climate change. Oxford University Press, 193p

McClanahan, T. R., Ateweberhan, M., & Omukoto, J. (2008). Long-term changes in coral colony size distributions on Kenyan reefs under different management regimes and across the 1998 bleaching event. *Marine Biology, 153*(5), 755-768. http://dx.doi.org/10.1007/s00227-007-0844-4

McClanahan, T. R., Starger, C. J., & Baker, A. C. (2014). Decadal changes in common reef coral populations and their associations with algal symbionts (Symbiodinium spp.) *Marine Ecology,* 1-15. http://dx.doi.org/10.1111/maec.12225.

McKibben, B. (2012). The global warming reader: a century of writing about climate change. Penguin Books, New York, NY.

Motai, S., Nagai, T., Sowa, K., Watanabe, T., Sakamoto, N., ... Kawano, J. (2012). Needle-like grains across growth lines in the coral skeleton of *Porites lobata. Journal of Structural Biology, 180*(3), 389-393. http://dx.doi.org/10.1016/j.jsb.2012.09.009

Obura, D. O. (1995). *Environmental stress and life history strategies, a case study of corals and river sediment from Malindi Kenya*. PhD Thesis, University of Miami 326p.

Ortiz, J. C., Bozec, Y. M., Wolff, N. H., Doropoulos, C., & Mumby, P. J. (2014). Global disparity in the ecological benefits of reducing carbon emissions for coral reefs. *Nature Climate Change, 4*(12), 1090-1094.

Ortiz, J. C., González-Rivero, M., & Mumby, P. J. (2013). Can a thermally tolerant symbiont improve the future of Caribbean coral reefs?. *Global change biology, 19*(1), 273-281.

Osinga, R., Schutter, M., Griffioen, B., Wijffels, R. H., Verreth, J. A. J., Shafir, ... Lavorano, S. (2011). The biology and economics of coral growth. *Marine Biotechnology, 13*(4), 658-671. http://dx.doi.org/10.1007/s10126-011-9382-7

Perry, C. T., & Harborne, A. R. (2016). Bioerosion on Modern Reefs: Impacts and responses under changing ecological and environmental conditions. In *Coral Reefs at the Crossroads* (pp. 69-101). Springer Netherlands. http://dx.doi.org/10.1007/978-94-017-7567-0_4

Perry, C. T., Murphy, G. N., Kench, P. S., Edinger, E. N., Smithers, S. G., … Mumby, P. J. (2014). Changing dynamics of Caribbean reef carbonate budgets: emergence of reef bioeroders as critical controls on present and future reef growth potential. In *Proceedings of the Royal Society B-Biological Science, 281*(1796), 20142018. The Royal Society. http://dx.doi.org/10.1098/rspb.2014.2018

Pisapia, C., Anderson, K., & Pratchett, M. S. (2014). Intraspecific variation in physiological condition of reef-building corals associated with differential levels of chronic disturbance. *PloS one, 9*(3), e91529. http://dx.doi.org/10.1371/journal.pone.0091529

Prouty, N. G., Goodkin, N. F., Jones, R., Lamborg, C. H., Storlazzi, C. D., & Hughen, K. A. (2013). Environmental assessment of metal exposure to corals living in Castle Harbour, Bermuda. *Marine Chemistry, 154*, 55-66. http://dx.doi.org/10.1016/j.marchem.2013.05.002

Ramos-Scharrón, C. E., Torres-Pulliza, D., & Hernández-Delgado, E. A. (2015). Watershed-and island wide-scale land cover changes in Puerto Rico (1930s–2004) and their potential effects on coral reef ecosystems. *Science of the Total Environment, 506*, 241-251. http://dx.doi.org/10.1016/j.scitotenv.2014. 11.016

Reyes-Nivia, C., Diaz-Pulido, G., Kline, D., Guldberg, O. H., & Dove, S. (2013). Ocean acidification and warming scenarios increase microbioerosion of coral skeletons. *Global Change Biology, 19*(6), 1919-1929. http://dx.doi.org/10.1111/gcb.12158

Risk, M. J. (2014). Assessing the effects of sediments and nutrients on coral reefs. *Current Opinions in Environmental Sustainability, 7*, 108-117. http://dx.doi.org/10.1016/j.cosust.2014.01.003

Robinson, L. F., Adkins, J. F., Frank, N., Gagnon, A. C., Prouty, N. G., ... van de Flierdt, T. (2014). The geochemistry of deep-sea coral skeletons: A review of vital effects and applications for palaeoceanography. *Deep Sea Research Part II, 99*, 184-198. http://dx.doi.org/10.1016/j.dsr2.2013.06.005

Roche, R. C., Abel, R. A., Johnson, K. G., & Perry, C. T. (2011). Spatial variation in porosity and skeletal element characteristics in apical tips of the branching coral *Acropora pulchra* (Brook 1891). *Coral Reefs, 30*, 195-201. http://dx.doi.org/10.1007/s00338-010-0679-1

Roughael, A. B., & Inglis, G. J. (1997). Impacts of recreational SCUBA diving on sites with different reef topographies. *Biological Conservation, 82*, 329-336. http://dx.doi.org/10.1016/S0006-3207(97)00047-5

Ryshkewitch, E. (1953). Compression strength of porous sintered alumina and zirconia. *Journal of American Ceramic Society, 36*(2), 65-68. http://dx.doi.org/10.1111/j.1151-2916.1953.tb12837.x

Samuel, L. C., & Monismith, S. G. (2013). Drag coefficients for single coral colonies and related spherical objects. *Limnology and Oceanography, 3*(1), 173-181. http://dx.doi.org/10.1215/21573689-2378401

Sawall, Y., Teichberg, M.C., Seemann, J., Litaay, M., Jompa, J., & Richter, C. (2011). Nutritional status and metabolism of the coral *Stylophora subseriata* along a eutrophication gradient in Spermonde Archipelago (Indonesia). *Coral Reefs, 30*(3), 841-853.

Schuhmacher, H. (1984). Reef building properties of *Tubastraea micranthus* (Scleractinia, Dendrpphyliidae), a coral without zooxanthellae. *Marine Ecology Progress Series, 20*, 93-99.

Schuhmacher, H., & Plewka, M. (1981a). Mechanical resistance of reefbuilders through time. *Oecologia, 49*, 279-282

Schuhmacher, H., & Plewka, M. (1981b). The adaptive significance of mechanical properties versus morphological adjustments in skeletons of *Acropora palmata* and *Acropora cervicornis* (Cnidaria, Scleractinia). *Proc 4th International Coral Reef Symposium Manila, 2*, 121-128

Selig, E. R., Casey, K. S., & Bruno, J. F. (2012). Temperature-driven coral decline: the role of marine protected areas. *Global Change Biology, 18*(5), 1561-1570. http://dx.doi.org/10.1111/j.1365-2486.2012.02658.x

Shirai, K., Schöne, B. R., Miyaji, T., Radarmacher, P., Krause, R. A., & Tanabe, K. (2014). Assessment of the mechanism of elemental incorporation into bivalve shells (Arctica islandica) based on elemental distribution at the microstructural scale. *Geochimica et Cosmochimica Acta, 126*, 307-320. http://dx.doi.org/10.1016/j.gca.2013.10.050

Silbiger, N. J., Guadayol, Ò., Thomas, F. I., & Donahue, M. J. (2014). Reefs shift from net accretion to net erosion along a natural environmental gradient. *Marine Ecology Progress Series, 515*, 33-44. http://dx.doi.org/10.3354/meps 10999

Spiske, M., Böröcz, Z., & Bahlburg, H. (2008). The role of porosity in discriminating between tsunami and hurricane emplacement of boulders – A case study from the Lesser Antilles, Southern Caribbean. *Earth and Planet Science Letters, 268*, 384-396. http://dx.doi.org/10.1016/j.epsl.2008.01.030

Stambler, N., Popper, N., Dubinsky, Z., & Stimson, J. (1991). Effects of nutrient enrichment and water motion on the coral, *Pocillopora damicornis*. *Pacific Science, 45*(3), 299-307.http://hdl.handle.net/10125/1396

Tambutté. E., Venn, A. A., Holcomb, M., Segonds, N., Techer, N., Zoccola, D., ... Tambutté, S. (2015). Morphological plasticity of the coral skeleton under CO^2-driven seawater acidification. *Nature Communications, 6*. http://dx.doi.org/10.1038/ncomms8368

Tanaka, Y., Inoue, M., Nakamura, T., Suzuki, A., & Sakai, K. (2014). Loss of zooxanthellae in a coral under high seawater temperature and nutrient enrichment. *Journal of Experimental Marine Biology and Ecology, 457*, 220-225. http://dx.doi.org/10.1016/j.jembe.2014.04.019

Tomiak, P. J., Penkman, K. E. H., Hendy, E. J., Demarchi, B., Murrells, S., Davis, S. A.,....Collins, M. J. (2013). Testing the limitations of artificial protein degradation kinetics using known-age massive *Porites* coral skeletons. *Quaternary Geochronology, 16*, 87-109. http://dx.doi.org/10.1016/j.quageo.2012.07.001

Tribollet, A., & Golubic, S. (2011). Reef Bioerosion: Agents and Processes In Z. Dubinsky & N. Stambler (eds) Coral Reefs: An Ecosystem in Transition, Springer Science + Business Media BV. http://dx.doi.org/10.1007/978-94-007-0114-4_25

van de Kerkhof, K., & Giesen W. (1994). The impact of river discharge on a Kenyan coral reef ecosystem-the physical processes. Part 1: The Athi (Sabaki) river basin. Report No. 193. Laboratory of Aquatic Ecology, Catholic University of Nijmegen, The Netherlands.

van Katwijk, M. M., Meier, N. F., Van Loon, R., Van Hove, E. M., Giesen, W. B. J. T., van der Velde, G., & Den Hartog, C. (1993). Sabaki River sediment load and coral stress: correlation between sediments and condition of the Malindi-Watamu reefs in Kenya (Indian Ocean). *Marine Biology, 117*(4), 675-683. http://dx.doi.org/10.1007/BF00349780

Venn, A. A., Tambutté, E., Holcomb, M., Laurent, J., Allemand, D., & Tambutté, S. (2013). Impact of seawater acidification on *p*H at the tissue–skeleton interface and calcification in reef corals. *Proceedings National Academy Science, 110*(5), 1634-1639. http://dx.doi.org/10.1073/pnas.1216153110

Walsh, S-J., Brading, P, Suggett, D. J., & Smith, D. J. (2012). Working with Nature to Identify Coral Reefs with Increased Environmental Tolerance. *Proceedings 12th International Coral Reef Symposium, Cairns, Australia, 9A*, 1-5

Wiedenmann, J., D'Angelo, C., Smith, E. G., Hunt, A. N., Legiret, F. E., Postle, A. D., & Achterberg, E. P. (2013). Nutrient enrichment can increase the susceptibility of reef corals to bleaching. *Nature Climate Change, 3*(2), 160-164. http://dx.doi.org/10.1038/nclimate1661

Wooldridge, S. A., Done, T. J., Thomas, C. R., Gordon, I. I., Marshall, P. A., & Jones, R. N. (2012). Safeguarding coastal coral communities on the central Great Barrier Reef (Australia) against climate change: realizable local and global actions. *Climatic Change, 112*(3-4), 945-961. http://dx.doi.org/10.1007/s10584 -011-0229-z

Yost, D. M., Wang, L. H., Fan, T. Y., Chen, C. S., Lee, R. W., ... Gates, R. D. (2013). Diversity in skeletal architecture influences biological heterogeneity and Symbiodinium habitat in corals. *Zoology, 116*(5), 262-269. http://dx.doi.org/10.1016/j.zool.2013.06.001

3

Integrated Use of Geological, Geophysical, Radiocarbon and Stable Isotopes Data for Tracing the Conduit Flow Paths in a Small Karstic Aquifer: Poocher Swamp Freshwater Lens, South Australia

Nara Somaratne[1] & Simon Mann[2]

[1] South Australian Water Corporation, 250 Victoria Square, Adelaide, SA 5000, Australia

[2] Zonge Engineering & Research Organization (Australia) Pty Ltd, 39 Raglan Avenue, Edwardstown, SA 5039, Australia

Correspondence: Nara Somaratne, South Australian Water Corporation, 250 Victoria Square, Adelaide, SA 5000, Australia. E-mail: nara.somaratnef@sawater.com.au

Abstract

An integrated approach combining lithological logs, downhole geophysics, electromagnetic survey and the distribution of radiocarbon (^{14}C) and the stable isotopes of water molecules (^{18}O) were used to identify the conduit flow paths of a small freshwater lens. Lost circulation zones, where drilling fluid flows into geological formation instead returning up the annulus recorded during water well drilling, were considered as the major fracture zones. The presence of high porosity zones within boreholes were identified using caliper, gamma and neutron logs. These methods were used to identify the depth intervals at which cavities and the existence of conduit porosity within the boreholes. Transient electromagnetic (TEM) method was used to investigate resistivity anomalies in the profiles along nine pre-determined lines across the freshwater lens. Resistivity anomalies were related to borehole information and other surface features such as sinkholes. Low resistivity zones of the TEM tomography sections had excellent correlation to fracture zones identified during well drilling, and downhole geophysical logs. Similarly, high resistivity zones in the profiles correlate well with the zones of cemented or recrystallized limestone identified from the lithological logs. The interpreted resistivity anomaly accurately depicts depth to watertable at 15-18 m, presence of main fracture zone at 20-25 m, presence of possible conduits flow paths and the cemented or recrystallized limestone below 35 m depth. The ^{14}C, ^{13}C and ^{18}O signatures of the groundwater confirm the presence of conduits and potential pathways of preferential flows. This investigation illustrates the effectiveness using an integrated approach to trace the conduit flow paths in karst aquifers. The information gained from the study is currently being used for the management of the freshwater lens.

Keywords: karst aquifer, groundwater, conduit flow, resistivity tomography, radiocarbon, stable isotopes

1. Introduction

When fresh water lenses are formed in association with high recharge in karstic aquifers, identification of conduit flow paths is critical for sustainable management of the resource. This is particularly important in arid and semi-arid countries as the demand for water is ever increasing for agriculture, industry and community water supplies (WWRD, 2015). Karst aquifers possess both diffuse and conduit flow and varying degrees of connectivity between surface and groundwater over spatial scales that results in complex hydrology (Edwards et al., 2013). The surface-groundwater connection and flow regime must be properly identified and characterized to improve management of karst aquifers. In most karst areas, the majority of groundwater is carried through a subsurface conduit system. Subsurface conduits often connect to land surface via sinkholes, and serve as major pathways for groundwater flow in karst aquifers. Locating conduit flow paths from the surface, however, is one of the most challenging tasks in karst research (Zhu et al., 2011). Exploitation of fresh water lenses that are formed due to point recharge from sinkholes is challenging. This involves identification of conduits pathways which serve groundwater flow in karst system (*ibid*). A number of direct and indirect techniques are used to identify conduit flows. Direct method of identifying conduit pathways is possible during drilling, as it is noticeable when intersecting a large cavity, but not so noticeable if intersecting a small conduit, unless core samples are collected

(Somaratne, 2015a). However, downhole geophysical techniques such as caliper, gamma and neutron logs provide additional information on fracture zones in a borehole. Collection of stratigraphic information by drilling large number of boreholes is prohibitively expensive and not always successful in intercepting conduit porosity zones, hence requires alternative techniques to identify major flow paths.

Geophysical methods are often deployed to help locate voids by mapping variations of physical properties of the subsurface (Zhu et al. 2011), as the conduits can cause significant contrasts of some physical properties that can be detected. Electrical resistivity is one of the commonly used geophysical methods for subsurface void detection, as a water-filled underground void has lower resistivity than most surrounding rocks whereas an air filled void has higher resistivity than most geologic material (Zhu et al., 2011). Geophysical techniques allow large areas to be covered in a short period of time and represent an efficient way of detecting subsurface heterogeneities including voids (Farooq et al., 2012). This provides much more effective and economic way to reduce the overall amount of drillings and to improve locating test drilling for verification (Metwaly & AlFouzan, 2013). The electrical resistivity method is based on assumption that various entities like minerals, solid bedrock, sediments, air and water filled structures have detectable electrical resistivity contrast relative to the host medium (Panek et al., 2010). The electromagnetic tomography involves determination of the subsurface distribution of electrical resistivity. The varying geoelectric response enables to obtain 2D profiles of the resistivity distribution underground to characterize possible discontinuities in the subsurface (Martinez-Lopez et al., 2013).

Geochemical and isotopic data have been used in numerous studies in different karst regions to characterize the flow in carbonate aquifers and degree of surface-groundwater connections (Edwards et al., 2013). Knierim et al. (2015) show that stable isotope analyses are valuable in karst settings. As groundwater flow in karst aquifers is shared between the granular porosity and karst conduits, an indirect way of identification of major flow paths are possible using salinity distributions (Somaratne, 2015a), tracer techniques such as radiocarbon activity (^{14}C) and stable isotope ($\delta^{18}O$) distribution (Somaratne et al., 2016). In a similar study, Einsiedl (2005) analyzed concentration breakthrough curves obtained from a tracer test and time series of environmental tracers to characterize slow and preferential flow in a karst aquifer. Tritium (^{3}H) and chemical tracers (bromide, strontium) have been used in this study. For tracing the young fraction of groundwater mixture, Plummert et al. (1998) used transient tracers and dissolved substances including: chloride, ^{3}H, tritogenic helium-3, chlorofluorocarbons, organic carbon, $\delta^{18}O$. Plummer et al.'s study revealed that concentrations have been affected by dilution and mixing; groundwater mixing fractions have been determined by using dissolved chloride and $\delta^{18}O$. Groundwater chemistry has been used to distinguish groundwater flow paths in different fractured-rock aquifers by Tweed et al. (2005). In their study, differences in groundwater chemistry between aquifers and between shallow and deeper flow systems, highlights inter-aquifer mixing. Nativ et al. (1999) used radioactive and stable isotope to identify and separate groundwater flow components in a karstified aquifer, with the use of ^{14}C to identify mixing of recent and older waters.

Problems arise when attempting to trace conduit flowpaths using a single technique given limitations in the each resulting in large uncertainty. Therefore, in this study, we use an inter-disciplinary integrated approach to study distribution of conduit flow paths in a small fresh water lens in the south-east of South Australia. The study uses geological information obtained by well drilling, downhole geophysical logs, resistivity data from high resolution electromagnetic (EM) in conjunction with radiocarbon and stable isotope distribution within the fresh water lens.

2. Study Site: Poocher Swamp Freshwater Lens

The study site, Poocher Swamp fresh water lens, is located 275 km south-east from Adelaide (Figure 1) and is within the Tatiara catchment. The catchment area extends across the South Australian border into western Victoria and previously described by Somaratne (2014). For brevity, a succinct summary of the study freshwater lens is provided here. The Tatiara catchment area is approximately 500 km^2 (Herczeg et al., 1997) and the Creek flows in a westerly direction through to Poocher Swamp and disappear via sinkholes. The catchment features average annual rainfall ranging from 400 to 500 mm and pan evaporation of 2000 mm. The unconfined aquifer is Murray Group Limestone and contains brackish water with average total dissolved solids (TDS) > 1400 mg.L^{-1}. Freshwater with TDS < 1000 mg L^{-1} occurs at locations where point recharge takes place through sinkholes. The Poocher Swamp freshwater lens, which is the largest of these freshwater plumes that float on brackish water, is a result of flows from Tatiara Creek that enter Poocher Swamp. The major recharge is through two sinkholes located in the northwest section of the swamp (Herczeg et al., 1997). The area encompassed by the 1000 mg.L^{-1} salinity contour comprises approximately 9 km^2 in 2016. Surface water flows in Tatiara Creek is highly irregular, with annual volumes of freshwater (0.05-2)$\times 10^6$ m^3 per year, but on rare occasions up to 19$\times 10^6$ m^3 per year (Figure 2). Groundwater flow direction of the lens is from east to west. The study area is a relatively flat terrain with ground elevations at WRG032 well near the Poocher Swamp (Figure 1) at 69.83 m AHD (Australian Height Datum) while the most down gradient well WRG046 is at 66.24 m AHD. Depth to water varies, about 19 m at WRG032 and 15.6 m at WRG 046, with overall average of the freshwater lens area about 17 m. Currently, an annual volume of 0.6×10^6 m^3 of groundwater is extracted from the freshwater lens for town water supply to Bordertown, approximately 7 km east from the lens.

The unconfined aquifer is part of the Murray Group Limestone aquifer, which comprises a consolidated, highly fossiliferous and fine to coarse bioclastic limestone. Saturated thickness of the limestone unconfined aquifer is approximately 50-60 m. During high flows in the Tatiara Creek, Poocher Swamp overflows through the outlet channel to Scowns sinkholes, about 4 km further west (Figure 1). Based on average annual flow data of Tatiara Creek near Bordertown for 1980-2010 periods, Somaratne (2014) suggested annual recharge to the lens through sinkholes at Poocher Swamp was about 2.5×10^6 m^3, but could not be quantified with any reliable precision.

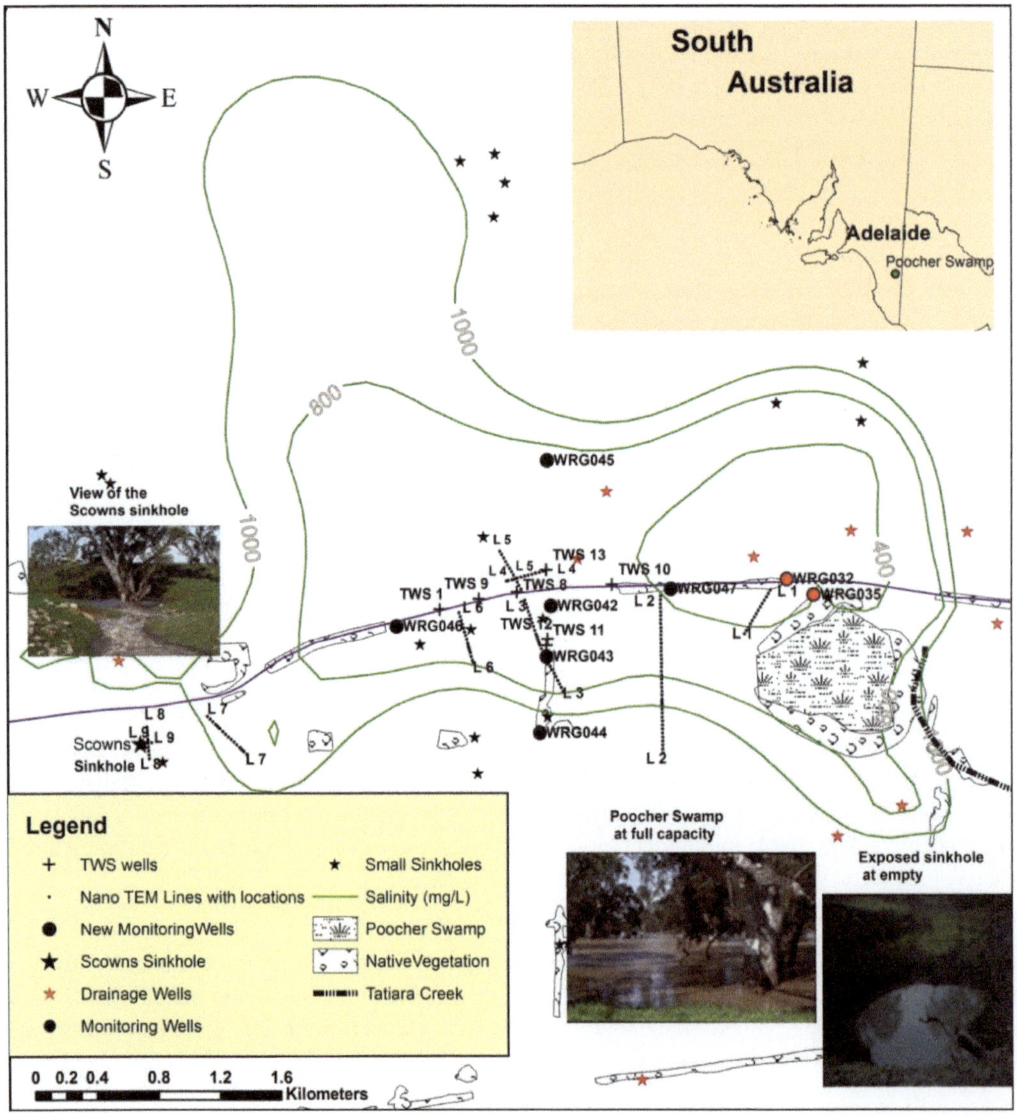

Figure 1. Poocher Swamp freshwater lens

In addition to town water supply, groundwater is also extracted within and from surrounding areas for irrigated agriculture, mainly for pasture, legumes and for viticulture. Dryland grazing is common within the lens area and the surroundings. As such, prudent management of the Poocher Swamp fresh water lens is vital for sustaining local economic activities.

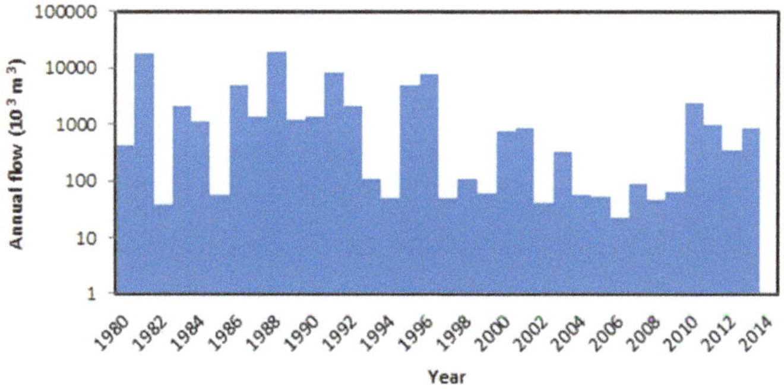

Figure 2. Annual flow of Tatiara Creek measured at Bordertown (in 2014 and 2015, flow measurements were not undertaken)

3. Method

Water samples were collected from Tatiara Creek, Poocher Swamp, monitoring wells and town water supply wells in 2012, 2014 and 2016. For monitoring wells, groundwater samples were collected using micro-purge (low-flow) sampling procedure (Vail, 2011) and for town water supply wells, samples were collected after allowing three-bore volume discharge to ensure samples are representative of the aquifer water. Water samples were preserved according to the standard methods specified in the Australia/New Zealand standard, AS/NZS 5667.5 (1998) guidelines. Water samples for stable isotope analysis were collected in 30 mL McArtney glass bottles with metal screw top lids with rubber insert. In order to reduce evaporation, samples were stored and transported inverted. Stable isotope (δ^2H and $\delta^{18}O$) samples were analysed at Flinders University of South Australia isotope analytical laboratory in Adelaide. The standard procedure for analysis of the Deuterium and Oxygen-18 composition of water is that described by PICARRO (2015). The water samples for ^{14}C were collected using 1.25 L polyethylene terephthalate (PET) bottles. A total of 5.0 L of groundwater was collected from each well. The sample bottles were sealed and securely capped. Radiocarbon was analysed using radiometric counting and accelerator mass spectrometry at the Rafter Radio Carbon Laboratory, National Isotope Centre, New Zealand (Rafter Radiocarbon Laboratory, 2016).

The downhole geophysical logs were obtained using borehole logging unit of the South Australian Department of Environment, Water and Natural Resources. The unit consists of data processing unit, winch, cable and logging tools installed in a van. Gamma and neutron probes were run simultaneously and 3-arm caliper tool was run separately. The surface unit controls the logging and movement of the probe and displays and stores data. The probe is connected to the surface unit with the cable that conveys the electrical signals (Wonik, 1990).

Nine high resolution EM lines of varying orientation (Figure 1) with station spacing of 25 m with infill to higher resolution (12.5 m) were used over areas of surface features such as sinkholes. A Zonge International GDP-32ii receiver equipped with NanoTEM cards were used to acquire all data for this study. Data were sensed using a single turn 5×5 m wire loop placed at the centre of each transmitter loop. Data were recorded over two channels, with the second gained up to improve late time decay resolution. Decay data were recorded over 31 time windows to approximately 2 milliseconds after transmitter turnoff. Transmitted fields were generated with a Zonge NT-20 geophysical transmitter energizing a single turn 25×25 m loop. This transmitter provides an output of 3 Amps, at 32 Hertz with a turn-off ramp time of 1.5 µs. Synchronisation was controlled directly using the GDP-32ii receiver. Data were then used to produce 1D inversion model sections of resistivity using Zonge International's STEMINV modelling program (Mann, 2016).

4. Results and Discussion

Direct information about the structure of the subsurface geology can be obtained by drilling boreholes. Downhole geophysical logging provides additional in-situ information to supplement the direct geological information. This is particularly useful where borehole intercept cavities but insufficient recovery of core sample (lost-circulation). Downhole geophysical logs provide continuous depth record of formation properties (Wonik, 2007) such as: caliper logs provide information on drillhole diameter, fractures and are essential in interpreting other logs; gamma logs provide a record of natural radioactivity of the rock penetrated by a drillhole and provides information on lithology and stratigraphic correlation by relating to clays and silt content and permeability; neutron log measures neutrons from an isotopic source at one or several detectors after they migrate through material in, and adjacent to, the borehole. Log response results primarily from hydrogen content, but it can be related to saturated porosity and moisture content (Keys, 1990).

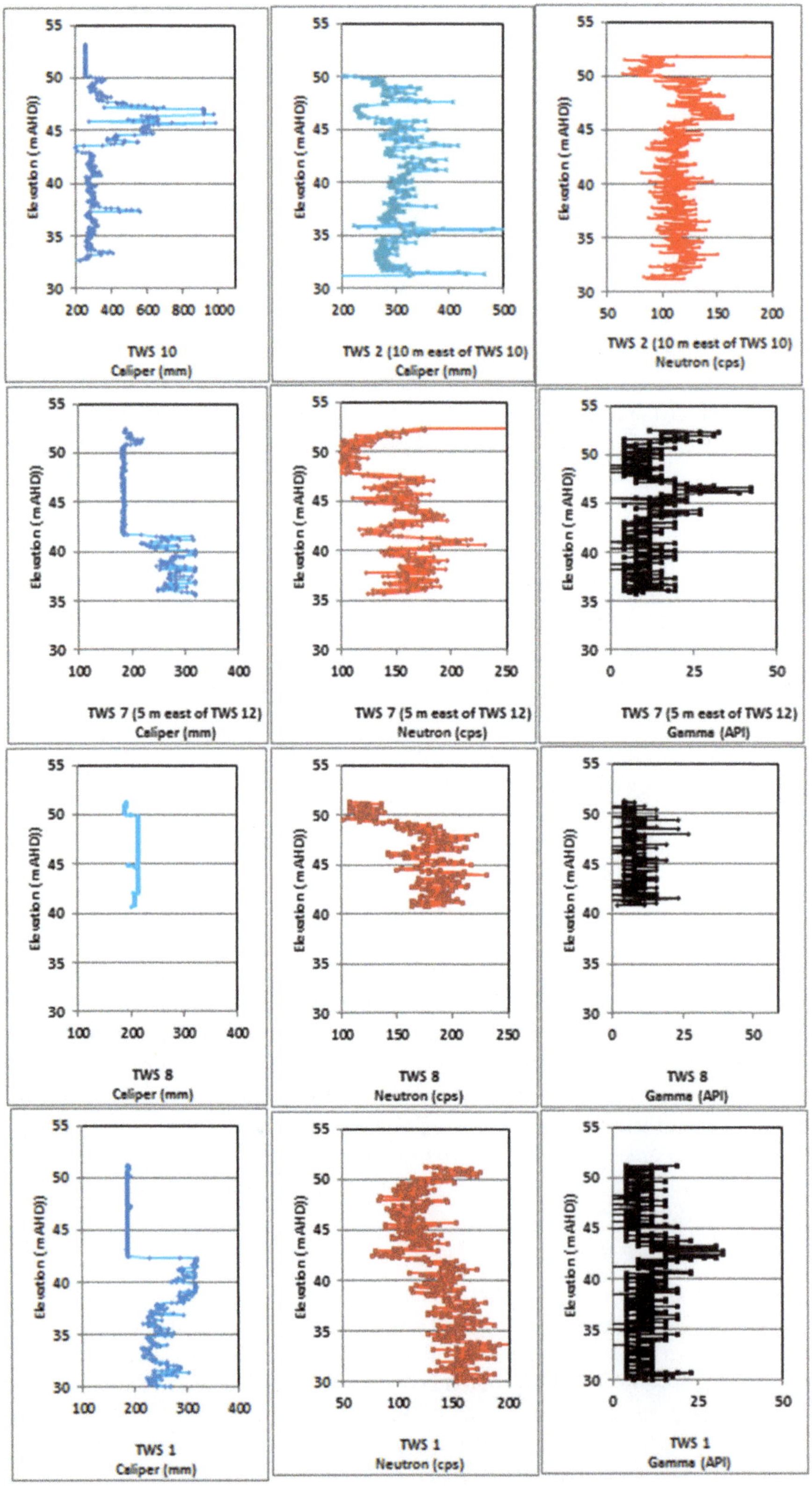

Figure 3. Caliper, gamma and neutron logs for water supply wells

4.1 Evidence of Conduit Porosity from Lithological and Downhole Geophysical Logs

Since geophysical logs have no unique response, logs need to be interpreted with geological information obtained from boreholes. Stratigraphic correlation using caliper, gamma and neutron logs are shown in Figure 3. Bordertown town water supply wells (TWS) 1, 2, 7, 8 and 10 wells have been logged for depth, gamma, neutron and caliper. Old wells TWS 2 and TWS 7 have been replaced with TWS 10 and TWS 12. As the new well is located within 10 m from the old well, geophysics for the older wells were used as the replacement well of TWS 12. The six monitoring wells (WRG042-WRG047) along with TWS 10 to TWS 13 were drilled using the rotary drilling method, with resultant loss-circulation occurring in 7 wells. These were encountered; WRG042 at 10-16 m (58 to 52 mAHD), WRG043 at 18-23 m (49 to 44 mAHD), WRG044 at 18-25 m (49 to 42 mAHD), WRG047 at 10-47 m (59 to 29 mAHD), TWS 10 at 18-25 m (50.7 to 43.7 mAHD), TWS 11 at 19-24 m (48.3 to 40.3 mAHD), TWS 12 at 21-26 m (46.3 to 41.3 mAHD), TWS 13 at 12-18 m (54.5 to 48.5 mAHD) depths (Figure 4). In TWS 10, the caliper log indicated full arm extension between 21 m to 24m (47 to 44 mAHD). The results indicate that loss-circulation (open cavities) occur at the Bridgewater Formation and the upper part of the Murray Group Limestone Formation.

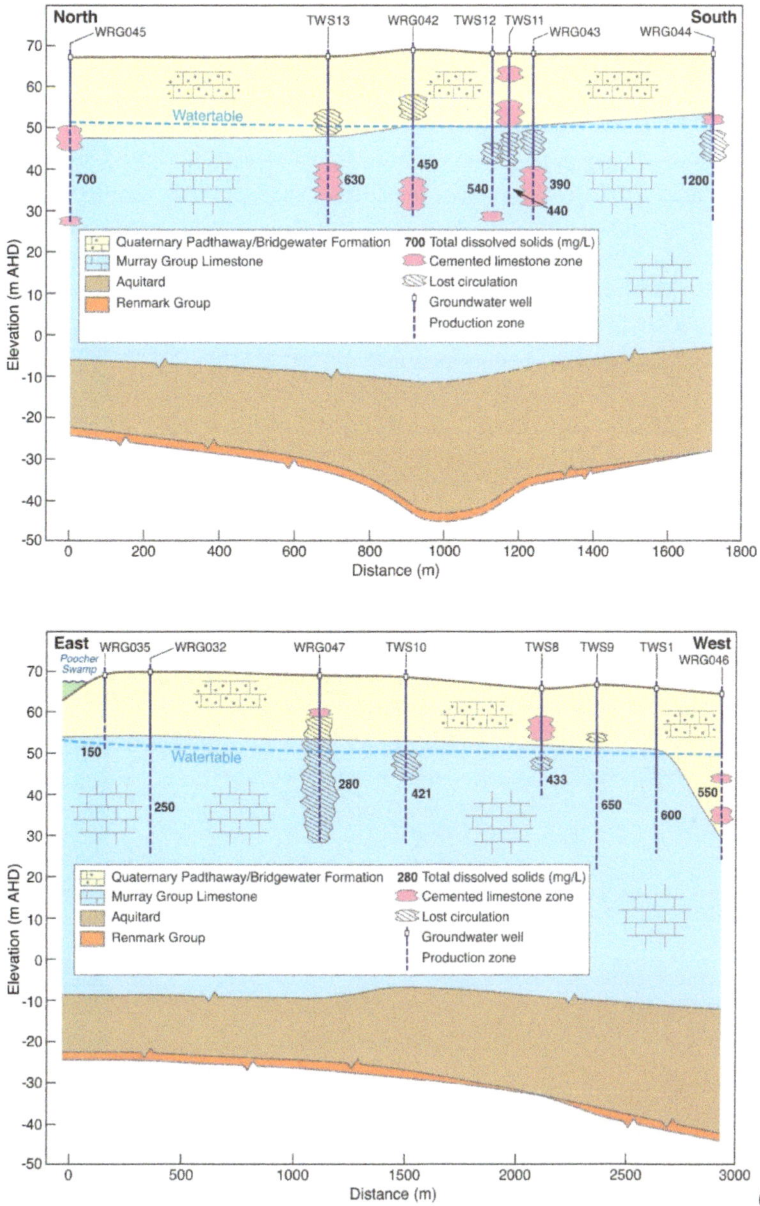

(a)

(b)

Figure 4. Geological cross sections across the study area (a) north-south (b) east-west

The older wells TWS 1, TWS 2, TWS 7 and TWS 8 have been drilled using cable tool method and therefore there are no records of cavity development. Gamma log of TWS 1 show at 43 mAHD, higher counts up to 33 API (American Petroleum Institute) indicating a possible clay rich zone. Lithological description at this depth of the TWS 1 is 'white limestone with yellow clay'. Immediately below this, 24-29 m (42 to 37 mAHD) depth interval lithology is described as 'limestone with yellow, sandy clay and gravel layer', stretching caliper arm to 317 mm. Below 41 mAHD, neutron reading slightly increased indicating higher porosity zone of the profile. Higher counts of neutron logs of TWS 1, below 40 mAHD indicate more open limestone. There can be a mismatch between lithological logs when directly compared to downhole geophysical logs. This is because lithological samples are collected at certain depth interval (in this case 3 m interval) and hence represent 'average of the samples collected' within the depth interval. Therefore, combining geophysical logs with lithological logs is more useful for interpretation than relying on one set of logs. Similarities are observed between caliper log of TWS 10 and neutron log of TWS 2.

The caliper log of TWS 2 shows presence of cavity or wash-out zones at different depths in the profile. It is apparent that TWS 10 has intercepted a major cavity than that present in TWS 2, even though two sites are about 10 m apart, indicating complexity of tracking conduits pathways using boreholes. The old well TWS 7 was replaced by TWS 12, and the neutron log of TWS 7 indicates high porosity zone at 41 mAHD (softer limestone with shelly and slightly marly) but 2 m above this at 43 mAHD gamma counts increased to 42 API indicating a clay or shale rich zone (marl with limestone). Geophysical logs of TWS 8 well shows more or less uniform gamma and caliper logs throughout the profile indicating homogeneous nature of limestone at this location.

4.2 Evidence from EM Survey

EM data were examined before being averaged to create a single record; for each channel, for each soundings using TEMAVGW (MacInnes, 2010) software (Mann, 2016). Channels that were considered of poor quality were skipped before averaging each station data. Ungained and gained data were merged to optimize early and late data to create a single sounding for each station. Data were then used to produce 1D inversion models of resistivity using STEMINV (MacInnes and Raymond, 2001) modelling program. Topographic information was used in modelling so sections are presented relative to SRTM (2000) elevations, comparable to m AHD (Mann, 2016).

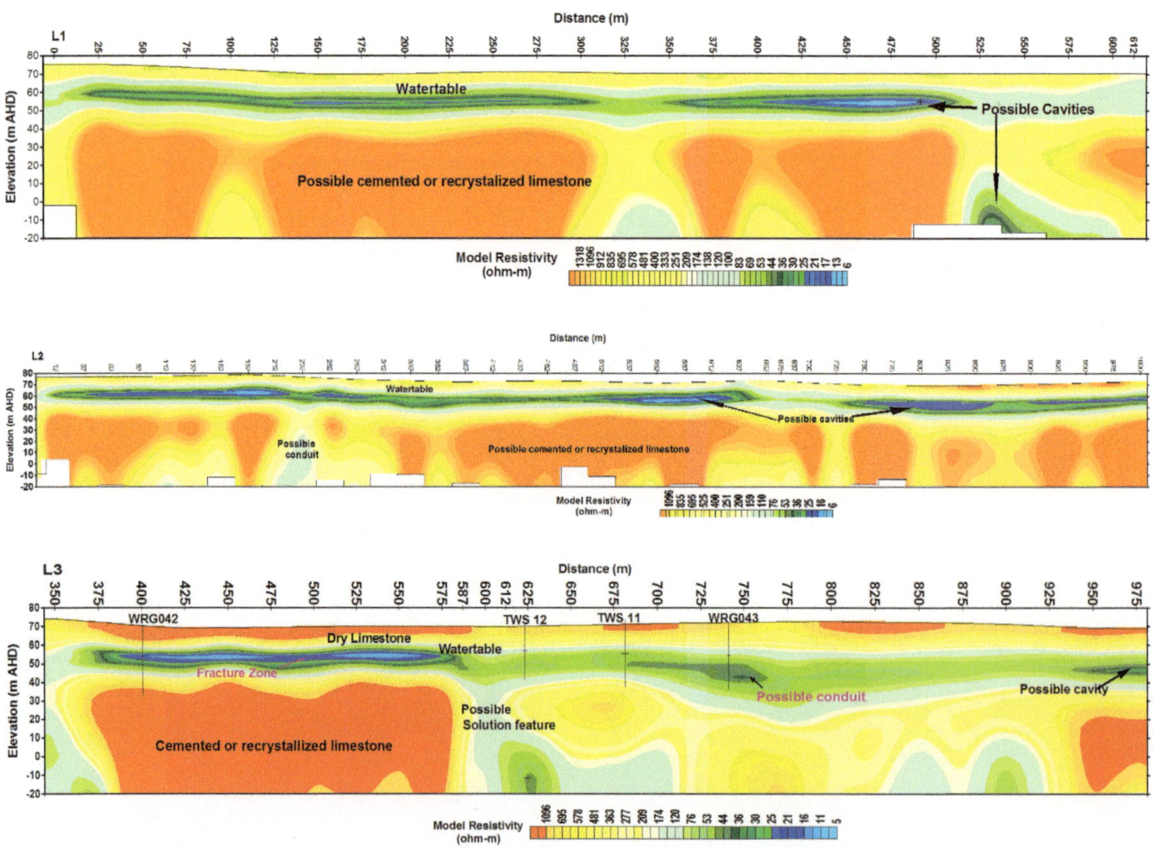

Figure 5. Inverted resistivity sections along lines 1 to 3 (L1 to L3)

The variation in resistivity of particular minerals can be enormous, for example limestone: 50-10^7 Ωm; Marls 3-70 Ωm; Clays: 1-100 Ωm (Telford et al., 1990). Other factors such as dissolved minerals in groundwater, fracture zone filled with air or water, consolidation of geologic materials influence the local changes to resistivity. In most cases, a small change in the percentage of water affects the resistivity (Telford et al., 1990). The resistivity of fresh water is about 10-100 Ωm (Palacky, 1987), overlapping of resistivity values in the saturated zone requires known features for interpretation of TEM sections. In the shallow substructures, presence of water controls much of the resistivity variations. Measurement of resistivity in general, is a measure of water saturation and connectivity of pore spaces (Cardimona, 2002). Therefore, increasing number of fractures or porosity tends to decrease the resistivity (Cardimona, 2002).

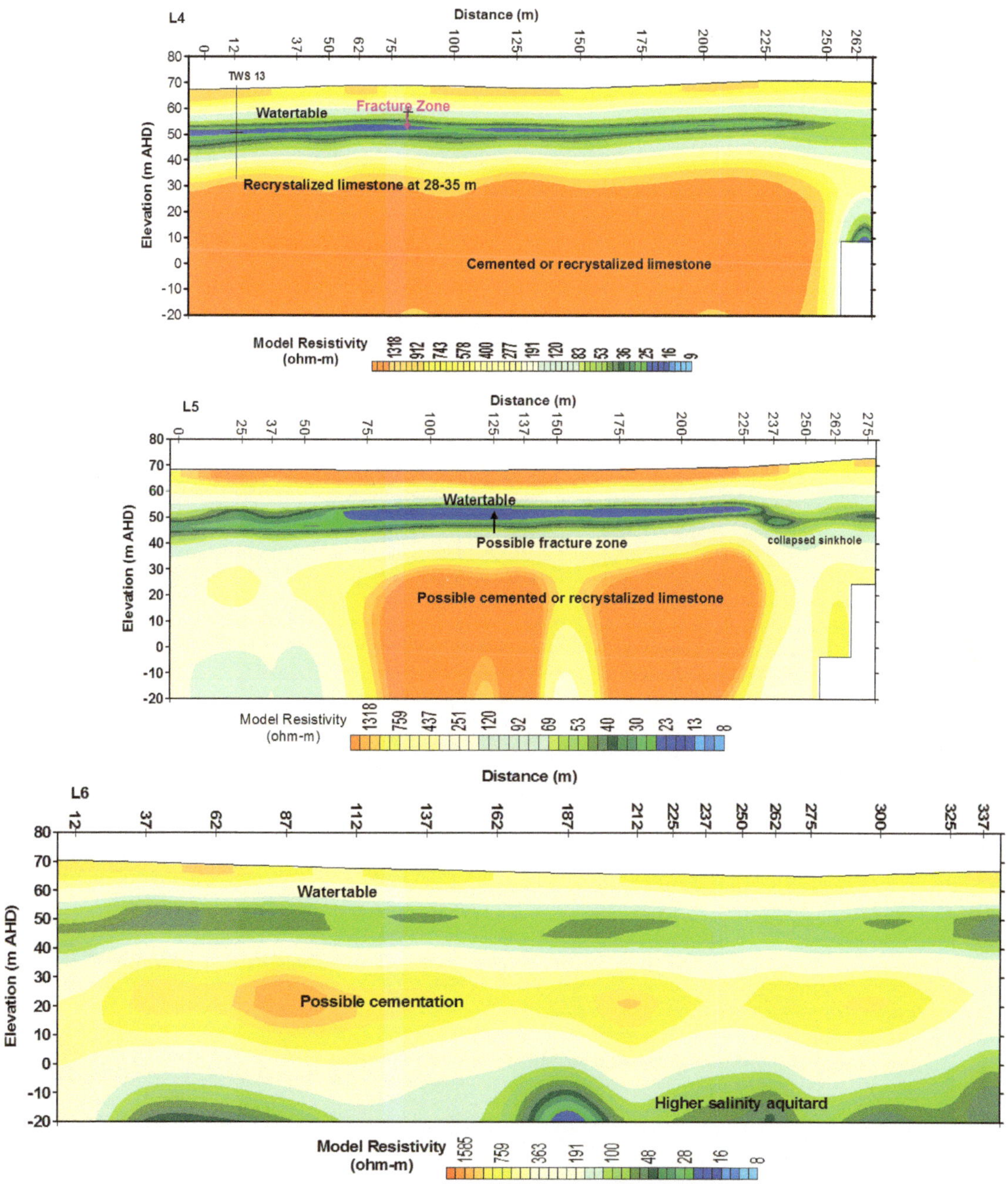

Figure 6. Inverted resistivity sections along lines 4 to 6 (L4 to L6)

Figure 5 shows the inverted resistivity sections for survey lines L1 to L3. Four distinct resistivity contrast areas are found. The upper high resistivity zones up to 15 m depth represent the unsaturated zone as the dry limestone is a poor electrical conductor. Marked resistivity anomaly begins about 18 m depth, which represent the watertable. The lowest resistivity zone (5 to 25 Ωm) from about 20 m to 30 m represents the main fracture zones. For interpretation purpose, wells drilled close to the line L3 are projected on to section L3. For example, fracture zones of wells: WRG042 at 10-16 m; TWS 12 at 21-26 m; TWS 11 at 19-27 m; and WRG043 at 18-23 m are within or at close range to the low resistivity zones of TEM section. In addition, pockets of low resistivity zones were identified as possible subsurface conduits or cavities. The higher resistivity >250 Ωm is considered as the cemented or recrystallized limestone. As the degree of cementation varies, the resistivity varies accordingly. The wells projected on to line L3 found cemented limestone: WRG042 at 30-38 m; TWS 12 at 38-40 m; TWS 11 at 3-17 m; and WRG043 at 27-36 m depth intervals, and are compatible with higher resistivity. Groundwater flow may be dominated through fracture zone located at 20-30 m below ground level.

Resistivity sections of TEM lines 4 to 6 (L4 to L6) are given in Figure 6. The line L4 begins at TWS 13 where depth to water is found at 16.2 m, the lost circulation zone at 12-18 m depth and the cemented limestone at 28-35 m depth interval. These observations are in good agreement with resistivity data interpretation and are comparable to range of resistivity values described previously. Resistivity contrast in inverted section L6 is minor (Figure 6), indicating the presence of more homogeneous limestone (no dominant fracture zone or cemented limestone). This is similar to the homogeneity found in geophysical logs of TWS 8 well. Hence groundwater flow in this part of the aquifer may be considered as the matrix flow. Presence of deeper aquitard is evidenced in L6 as indicated by low resistivity at about -10 m AHD.

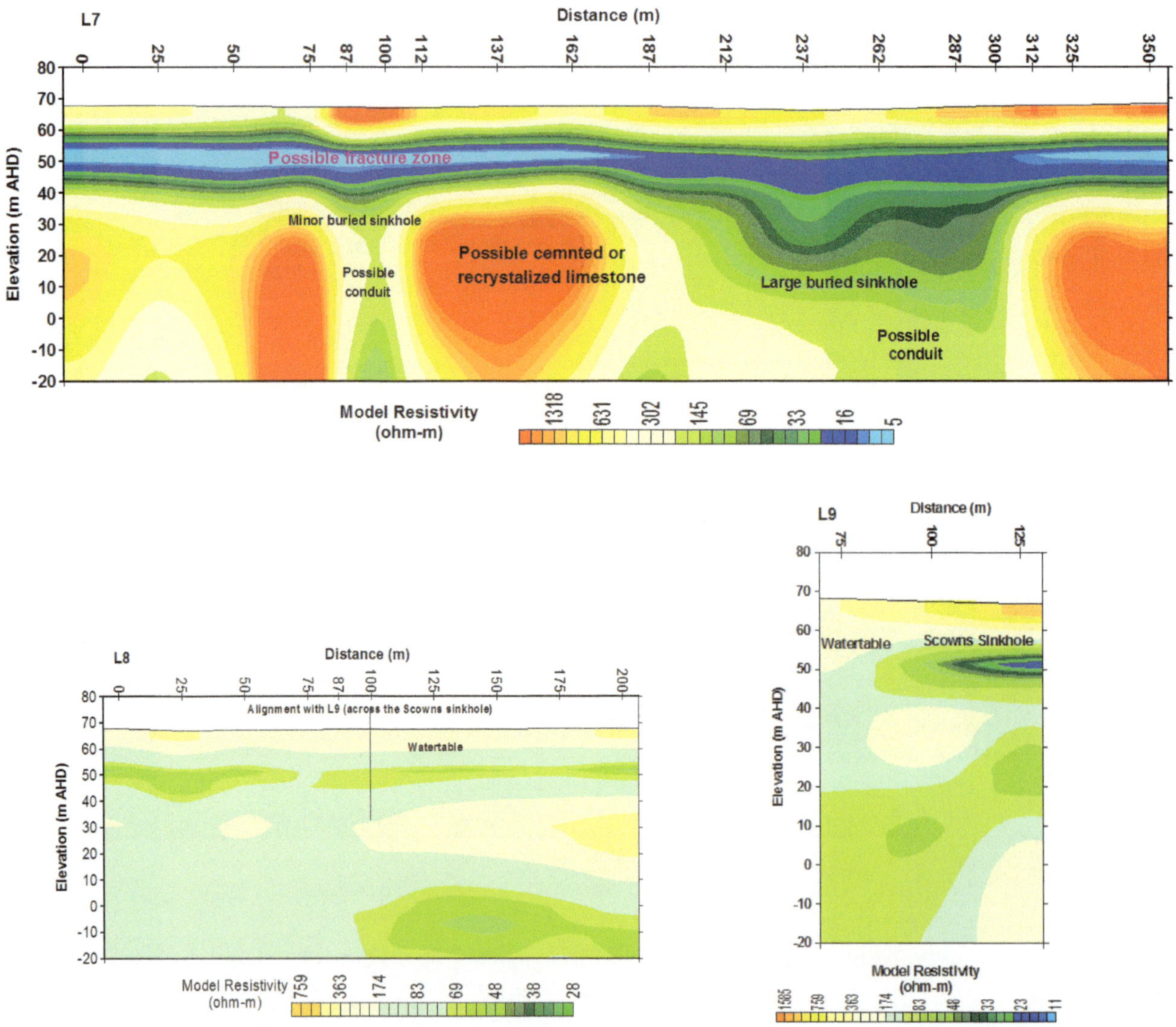

Figure 7. Inverted resistivity sections along lines 7 to 9 (L7 to L9)

Resistivity sections L7 (Figure 7) runs across two buried sinkholes, as evidenced in circular shaped and minor ground surface depressions at about 87 m and 237 m along the line. Resistivity contrasts reappear in this section as lower resistivity, possible fracture zone, occurs at shallow depth of 10 to 20 m depth (56-46 m AHD), and pockets of high resistivity (>750 Ωm) limestones occurring at below 8 m depth (at 30 m AHD), indicating possible layers of cemented limestone. At buried sinkhole locations, near vertical low resistivity portions in the profile may be the possible conduits, could probably be filled with sediment and water. No borehole information is available for this line. Thus, buried or subsidence sinkholes (Waltham et al. 2005) or subsurface cavity locations can be identified based on the characteristics of low resistivity readings. In the saturated zone, when the cavities are filled with fine, loose materials, the electrical resistance is lower as both water and clay are good conductors (Panek et al. 2010). In general, water filled cavities are abundant in karstic limestone aquifers.

The line L8 is about 100 m upgradient (east) to the Scowns sinkhole and is 200 m long. The station at 100 m along L8 (Figure 7) is aligned with the centre of the sinkhole and is intentionally placed to intercept any conduits flow paths that connect to the sinkhole. However, resistivity values indicate presence of homogeneous limestone. This indicates, either conduits connecting the Scowns sinkhole begins at this location or interconnection to existing conduits occur at about 50 m further south from L8, as occurrence of another sinkhole is found about 100 m east of L8 (Figure 1). This conjecture was not investigated. The line L9 is positioned across the Scowns sinkhole. The station 125 m (on L9) is at the centre of the sinkhole. Down to watertable depth, presence of dry soil is indicated by higher resistivity (>174 Ωm). The low resistivity about 10-25 Ωm, is the water filled gallery of the sinkhole and the conduit system is indicated by low resistivity zones of 25-83 Ωm.

In arid and semi-arid regions, it is common that some of the active sinkholes are capped with a layer of sediment and become dormant during dry summer months. This is evidence in both Poocher Swamp and Scowns sinkholes (Figure 8). With the onset of receiving surface runoff, these sinkholes become active and would act as the sources of point recharge (Figure 9). Somaratne (2015b) reports that during the 1986 flood event, discharge to the Scowns sinkhole was measured at the rate of 500-2000 $m^3.hr^{-1}$ during 17th to 28th August 1986.

(a) (b) (c)

Figure 8. Sinkhole surface during dry summer (a) Poocher Swamp sinkhole during dry summer covered with sediment (b) Scowns sinkhole during dry summer covered with sediment (c) Scowns sinkhole during August 1986 high rainfall events (Somaratne, 2015b) showing point recharge through the sinkhole

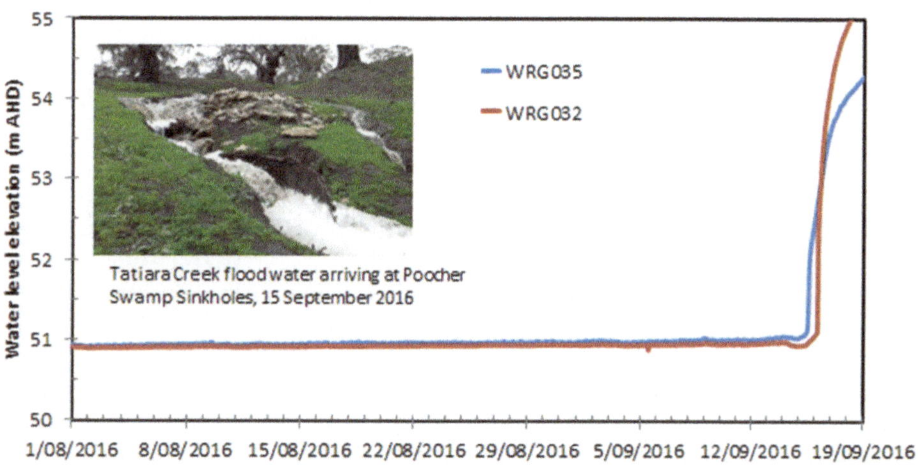

Figure 9. Hydrographs for WRG032 and WRG035 wells in response to recharge event on 15th September 2016

4.3 Stable Isotope and Radiocarbon Distribution and Possible Conduit Flow Paths

Chloride (Cl^-) is a conservative tracer, and in the absence of other sources and sinks, can be used to compare recharge sources with stable isotope $\delta^{18}O$. The Cl^- and $\delta^{18}O$ data were used to infer the degree of mixing along flow paths from the primary recharge source (Poocher Swamp) to the sampling point. In a karsitic aquifer composition of groundwater chemistry at a sampling point represents a contribution from matrix and conduit flow components. The matrix flow has a longer residence time than the conduit flow. It is therefore possible to compare degree of mixing of matrix and conduit flows relative to two end members, the Poocher Swamp and the pristine limestone aquifer water in WRG044 well (Figure 1) which is located outside the freshwater lens. The degree of mixing between two end members within the freshwater lens is shown in Figure 10 by plotting the values of $\delta^{18}O$ *versus* Cl^- concentrations. For comparison, available long-term observed rain water $\delta^{18}O$ and Cl^- from Adelaide is also included. It is noted that Poocher Swamp is about 100 km inland from the nearest coastline; therefore rain water $\delta^{18}O$ may be more depleted than the Adelaide (coastal) winter average of -4.19‰ VSMOW. Radiocarbon (^{14}C) data was used to identify relatively younger water indicated by higher ^{14}C activity in the groundwater mixtures along flow paths. Somaratne et al. (2016) used the relative presence of measured ^{14}C activities, in preference to the corrected ^{14}C or absolute residence time of groundwater recharge to compare residence time between study sites. In this study, correction to measured ^{14}C activity was not made rather comparison between sampled points within the lens was done using measured ^{14}C activity. The Cl^-, $\delta^{18}O$ and ^{14}C data provide degree of mixing along inferred flow paths. Stable isotope, radiocarbon, Cl^-, EC, pH data with well details are presented in Table 1.

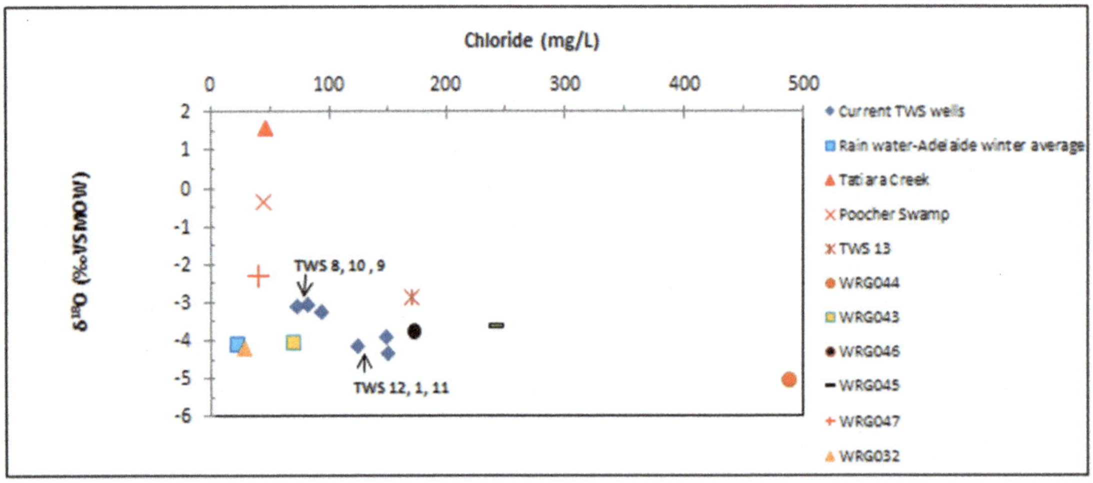

Figure 10. $\delta^{18}O$ *versus* Cl^- concentrations of groundwater between two end members; Poocher Swamp and WRG044

Both Tatiara Creek and Poocher Swamp water are isotopically enriched relative to regional groundwater (WRG044) and the local rainfall. The enrichment of $\delta^{18}O$ (1.58‰) in the Tatiara Creek indicates creek water has undergone evaporation during sampling period in October 2012, due to relatively high temperature in spring period (Figure 2). The Poocher Swamp receives local surface runoff, and therefore, the enrichment reduced to -0.34‰. The well WRG032 is located 150 m downgradient to the Poocher Swamp and has $\delta^{18}O$ (-4.19‰) and Cl^- (29-35 mg.L^{-1}) indicating the well water is reflective of more recent winter (June-August) recharge received from the Poocher Swamp. The well was sampled in May, 2016. Tatiara Creek and Poocher Swamp are highly dynamic systems, and as such, water chemistry varies significantly over the year, depending on wet-dry cycles. Herczeg et al. (1997) reports annual variation of ^2H (10 to -40 ‰ VSMOW) with variation of chloride concentration in Poocher at about 10-40 mg.L^{-1}. Thus, wells located closer to the Poocher Swamp (WRG035 and WRG032) tend to mirror water quality of Poocher Swamp. Further downgradient (1113 m) well, WRG047 recorded lost-circulation throughout the 10-40 m depth interval indicating the presence of large cavity. The $\delta^{18}O$ (-2.31‰) and Cl^- (40 mg.L^{-1}) indicate the well has intercepted a major conduit flow path that originates from the Poocher Swamp. The characteristic of regional water quality in the limestone aquifer can be found in the WRG044 well with higher Cl^{-1} of 487 mg.L^{-1} and $\delta^{18}O$ of -5.06‰ indicating older groundwater in Murray Group Limestone.

Table 1. Source details and radiocarbon and stable isotope data

Source	Depth (m)	Prod. Interval (m)	SWL (m)	Sample Depth (m)	δ^2H (‰VSMOW)	$\delta^{18}O$ (‰VSMOW)	$\delta^{13}C$ (‰)	^{14}C (pmC)	Cl^{-1} ($mg.L^{-1}$)	EC ($\mu S.cm^{-1}$)	pH	Temp. (^0C)
Tatiara Creek	n/a	n/a	n/a	n/a	11	1.58	-	-	46	525	8.3	-
Poocher Swamp	n/a	n/a	n/a	n/a	-4.5	-0.34	-	-	27-44	421	8.2	-
WRG035	18	12-18	dry	dry	-	-	-		40	699	7.4	-
WRG032	38.7	19.3-38.7	19.1	21	-28.2	-4.19	-14.3	88	29-35	612	6.7	17.1
WRG047	40	16-40	18.3	21	-14.05	--2.31	-14.6	74.55	35-40	670	6.8	17
TWS 10	40	16-40	-	pump	-9.4	-3.05	-11,5	45.77	81	962	7.1	18
TWS 8	26.3	16-26.3	-	pump	-17.5	-3.08	-12	61.53	73	739	7.1	17.9
TWS 9	45	20-45	-	pump	-18.8	-3.23	-11.3	60.46	93.3	1054	7.01	18
TWS 1	40	19-40	-	pump	-22.4	-3.89	-10.6	60.29	149	805	7	18.1
WRG046	40	16-40	15.5	20	-23.22	-3.76	-	-	173	1326	11.6	17.8
WRG045	40	16-40	15.3	20	-22.68	-3.64	-12.2	60.59	242	2032	6.7	18.1
TWS 13	40	25-40	16.7	20	-17.76	-2.86	-12.6	53.62	171	808	7.1	17.6
WRG042	40	16-40	17.8	21	-21.64	-3.49	-12.5	59.95	-	888	6.9	17.4
TWS 12	37	25-37	-	pump	-24.3	-4.13	-10.4	48.83	124	907	7	18
TWS 11	37	25-37	-	pump	-22	-4.31	-10.9	55.15	150	805	7	17.8
WRG043	36	28.8-36	16.6	30	-23.58	-4.05	-12.4	67.06	70	733	7.1	16.5
WRG044	40	16-40	16.3	20	-31.29	-5.06	-9.7	56.49	487	2849	6.8	17.3

Note: *Tatiara Creek, Poocher Swamp and WRG035 well were dry during May 2016 sampling round. Stable isotope, Cl, EC and pH data were collected in 2012, 2014 and 2016. Prod. Interval = Production zone of the well; SWL=Standing Water Level, EC=Electrical Conductivity.*

Gradual depletion of $\delta^{18}O$ from -0.34‰ at Poocher Swamp to -2.31‰ at WRG047 may be due to minor mixing of conduit flow with ambient groundwater. The highest ^{14}C activity, 88 pmC, is at the WRG032 well is the closest location to the point recharge source (Figure 11). The ambient groundwater just outside the freshwater lens, 56.49 pmC, is found in WRG044 well confirming the presence of 'older' groundwater. The WRG045 located closer to the boundary records 60.59 pmC, presumably due to local recharge. Along the flowpaths, ^{14}C activity decreases due to mixing of recharge water with ambient groundwater: WRG032 (88 pmC), WRG047 (74.55 pmC), TWS 10 (45.77 pmC). Higher radiocarbon values were observed from TWS 8 to TWS 1, which may be due to local diffuse recharge; WRG042 (59.95 pmC), TWS 8 (61.53 pmC), TWS 9 (60.46 pmC), TWS 1 (60.29 pmC). The lower radiocarbon values in TWS 10 and TWS 13 indicates the wells may intercept older groundwater from the north-east direction. In WRG045, depleted $^{18}\delta O$ (-3.64‰), higher Cl^{-1}(242 $mg.L^{-1}$) and relatively younger water (60.59 pmC) can be interpreted as the well received recent vertical recharge that has undergone evaporation. Observed lack of significant variation of $\delta^{18}O$ as Cl^{-1} concentration increases in wells WRG043, TWS 12 and TWS 11 can be interpreted as mixing of inflow from the Poocher Swamp with some vertical recharge received in this part of the aquifer. The $\delta^{18}O$ values in these wells are comparable to that of Adelaide rain water, enrichment is minor. Lost-circulation zones (Figure 4a) and TEM tomography section (Figure 5-L3) show possible cavity zones in WRG043, TWS 12 and TWS 11, but $\delta^{18}O$ values are more depleted than TWS 8 to TWS 10 (Table 1). However, ^{14}C data indicates older water in TWS 11 (55 pmC), TWS 12 (48 pmC) and younger water in WRG043 (67 pmC) suggesting that only WRG043 well has a direct link to the Poocher Swamp.

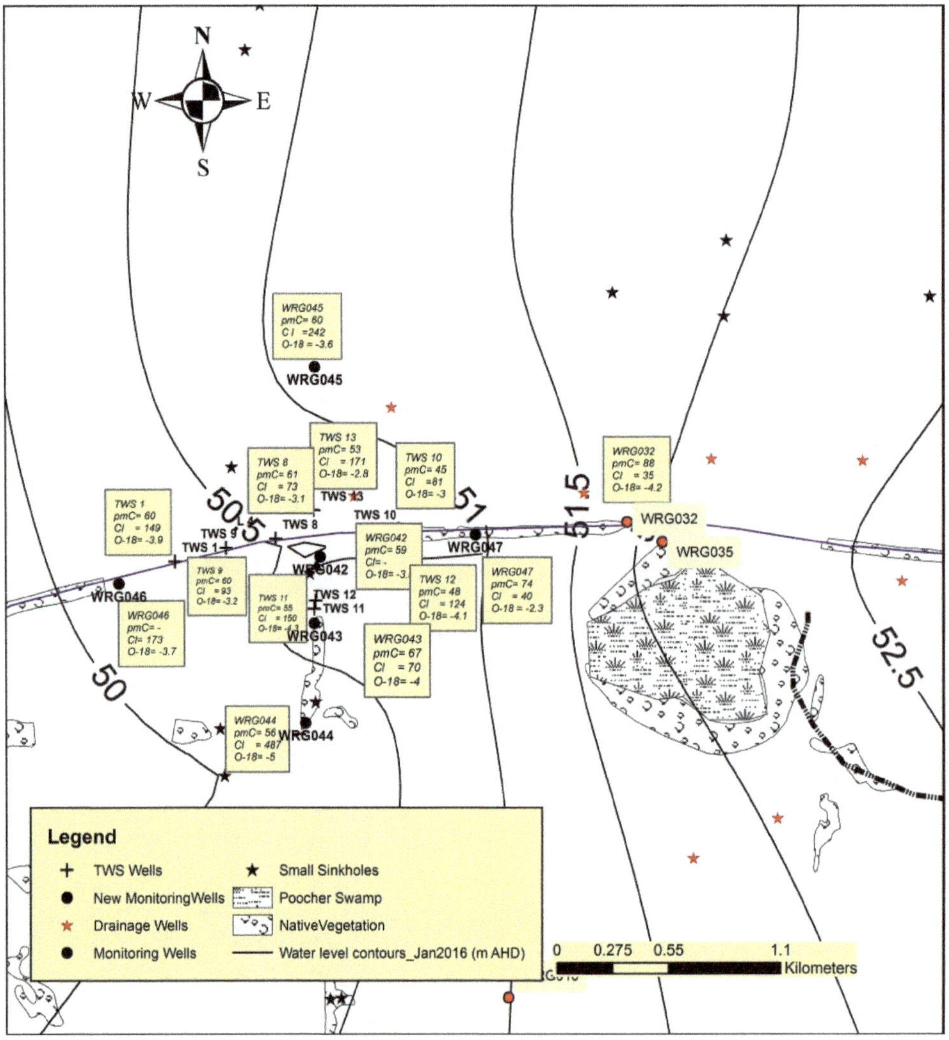

Figure 11. Radiocarbon and stable isotope distribution with inferred flow paths

4. Conclusion

Inherent difficulties arise when attempting to trace conduit flowpaths using a single technique given limitations in the each and the large variation in scale of influencing structures. Well drilling is effective but unlikely to intercept discrete structures; geophysical methods are commonly used to compliment well drilling however interpretation of responses from both geologic and hydrologic sequences can be complex; tracer techniques are widely used but defining flowpaths become problematic due to dilution and mixing. We used geological, geophysical, stable isotope and radiocarbon data for assessing the conduit flow paths in the Poocher Swamp freshwater lens. The drillhole data was used to validate and interpret the geophysical results. Low resistivity zones of the TEM tomography sections had excellent correlation to fracture zones identified during well drilling, and dowhnhole geophysical logs. In the saturated zone, conduits may be filled with both water and clay and therefore, is difficult to identify by resistivity survey alone as both water and clay are low resistivity materials. Therefore use of stable isotopes and radiocarbon is a useful tool in identifying preferential groundwater flow paths. Wide range (-0.34‰ at Poocher Swamp to -5.06‰ at WRG044) of ^{18}O and ^{14}C activity (88 pmC at WRG032 to 56.49 pmC at WRG044) indicate spatial inhomogeneity of isotopes and groundwater age. The presence of relatively enriched ^{18}O signatures and higher radiocarbon activity is an indication of presence of conduit flows that carry recharged water from the Poocher Swamp.

Acknowledgments

This study was funded by the South Australian Water Corporation under resource assessment of the Poocher Swamp freshwater lens. Authors thank Nardine Habner for arranging water sampling and laboratory analyses in 2016, and Jeff Lawson for providing lithological and downhole geophysical data.

References

AS/NZ 5667.5. (1998). Water Quality Sampling-Guidance on the Design of Sampling Programs, Sampling Techniques and the Preservation and Handling of Samples; AS/NZS 5667.5; Australia/New Zealand Standard: Homebush, Australia.

Cardimona, S. (2002). *Electrical resistivity techniques for subsurface investigation*. Department of Geology and Geophysics, University of Missouri-Rolla, Rolla, USA. Retrieved April 21, 2016 from https://www.researchgate.net/publication/242692638_ELECTRICAL_RESISTIVITY_TECHNIQUES_FOR_SUBSURFACE_INVESTIGATION

Edwards, A. E., Amatiya, D. M., Williams, T. M., Hitchcock, D. R., & James, A. L. (2013). Flow characterization in the Santee Cave system in the Chapel Branch Creek watershed, upper coastal plain of South Carolina, USA. *Journal of Cave and Karst Studies, 75*(2), 136-145. http://dx.doi.org/10.431/2011ES0262.

Einsiedl, F. (2005). Flow system dynamics and water storage of a fissured porous karst aquifer characterized by artificial and environmental tracers. *Journal of Hydrology, 312*(1-4), 312-321. http://dx.doi.org/10.1016/j.jhydrol.2005.03.031.

Farooq, M., Park, S., Song, Y.S., Kim, J.H., Tariq, M., & Abraham, A. A. (2012). Subsurface cavity detection in a karst environment using electrical resistivity (er): a case study from yongweol-ri, South Korea. *Earth Sci. Res. SJ., 16*(1), 75-82.

Herczeg, A. L., Leaney, F. W. J., Stadter, M. F., Allan, G. L., & Fifeld, L. K. (1997). Chemical and isotope indicators of point source recharge to karst aquifers, South Australia. *J. Hydrol., 192*, 271-299.

Keys, W. S. (1990) Borehole geophysics applied to groundwater investigations, Chapter E2, Techniques of water resources investigations of the United States Geological Survey, U.S. Geological Survey, Denver, USA.

Knierim, K. J., Pollock, E. D., Hays, P. D., & Khojasteh, J. (2015). Using stable isotopes of carbon to investigate the seasonal variation of carbon transfer in a northwestern Arkansas cave. *Journal of Cave and Karst Studies, 77*(1), 12-27. http://dx.doi.org/10.4311/2011ES0264

MacInnes, S. (2010). TEMAVGW Documentation v1.10 Documentation Zonge Engineering and Research Organization, Inc., Tucson, USA.

MacInnes, S., & Raymond, M. (2001). STEMINV: Smooth-Model TEM Inversion, ZONGE Data Processing, Version 3.0, Zonge Engineering and Research Organization, Inc., Tucson, USA.

Mann, S. (2016). NanoTEM Electro Magnetic Survey, Logistic Summary. Zonge Enginnering and research Organization (Australia) Pty Ltd, Report no: 150265, 39 Raglan Avenue, Edwardstown, SA 5039, Australia.

Martinez-Lopez, J., Rey, J., Duenas, J., Hidalgo, C., & Benavente, J. (2013). Electrical tomography applied to the detection of subsurface cavities. *Journal of Caves and Karst Studies, 75*(1), 28-37. http://dx.doi.org/10.4311/2011ES0242.

Metwaly, M., & AlFouzan, F. (2013). Application of 2-D geoelectrical resistivity tomography for subsurface cavity detection in the eastern part of Saudi Arabia. *Geoscience Frontiers, 4*(2013), 469-476.

Nativ, R., Gunay, G., Hotzl, H., Reichert, B., Solomon, D. K., & Tezcan, L. (1999). Separation of groundwater flow components in a karstified aquifer using environmental tracers. *Applied Geochemistry, 14*(8), 1001-1014. http://dx.doi.org/10.1016/so883-2927(99)00012-8

Palacky, G. V. (1987). Resistivity characteristics of geologic targets. In Nabighian, M.N (ed.), Electromagnetics Methods in Applied Geophysics Theory: Tulsa, Okla.. *Society of Exploration Geophysicists, 1*, 53-129.

Panek, T., Margielewski., Taborik, P., Urban, J., Hradecky, J., & Szura, C. (2010). Gravitationally induced caves and other discontinuities detected by 2D electrical resistivity tomography: case studies from the Polish Flysch Carpathians. *Geomorphology, 123*, 165-180.

PICARRO. (2015). PICARRO L2130-i δD and $\delta^{18}O$ High Precision Isotopic Water Analyser, PICARRO Inc, Santa Clara, USA. Retrieved April 28, 2016, from http://www.picarro.com/products_solutions/peripherals/for_h2o/ih2o_solids_liquids_and_vapor

Plummer, L. N., Busenberg, E., McConnell, J. B., Drenkard, S., Schlosser, P., & Michel, R. L. (1998a). Flow of river water into a karstic limestone aquifer-1. Tracing the young fraction in groundwater mixtures in the Upper Floridan aquifer near Valdosta, Georgia. *Applied Geochemistry, 13*(8), 995-1015.

Rafter Radiocarbon Laboratory. (2016). GNS Science, National Isotope Centre, New Zealand. Retrieved April 28, 2016, from http://www.gns.cri.nz/Home/Services/Laboratories-Facilities/Rafter-Radiocarbon-Laboratory

Somaratne, N. (2014). Characteristics of point recharge in karst aquifers. *Water, 2014*(6), 2782-2807. http://dx.doi.org/10.3390/w6092782.

Somaratne, N. (2015a). Karst aquifer recharge: A case history of over simplication from the Uley South Basin, South Australia. *Water,* 2015(7), 464-479. http://dx.doi.org/10.3390/w7020464

Somaratne, N. (2015b). Pitfalls in application of the conventional chloride mass balance (CMB) in karst aquifers and use of the generalized CMB method. *Environ. Earth Sci.* http://dx.doi.org/10.1007/s12665-015-4038-y

Somaratne, N., Mustafa, S., & Lawson, J. (2016). Use of hydrochemistry, stable isotope, radiocarbon, ^{222}Rn and terrigenic ^{4}He to study the geochemical processes and the mode of vertical leakage to the Gambier Basin Tertiary confined sand aquifer, South Australia. *Water, 2016*(8), 180. http://dx.doi.org/10.390/w8050180

SRTM. (2000). Shuttle Radar Topography Mission. The National Space Administration (NASA) and the National Geospatial Intelligence Agency (NGA). U.S. Geological Survey, USA. Retrieved February 2, 2016, from https://ita.cr.usgs.gov/SRTM

Telford, W. M., Geldart, L. P., & Sheriff, R. E. (1990). *Applied Geophysics* (2nd Ed.). Cambridge University Press.

Tweed, S. O., Weaver, T. R., & Cartwright, I. (2005). Distinguishing groundwater flow paths in different fractured rock aquifers using groundwater chemistry: Dandenong Ranges, southeast Australia. *Hydrogeology Journal, 13*(5), 771-786.

Vail, J. (2011). *Groundwater Sampling*; SESDRPROC-301-R3; U.S. Environmental Protection Agency, Science and Ecosystem Support Division: Athens, GA, USA.

Waltham, T., Bell, F. D., & Culshaw, F. D. (2005). *Sinkholes and Subsidence: Karst and Cavernous Rocks in Engineering and Construction.* Springer-Verlag, Chichester, UK.

Wonik, T. (2007). Borehole logging. In K. Knodel, G. Lange, & H. J. Voigt (Eds.), *Environmental Geology: Handbook of Field Methods and Case Studies* (pp. 431-474). Springer Press, Berlin.

WWRD. (2015). Water for a sustainable world, The United nations World water Development Report 2015. UNESCO, Paris, France.

Zhu, J., Currens, J. C., & Dinger, J. S. (2011). Challenges of using electrical resistivity method to locate karst conduits-A field case in the Inner Bluegrass Region, Kentucky. *Journal of Applied Geophysics, 75*, 523-530.

4

Trajectory Modelling of Atmospheric Pollutants Associated with Aircraft Emissions for Air Quality Assessment

Marilia Mitidieri Fernandes de Oliveira[1*], Nelson Francisco Favilla Ebecken[1], Jorge Luiz Fernandes de Oliveira[2], Marina Aires[2] & José Maria de Castro Jr.[3]

[1] Center of Technology, Federal University of Rio de Janeiro-Civil Engineering Postgraduate Program–COPPE/UFRJ, Rio de Janeiro, Brazil

[2] Geography Postgraduate Program, Geoscience Institute of Fluminense Federal University-UFF, Niterói, Brazil

[3] Defense and Civil Security Postgraduate Program of Fluminense Federal University-UFF, Niterói, Brazil

Correspondence: Marilia Mitidieri Fernandes de Oliveira, Center of Technology, Federal University of Rio de Janeiro, Rio de Janeiro State. E-mail: marilia@coc.ufrj.br

Abstract

Aviation is suffering a large increase in recent years and as result there is a consequent increase in emissions of air pollutants. The highest exposure levels of air pollution in airports and near them are found in the Landing and Take-off (LTO) cycle responsible for negative impacts on air quality. The city of Rio de Janeiro, in the coastal region of the South East of Brazil, has two major airports; the International airport and one for domestic flights. Both are near the Guanabara Bay (GB), in Metropolitan Region of Rio de Janeiro (MRRJ). In this research, it was verified the pollutant trajectories emitted by these two airports using the Brazilian Regional Atmospheric Modelling System (BRAMS) to generate wind fields in the MRRJ. From the wind fields, 3D kinematic trajectories Lagrangian model was used to calculate the pollutant trajectories. Atmospheric instability indices were tested from the BRAMS simulations to verify the days with more stable atmospheric conditions in winter month. The Total Totals index (TT) was selected to be more appropriate for the purpose of this study. Obtained results showed that the cities of Rio de Janeiro and Duque de Caxias, in the western of the MRRJ, are critical points to be affected continually by pollutant transport due to the prevailing meteorological conditions, harming the air quality in the region. These models captured the influence of mesoscale and large-scale systems, showing the dependence of the trajectories of both systems related to the season and time of day. They are important tools in decision-making for the control of emissions, contributing to better management of urban air quality.

Keywords: aircraft transportation, aircraft emissions, air pollution, meteorological models

1. Introdution

Civil aviation is a strongly growing transportation sector and has been steadily growing since the late 1960s when air travel started to become more widespread and economically accessible (Lee et al., 2010) and therefore pollutant gas emissions from air transportation play a key role in reduce air quality. It is an important contributor to global emissions of many different gases and aerosols that can have an impact on climate, either directly or indirectly (Fuglestvedt et al., 1999; Ambrosino, Sassoli, Bielli, Carotenuto, & Romanazzo, 1999; Eyring, Kohler, van Aardenne, & Lauer, 2005; Fuglestvedt et al., 2010). Lately it has been investigated aviation's effects on the global atmosphere in national and international research programmes, including the IPCC (1999; 2000; 2001; 2007a; 2007b; 2014) report.

Photochemical air pollution is recognized to be a world–wide problem in areas where emissions from mobile and stationary sources occur, affecting the air quality due to meteorological phenomena. Nowadays, climate impacts of emissions from aviation, maritime and road transports have been investigated in parallel by applying consistent methodologies, which allow direct comparison of their impact on global climate (Sausen, 2010; Heal, Kumarbc, & Harrisonde, 2012).

Kerosene is a conventional jet fuel, being a derivative of crude oil (Wood, Long, & Morehouse, 2004) with a great specific energy content, allowing fly on longer distances, carrying passengers and cargo (Janic, 2010). According to the International Civil Aviation Organization (ICAO), Volatile Organic Compounds (VOC), Nitogen-Oxides (NO_x),

Carbon-Monoxide (CO), Carbon-Dioxide (CO_2), Sulfhur-Dioxide (SO_2), Methane (CH_4), Nitrous-Oxide (N_2O), water vapour (H_2O) and soot particles are considered typical aircraft emissions related to burning conventional jet fuel, contributing in various ways to air pollution. As aviation is suffering a large increase there is a consequent increase of greenhouse gases (GHG) (CO_2, CH_4 and N_2O) and ozone (O_3) depletion, puting this modal on the list of priorities (Brüning, 2010). Aircraft emissions of CO_2, SO_2, NO_x and O_3 precursors change atmospheric concentrations of GHG and aerosol mainly Sulphate (SO_4), causing both positive and negative contributions to direct radiative forcing (RF) of climate and, by convention, these impacts are quantified using this metric (Lee et al., 2010). The particles can also have an indirect effect on climate through their ability to change the properties of clouds, mainly in formation of contrail-cirrus cloud from spreading contrails (Eyring, 2010).

The incremental contribution of civil air transport system occurs at ground level during landing and take-off (LTO) cycles (below 3000 ft) and when the plane is at cruise altitude according to Aviation Environmental Design Tool (AEDT FAA).Therefore is not clear whether aircraft would be more like mobile sources or like power plants in the magnitude and spatial distribution of their impacts, especially for emissions below cruise altitude and during LTO cycles at airports considered so as stationary sources.

Airports represent a source of growing interest for health impact analysis related to air pollution and may represent an increasingly large fraction of the emissions inventory (Arunachalam, Wang, Davis, Baek, & Levy, 2011; Ashok et al., 2013). Some studies have contributed to the understanding spatial patterns of air pollution near airports, quantifying the air quality related to aircraft contributions but lack information about spatial extent of these effects, specially on population health (Moussiopoulos, Sahm, Karatzas, Papalexiou, & Karagiannidis, 1997; Pison & Menut, 2004; Yu, Cheung, & Henry, 2004, Borrego et al., 2010, Koo, Wang, Henze, Waitz, & Barrett, 2013).

In general, the total air pollution is proportional to the number of polluting activities, being its intensity related per activity which in turn is related to the volume of traffic, usually expressed as the number of passengers and freight shipments during the LTO cycle. The intensity of air pollution per activity depends on the quantity and type of the energy consumed and, in addition, must be consider the movement of vehicles, aircraft servicing and the heating and lighting of the passenger and cargo terminals (Janic, 2010).

From the 2000s, in order to reduce the effects of air pollutant emissions by ships and aircraft, several measures are imposed from international, national and local legislators such as the United Nations Framework Convention on Climate Change (UNFCCC), International Maritime Organization (IMO) and the International Civil Aviation Organization (ICAO) (Dessens, Anger, Barker, & Pyle, 2014).

Brazil follows the United States Environmental Protection Agency (US-EPA) and the World Health Organization (WHO) air pollution guidelines since 1976. In 1989, the National Program for the Air Quality Control (PRONAR) was launched as baseline for the environmental management and human health protection. The first legal instrument connected to the PRONAR was released in 1990 where new air quality patterns were determined. From 2000 was represented for the first time in the Reference Report of Greenhouse Gas Emissions as part of the 2nd National Communication to the United Nations Framework Convention on Climate Change (MCT, 2010).

The Brazilian Action Plan (2013) aims to share information with ICAO for contributing with the global effort of reducing GHG from international civil aviation. Therefore, between 1990 and 2012 was verified a rapid growth in domestic operations. Now, the publication of the National Inventory of Civil Aviation Air Emissions (ANAC, 2013; 2014) enables various actions such as evaluation of emissions and better knowledge about local pollutant profiles as well as direct GHG.

The State Environmental Institute (INEA) is the agency responsible for monitoring of air quality in the state of Rio de Janeiro since 1967, when the first measurement stations were installed. The resolution of the National Environment Council (CONAMA n° 03/1990) indicates the air quality patterns by pollutants levels, which cannot be exceeded in a given time and region. These parameters must be regarded for the health protection of the people, buildings and environmental resources. This Council classifies these patterns according to the exposure time, as short or long period, depending on the damage.

The concentration of pollutants is strongly related to meteorological conditions and some parameters favorable to high levels of pollution are calm, light winds and temperature inversions at low altitude. Temperature inversions are conditions that occur with regular frequency, being more frequent in the dry season. However, the inversion of subsidence may occur with the passage of cold mass over the region or by blocking condition caused due to an intense high center pressure. The temperature inversion is extremely unfavorable to the health and can stay for a few days, concentrating pollution near the ground. This behavior of the air temperature limits lateral and vertical dispersion of the pollutants, making the concentrations abnormally high if the emissions continue (Oliveira, 2004).

The MRRJ is the region with the second largest concentration of population, vehicles, industries and emission sources of pollutants, generating serious problems of air pollution. The dry season is under weak synoptic northeasterly wind and low precipitation, potentially leading to serious air pollution in MRRJ. This region, considered the second most polluted region of the country and is located in Southeast Brazilian coastal where the Guanabara Bay (GB) is inserted, undergoing the effects of the breezes.

Eulerian and Lagrangian modelling are usually used to compute the pollutant plumes which follow the atmospheric flow field (Oliveira, 2004). In the Eulerian method is traced tangent lines at speeds observed simultaneously at different points of the flow, called current lines, allowing an overview of the movement of particles in the flow. Otherwise, the Lagrangian method allows viewing the successive positions of the same particle over time (Draxler & Hess, 1998). According Doty and Perkey (1993) the path taken by a given air parcel for a period of time in a balanced flow is called trajectory but the knowledge of wind at the study area is not sufficient to determine the source of air masses. Only through the knowledge of dynamic of atmosphere is that we know the path followed by them. The calculation of air mass trajectories is very important because it enables the knowledge of pollutant emission sources transported in the atmosphere.

The above is an overview of the pollutant emissions related to air transport sector and its impact on air quality. This study is part of the research in analysis of the pollutant concentrations and trajectories in the MRRJ to verify the impacts of air pollution related to the transport sector in an urban area, densely populated and industrialized that has the air quality reduced. In addition to industries (stationary sources) and road transports (mobiles), this region is also affected by other sources such as maritime and aviation transport sectors (Oliveira, Ebecken, Oliveira, & Aires, 2016). In this context, the purpose of this research is to use the BRAMS to generate the wind fields in the MRRJ and a 3D kinematic trajectories Lagrangian model to compute the trajectories of pollutants generated by the aviation sector at ground level during LTO flight phases from two airports in Rio de Janeiro: Antônio Carlos Jobim International Airport (Galeão) and Santos Dumont Airport (Domestic flights) considered as stationary sources. These two airports are located in urban and central city of Rio de Janeiro with an intense traffic, contributing to the pollutant emissions in MRRJ. The contribution of this modal is analysed not only in MRRJ but also in other regions for the wintertime on June, 2016, considering meteorological conditions favourable to pollutant concentrations.

2. Study Area

In order to investigate the contribution of the civil aviation transport sector in Rio de Janeiro city (22°47'26"S, 43°09'20"W), South-east coast of Brazil, and its influence in the MRRJ and other regions (Figure 1), it was verified the spatial and temporal trajectories of pollutants under prevailing meteorological systems. This research is focused in two points considered as emissions from stationary sources (International Airport, one of the five largest airports in the country, and the Domestic Airport) both near to the margins of the GB. The GB is inserted in MRRJ surrounded by the cities of Rio de Janeiro, Duque de Caxias, São Gonçalo, Niterói and other smaller cities, where the breezes recirculate pollutants in the region. In addition to the meteorological mesoscale systems such as breezes, the study area is under influence by the South Atlantic semi-stationary subtropical anticyclone system, a large-scale circulation pattern. The high levels of air pollution depend on the region and topography relative to the general circulation of the atmosphere and this atmospheric system also determines the air quality in MRRJ according to its positioning along the Brazilian coastline, presenting a well-defined seasonal movement.

During the winter, this system becomes stronger than summer, moving northerly. On its western side the winds blow north-easterly towards the south-eastern coast of South America and they are less intense on the south-eastern coast of Brazil where the GB is located (Figure 2). This circulation is periodically disturbed by the passage of frontal systems caused by anticyclones that move from the south west towards the north-east on the south-eastern coast of Brazil. Therefore, there is a weakening of the northerly winds and an intensification of the south westerly winds which facilitates the track of the cold fronts that reach the south and south-east coast of Brazil (Ahrens, 2000).

Other important systems that affect the region are the mesoscale meteorological systems known as breezes. These wind systems are affected by local topography and the differential heating of the surface, favouring or harming the places where they move. The breezes are generated due to the differential heating between the GB and the continent, recirculating pollutants in MRRJ. Usually, the sea breeze follows the south-north direction carrying pollutants into the Metropolitan area during the day. Otherwise, during the night, the land breeze follows the north-south direction carrying pollutants to densely urbanized regions of the coast. The topography of the MRRJ with the hills located parallel to the coastline acts as a physical barrier to the winds from the sea, blocking the ventilation in areas located further inland. At the regional scale, the breeze circulation due to the temperature difference between the sea surface and the continent can change air quality by recirculation of pollutants (Luhar et al., 2004). Therefore, it is necessary to know the meteorological, geographical and social-economic variables and the chemistry of the troposphere to manage air quality in cities.

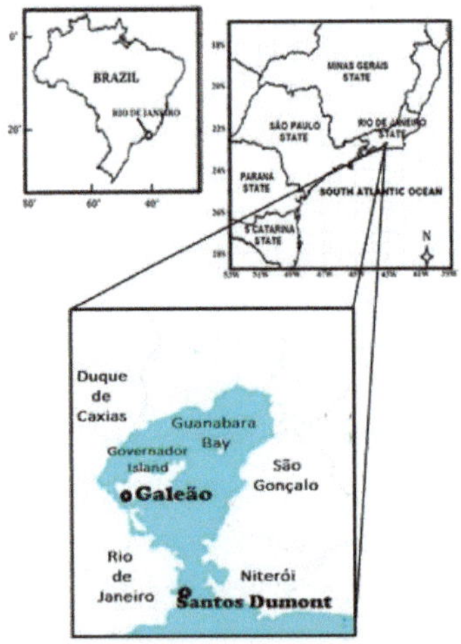

Figure 1. Study area with the localization of Galeão and Santos Dumont Airports. The International Airport is located on the Governador Island and the Domestic Airport near the city center, both close to the margins of the Guanabara Bay. The figure shows the major cities of MRRJ

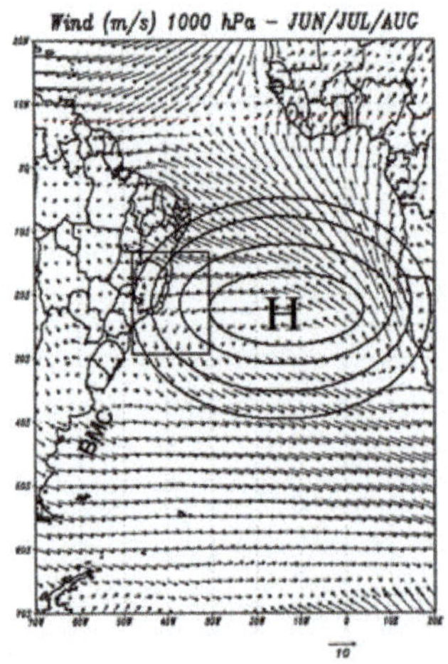

Source: Adapted from NCEP/NCAR reanalysis dataset.

Figure 2. Surface wind-flow pattern with a scheme of the South Atlantic high-pressure system for winter. The circles represent the isobars and the positions of the high center. The highlighted area shows the winds blowing over the study area; BMC - Brazil Malvinas Confluence region.

3. Materials and Methods

3.1 Meteorological Dataset

In this research were used the reanalysis from the National Centers for Environmental Prediction-NCEP and National Center for Atmospheric Research-NCAR (Kistler et al., 2001) for initialization of the Brazilian Regional Atmospheric Modeling System (BRAMS-http://www.brams.cptec.inpe.br) to generate the wind fields over the MRRJ on June, 2016. This model is a set of computer program that simulates the atmosphere for weather and climate research and for numerical weather prediction.

3.2 Atmospheric Simulations

The BRAMS was used with their parametrization standards and the initial condition and boundary to generate the wind fields over the MRRJ, using NCEP/NCAR reanalysis data set. The reanalysis data set of the climatological average of June, 2016 (dry season) were submitted to analysis package of this model, which treated and interpolated the data of wind components, temperature and relative humidity at several pressure levels for isentropic and sigma_Z levels. The initial condition and boundary for numerical integrations were obtained by applying objective analysis (Barnes, 1973) of these two levels.

The boundary layer was finely refined with stretch ratio of 1.2 up to 1 km. In the wind field simulations were used three nested grids (Clark, & Farley, 1984); the first with a resolution of 40 km x 40 km (thick grid), the second with 10 km x 10 km and the third, of most interest (high resolution grid), 2.5 km x 2, 5 km (thin grid).

The downscaling technique allowed the information of the largest scales passed to the smaller scales and vice-versa with the two-way nesting. For the model assimilate the data, in the center, on the sides and on top of the domain it was used the Newtonian relaxation (nudging) as an alternative form of dynamic initialization of the data set, with three nested grids centered between the two airports whose distance between them is approximately 20 km. The nudging allows the model to reach the desired fitting after a few hours of integration and was applied from the fifth grid point on the sides and until 15 km on top of the grid. The thick grid captured the large scale information and the thin grid covered the domain MRRJ. To view the wind field, it was used the Grids Analysis and Display System-GRADS (Doty & Perkey, 1993).

Atmospheric instability indices play an important role to verify synoptic and mesoscale features favorable for the development of thunderstorms. These indices were examined using the BRAMS to identify which days with atmospheric stability conditions occurred. These weather conditions are favorable to the concentration of pollutants over a region. Although these indices have been developed for mid-latitudes, it does not imply the non-use of them to Brazil; they can be parametrized and represent a useful tool to characterize the atmospheric stability conditions.

The LTO cycle includes all flight phases near the airport carried out by aircraft at altitudes less than 914.4 meters (or 3000 feet). Therefore, the trajectories of pollutants were calculated forward, considering this flight phase where there are the most intense emissions at low levels of the atmosphere and at ground. The trajectory of air pollutants was numerically investigated using a 3D kinematic trajectories Lagrangian mode (Freitas, Longo, Silva Dias, & Artaxo, 1996; Freitas, 1999; Freitas et al., 2000) to compute the trajectories of pollutants emitted by the two airports with the wind fields generated by the BRAMS atmospheric modeling. To verify the extent of the pollutants were used integration time of 24 hours for tracking the trajectories, both in the region and elsewhere outside the MRRJ. The trajectories calculated by the wind field generated in the high-resolution grid will not present herein because the results were not consistent with the flow of mesoscale systems type breezes due to the influence of GB. These local flows are affected by the topography and the differential heating of the surface, favoring or harming the places where they move. In some cases, the orientation and type of rocks influence the flow in the Planetary Boundary Layer (PBL), arising pollutants around 300-900 meters of height in MRRJ. After analysing the weather conditions in June 2016, the period from 26-29 at 0600, 1200 and 1800 local time (LT) was used in the simulations of the trajectories due to favourable conditions for air pollutant concentrations in the region.

3.3 Calculation of Air Parcel Trajectories

Air parcels trajectories were obtained at different times and heights in order to verify the influence of the meteorological systems in the trajectory of pollutants emitted in the region.

The calculation of trajectories is based on the Equation (1) (Petterssen, 1956):

$$\frac{dr(t)}{dt} = V(r,t) \tag{1}$$

where

r(t) is the position of the air parcel at time t;

V is the horizontal wind field (*u, v*) and the vertical wind (*w*).

Thereby is possible to know the output and arrival of pollutants emitted in a region. From the Equation (1), the position of the air parcel *r(t)* is defined by Equation (2) to the output trajectory (forward) after the time lag $\Delta t = (t_1 - t_0)$:

$$r(t_1) = r(t_0) + \int_{t_0}^{t_1} V[r(t), t]\, dt \tag{2}$$

The calculation of the arrival trajectory (backward) is also obtained from Equation (1).

The level (height) reached by a particle or pollutant, along the trajectory of air parcels, is represented by the color scale of the figure.

4. Results and Discussion

4.1 Atmospheric Instability Indices

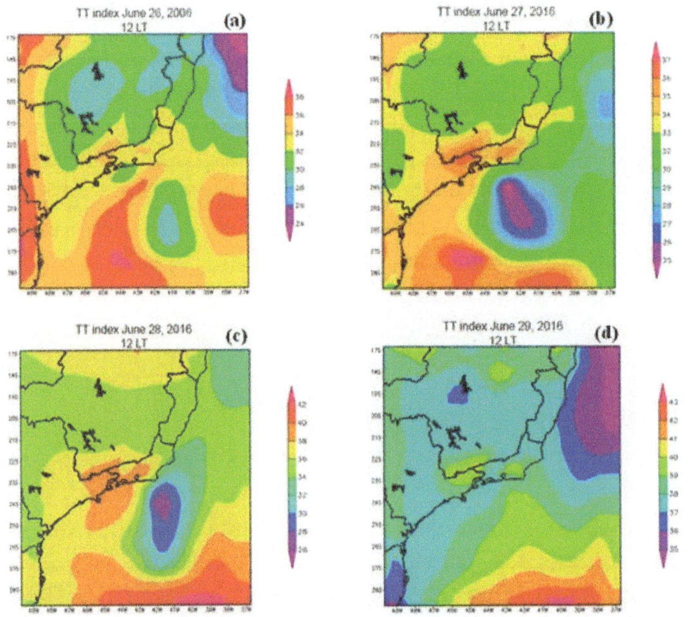

Figure 3 (a)-(d): TT Index simulation maps obtained with the BRAMS model for selected days in thick grid. TT values for the study area 12 LT, (a) June 26 around 32, (b) June 27 between 33-35, (c) June 28 around 38 and (d) June 29 around 38. All values were below 44, characterizing a condition close to atmospheric stability.

Table 1. Total Totals (TT) threshold values

<44	Convection not likely
44-50	Likely thunderstorms
51-52	Isolated severe storms
53-56	Widely scattered severe
>56	Scattered severe storms

Source: Miller, 1972; Severe weather índices page (www.theweatherprediction.com/severe/indices, accessed on October, 10 2016).

Table 2. Meteorological data obtained from automatic station. LT-local time; T- temperature; H- humidity; Pr-pressure and P-precipitation

Date (June, 2016)	LT	T (°C)	H (%)	Pr (hPa)	Wind spead (m/s)	Wind direction	P (mm)
26	12	22.9	68	1023.0	3.4	SE	0.0
26	18	19.5	88	1022.3	4..3	ESE	0.0
27	12	23.3	70	1022.9	2.9	SE	0.0
27	18	20.2	87	1021.9	6.4	ESE	0.0
28	12	24.0	72	1021.4	2.6	SE	0.0
28	18	21.2	84	1019.3	2.0	E	0.0
29	12	23.8	70	1019.6	1.6	ESE	0.0
29	18	21.3	76	1018.4	0.4	NE	0.0

Source: National Weather Institut (INMET) – Copacabana automatic station.

Atmospheric instability indices were tested from the BRAMS simulations showing the days with more stable atmospheric conditions The Total Totals index (TT) was more appropriate to evaluate stability categories of weather for the purpose of this study and the specific threshold values are presented in Table 1. Generally, values of TT greater than or equal to 44 are indicative of thunderstorms or high convective instability (Sioutas, Szilagyi, & Keul, 2013). The TT index calculated from the BRAMS simulations showed that June 26, 27, 28 and 29 had stable atmospheric conditions and therefore were selected for analysis. The meteorological data obtained from automatic station (Table 2) for selected days show a predominance of large-scale circulation pattern (Subtropical high pressure system) over the region with light winds in SE-E-NE direction and no precipitation. Results of TT index are presented in Figure 3.

4.2 Wind Field Simulations

The topography influences the speed and direction of winds in MRRJ. The atmospheric flow simulated for the selected days at 12 LT are presented in Figure 4 (a) to (d) and refer to the land breeze. Due to the heating of the surface in the morning, the land breeze is weak at 12 LT, but even so this mechanism is responsible for the transport of pollutants towards the coastline and ocean area. The wind field simulations on the coastline are consistent with the measured data in the automatic station (Table 2).

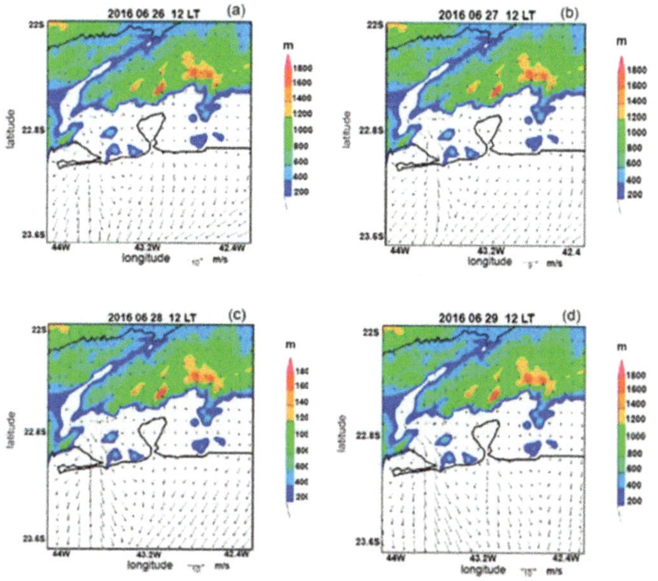

Figure 4 (a)-(d). Maps of wind field and topography on June 26, 27, 28, 29, 2016 at 1200 LT (a), (b), (c) and (d), respectively show weak land breeze in the MRRJ in high resolution grid

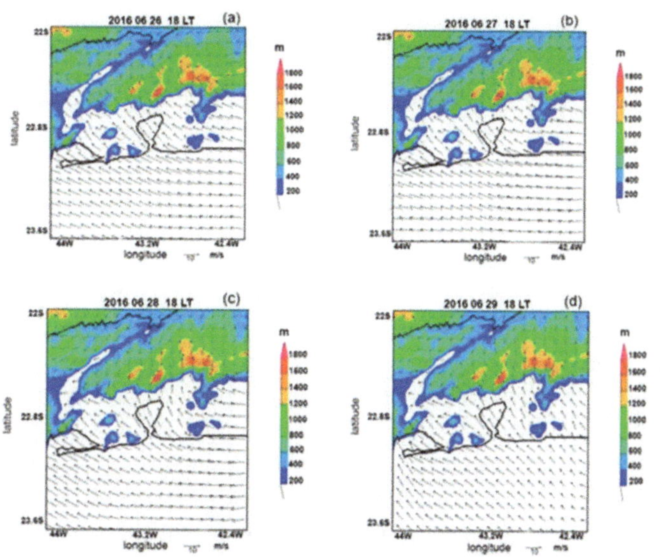

Figure 5 (a)-(d). Maps of wind field and topography on June 26, 27, 28, 29, 2016 at 1800 LT (a), (b), (c) and (d), respectively, show sea breeze more intense in the MRRJ in high resolution grid. Wind field shows E and SE wind direction with speed around 10 ms^{-1}

The atmospheric flow simulated for the selected days at 18 LT are presented in Figure 5 (a) to (d) and refer to the sea breeze. Due to the heating of the surface in the afternoon, the strong sea breeze at 18 LT is responsible for the transport of pollutants towards the MRRJ. It is also verified that the GB canalizes the flow over the region, facilitating the transport of pollutants. The sea breeze is undoubtedly the main natural mechanism of pollutant transport for densely populated area of MRRJ. The wind field simulations on the coastline are consistent with the measured data in the automatic station only on 26-27 (Table 2).

4.3 Trajectories of Air Pollutants

Obtained results of the 3D kinematic trajectories models during winter month (June 26 -29, 2016) from wind field simulations by means of the BRAMS showed the trajectories of pollutants emitted by two airports. Figure 5 and 6 show the trajectories calculated at 1200 LT and 1800 LT, respectively, using 40-km resolution outputs during 24 hours, within a domain surrounding each airport.

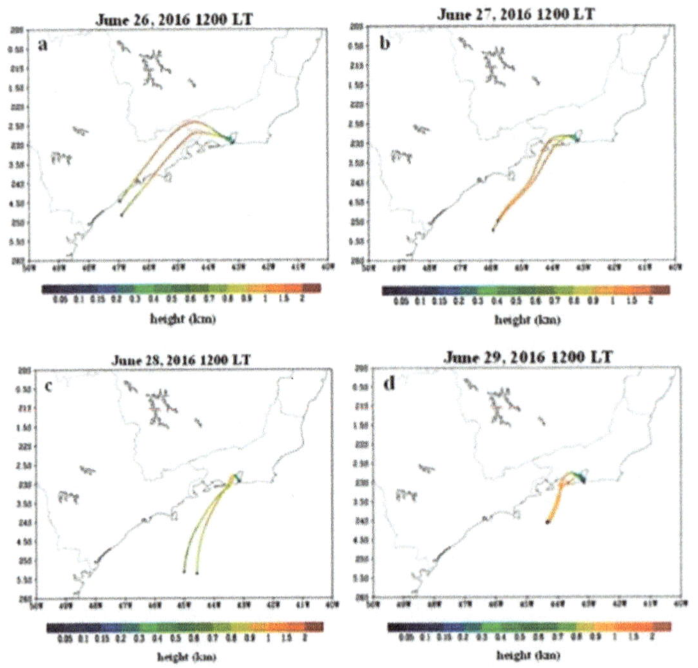

Figure 6 (a)-(d). Trajectories of pollutants simulated with 3D kinematic model on June 26-29, 2016 at 1200 LT. The heights of the pollutant trajectories are represented by the color scale of the figure

At 1200 LT (Figure 6) pollutants followed trajectories towards NW around 50 meters of height, on the Central and coastal areas of west zone of the city of Rio de Janeiro and then following the SW direction, reaching other states close to the state of Rio de Janeiro and the South Atlantic Ocean, remaining below 2 km of height. Figure 5a shows that on 26 pollutants initially moved to NW (from 50 to 500 meters, approximately) reaching the states of Minas Gerais and São Paulo that border with the state of Rio de Janeiro (mountain regions) around 1500 meters of height, returning towards SW and reaching the coast of the state of São Paulo. On 27 and 29 pollutants also moved to NW (from 50 to 500 meters) and to SW towards the Restinga da Marambaia and Sepetiba in the west zone of Rio de Janeiro, as well as Angra dos Reis Bay, lifting from 1000 to 1800 meters of height (Figures 5b and 5d), following towards the oceanic region. On 28 trajectories remained at low levels over the west zone of Rio de Janeiro, returning to SW towards the ocean, lifting from 400 to 900 meters of height. These simulations indicate that the trajectories not followed the land breeze that remained weak in the region in the analyzed period, being the trajectories associated with the positioning of South Atlantic subtropical high-pressure system.

Under the influence of sea breeze on 26-29 at 1800 LT pollutants were transported towards NW around 50 (near airports) and 500 (in the MRRJ and mountain region near the states of Rio de Janeiro and Minas Gerais) meters (Figure 7). On 27 pollutants moved to NW (around 50 meters), returning to SW also towards the Restinga da Marambaia and Sepetiba, as well as Angra dos Reis Bay, lifting from 150 to 800 meters (Figure 5b). On 28 pollutants also followed the NW direction in the MRRJ, rising around 400 meters and returning to SW towards ocean area, dropping between 150-100 meters. For the four analyzed days, it was verified that pollutants emitted by both airports remain at low levels in the west side of MRRJ due to meteorological conditions, harming the air quality in the region. This metropolitan region is densely populated and greatly affected continually by pollutant transport emitted from vehicles, industries and further the civil air transport system contributing to serious problems of air pollution, causing therefore degradation in human health.. In this scenario, the model captured well both meteorological systems, indicating a critical situation of the air quality in the study area in periods of atmospheric stability.These models are important tools in decision-making for the control of emissions to minimize the harmful effects of pollutants, contributing to better management of air quality.

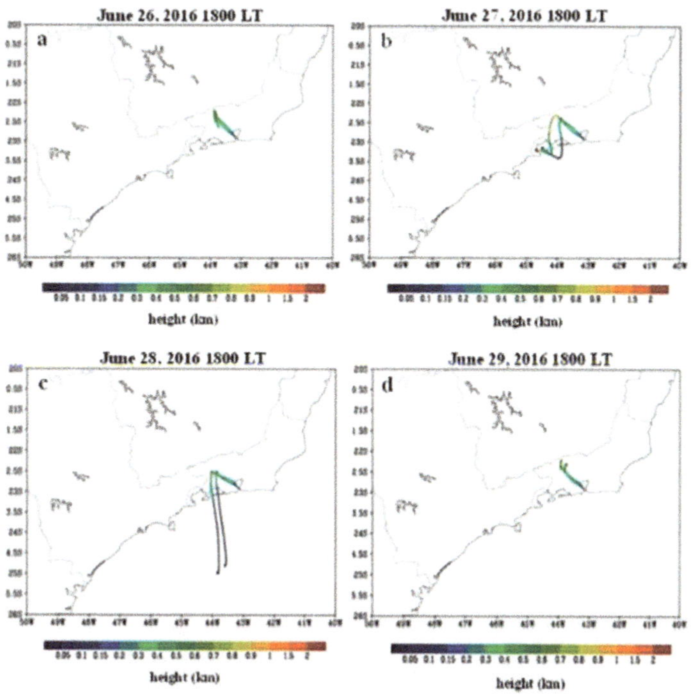

Figure 7 (a)-(d): Trajectories of pollutants simulated with 3D kinematic model on June 26-29, 2016 at 1800 LT. The height of the pollutant trajectories are represented by the color scale of the figure.

4. Conclusion

The contribution of air transport system in pollutant emissions that occurs at ground level during LTO cycles plays a key role, because is currently growing fast worldwide. Emissions of many different gases and aerosols induce an impact on air quality of the cities close to the airports, representing a source of growing interest for health impact analysis. The air transport influences the air quality in MRRJ and vicinity where the air quality has a strong influence of meteorological systems, especially type breeze, which recirculate pollutants in this region. The BRAMS and 3D kinematic trajectories used in this research showed the influence of these systems on the transport of pollutants generated by the two airports in the winter month. Usually during the occurrence of stability weather conditions as verified in this study, the trajectories follow westward of MRRJ, reaching the cities of the Rio de Janeiro and Duque de Caxias due to prevailing meteorological systems. In the analysed period pollutants followed trajectories on the central and coastal areas of west zone of the city of Rio de Janeiro during the morning and more intensely on the central regions in the afternoon keeping up at low heights, harming the air quality in the city. These areas of the city have air quality quite impaired not only because of the pollutants generated by the two airports but also from other sources mainly due to its concentration at low levels almost all day. Therefore the models captured the influence of mesoscale and large-scale systems, showing the dependence of the trajectories of both systems related to the season and time of day.

This study shows the need for better management of air transportation sector to reduce the harmful effects on air quality in the MRRJ generated by this sector. The complexity of the physical processes and their impacts must be analysed, as well as aspects related to legislation, regulatory systems and articulation between several plans and programs of different sectors of administration. It should consider a policy that favours the required patterns for urban development without environmental degradation, aiming to improve the air quality.

Acknowledgement

The authors acknowledge the financial support of the National Scientific and Technological Development Council (CNPq).

References

ANAC. (2013). National Civil Aviation Agency Brazil (ANAC). Brazil's Action Plan on the reduction of Greenhouse Gas Emissions from aviation. Retrieved from http://www.anac.gov.br/publicacoes/brazil_actionplan.pdf

ANAC. (2014). National Civil Aviation Agency Brazil (ANAC). National Inventory of Civil Aviation Air Emissions. Retrieved from http://www.anac.gov.br/assuntos/paginas-tematicas/meio-ambiente/inventario-nacional-de-emissoes-atmosfericas-da-aviacao-civil-2014

Ambrosino, G., Sassoli, P., Bielli, M., Carotenuto, P., & Romanazzo, M. (1999). A modeling framework for impact assessment of urban transport systems. *Transportation Research Part D: Transport and Environmental*, *4*(2), 73-79. http://dx.doi.org/10.1016/S1361-9209(98)00014-5

Ahrens, C. D. (2000). *Meteorology today: An introduction to weather climate and the environment* (6th ed.). London: Brooks/Cole.

Arunachalam, S., Wang, B., Davis, N., Baek, B. H., & Levy, J. I. (2011). Effect of chemistry-transport model scale and resolution on population exposure to PM2.5 from aircraft emissions during landing and takeoff. *Atmospheric Environment*, *45*(19), 3294-3300. h http://dx.doi.org/10.1016/j.atmosenv.2011.03.029

Ashok, A., Lee, I. H., Arunachalam, S., Waitz, I. A., Yim, S. H. L., & Barrett, S. R. H. (2013). Development of a response surface model of aviation's air quality impacts in the United States. *Atmosferic Environment*, *77*, 445-452. http://dx.doi.org/10.1016/j.atmosenv.2013.05.023

Barnes, S. L. (1973). Mesoscale objective map analysis using weighted time series observations. *NOAA Technical Memorandum NSSL-62*. Retrieved from http://docs.lib.noaa.gov/noaa_documents/OAR/NSSL/NOAA_TM_ERL_NSSL/ERL_NSSL_62.pdf

Borrego, C., Monteiro, A., Ferreira, J., Moraes, M. R., Carvalho, A., Ribeiro, I., Miranda, A. I., & Moreira, D. M. (2010). Modelling the photochemical pollution over the metropolitan área of Porto Alegre, Brazil. *Atmospheric Environment*, *44*, 370-380. http://dx.doi:10.1016/j.atmosenv.2009.10.027

Brüning, C. (2010). Transport impacts on atmosphere and climate. *Atmospheric Environment, 44*, 45-46. http://dx.doi.org/10.1016/j.atmosenv.2010.04.018

Clark, T. L., & Farley, R. D. (1984). Severe downslope windstorm calculations in two and three spatial dimensions using anelastic interactive grid nesting: a possible mechanism for gushiness. *Journal of Atmosphere Science*, *41*, 329-350. http://dx.doi.org/10.1175/1520-0469(1984)041<0329:SDWCIT>2.0.CO;2

Dessens, O., Anger, A., Barker, T., & Pyle, J. (2014). Effects of decarbonising international shipping and aviation on climate mitigation and air pollution *Environmental Science & Policy, 44*, 1–10. http://dx.doi.org/10.1016/j.envsci.2014.07.007

Doty, K. G., & Perkey, D. J. (1993). Sensitivity of trajectory calculations to the temporal frequency of wind data. *Monthly Weather Review*, *121*, 387-401. http://dx.doi.org/10.1175/1520-0493(1993)121<0387:SOTCTT>2.0.CO;2

Draxler, R. R., & Hess, G. D. (1998). An overview of the HYSPLIT_4 modelling system for trajectories dispersion and deposition. *Australian Meteorological Magazine, 47*, 295-308. Retrieved from http://www.arl.noaa.gov/documents/reports/MetMag.pdf

Freitas, S. R., Longo, K. M., Silva Dias, M. A. F., & Artaxo, P. (1996). Numerical modeling of air mass trajectories from the biomass burning areas of the Amazon Basin. *Anais da Academia Brasileira de Ciências, 68*. 193-206. Retrieved from https://www.researchgate.net/publication/257875020

Freitas, S. R. (1999). *Numerical Modeling of Transport and Emissions of Trace Gases and Aerosols from Fires in the Cerrado and Tropical Rainforest in South America* (Doctoral dissertation). Retrieved from http://bdtd.ibict.br/vufind/Record/USP_e372ff9fbd538f8c9100b8deae4049cc

Frcitas, S. R., Silva Dias, M. A. F., Silva Dias, P. L., Longo, K. M., Artaxo, P., Andreae, M. O., & Fisher, H. (2000). A convective kinematic trajectory technique for low-resolution atmospheric models. *Journal of Geophysical Research, 105*(19), 24, 375-24, 386. http://dx.doi.org/0148-0227/00/2000JD900217509

Fuglestvedt, J. S., Berntsen, T. K., Isaksen, I. S. A., Mao, H. T., Liang, X. Z., & Wang, W. C. (1999). Climatic forcing of nitrogen oxides through changes in tropospheric ozone and methane; global 3D model studies. *Atmospheric Environment, 33*, 961–977. http://dx.doi.org/10.1016/S1352-2310(98)00217-9

Fuglestvedt, J. S., Shine, K. P., Berntsen, T., Cook, J., Lee, D. S., Stenke, A., Skeie, R. B., Velders, G. J. M., & Waitz, I. A. (2010). Transport impacts on atmosphere and climate: metrics. *Atmospheric Environment, 44*, 4648–4677. http://dx.doi.org/10.1016/j.atmosenv.2009.04.044

Eyring, V., Kohler, H. W., van Aardenne, J., & Lauer, A. (2005). Emissions from international shipping: 1. The last 50 years. *Journal of Geophysical Research-Atmospheres, 110(D17)*, D17305. http://dx.doi.org/10.1029/2004JD005619

Eyring, V., Isaksen, I. S. A., Berntsen, T., Collins, W. J., Corbett, J. J., Endresen, O., Grainger, R. G., Moldanova, J., Schlager, H., & Stevenson, D.S. (2010). Transport impacts on atmosphere and climate: shipping. *Atmospheric Environment, 44*, 4735–4771. http://dx.doi.org/10.1016/j.atmosenv.2009.04.059

Heal, M. R., Kumarbc, P., & Harrisonde, R. M. (2012). Particles, air quality, policy and health. *Chemical Society Reviews, 41*, 6606–6630. http://dx.doi.org/10.1039/c2cs35076a

Federal Aviation Administration. (2016). *Aviation Environmental Design Tool (AEDT).* Retrieved June 8, 2016, from http://www.faa.gov/about/office_org/headquarters_offices/apl/research/models/aedt/

IPCC. (1999). Aviation and the global atmosphere. In E. Penner, J., D. H. Lister, D. J. Griggs, D. J. Dokken, M. McFarland (Eds.), *Intergovernmental Panel on Climate Change.* Cambridge University Press, Cambridge, UK. http://www.ipcc.ch/ipccreports/sres/aviation/

IPCC. (2000). *Emissions Scenarios. A Special Report of WG III of the Intergovernmental Panel on Climate Change.* Cambridge University Press, UK. https://www.ipcc.ch/pdf/special-reports/emissions_scenarios.pdf

IPCC. (2001). Climate change 2001: the scientific basis. In J. T. Houghton, Y. Ding, D. J. Griggs, M. Noguer, P. J. van der Linden, X. Dai, … C. A. Johnson (Eds.), *Contribution of Working Group I to the Third Assessment Report of the Intergovernmental Panel on Climate Change.* Cambridge University Press, UK. Retrieved from http://www.grida.no/climate/ipcc_tar/wg1/pdf/wg1_tar-front.pdf

IPCC. (2007a). Climate change 2007. The physical science basis. In S. Solomon, D. Qin, M. Manning, M. Marquis, K. Averyt, M. M. B. Tignor, … Z. Chen (Eds.), *Contribution of Working Group I to the Fourth Assessment Report of the Intergovernmental Panel on Climate Change.* Cambridge University Press, Cambridge, United Kingdom and New York, NY, USA. Retrieved from https://www.ipcc.ch/publications_and_data/publications_ipcc_fourth_assessment_report_wg1_report_the_physical_science_basis.htm

IPCC. (2007b). Climate change 2007. Mitigation of climate change. In B. Metz, O. R. Davidson, P. R. Bosch, R. Dave, & L. A. Meyer (Eds.), *Contribution of Working Group III to the Fourth Assessment Report of the Intergovernmental Panel on Climate Change.* Cambridge University Press, Cambridge, United Kingdom and New York, NY, USA. Retrieved from http://www.ipcc.ch/publications_and_data/ar4/wg3/en/contents.html

IPCC. (2014) *Climate change: Mitigation of Climate Change. Contribution of Working Group III to the Fifth Assessment Report of the Intergovernmental Panel on Climate Change. Chapter 8 (Transport).* Cambridge University Press, Cambridge, UK. Retrieved from https://www.ipcc.ch/pdf/assessment-report/ar5/wg3/ipcc_wg3_ar5_full.pdf

Janic, M. (2011). Is liquid hydrogen a solution for mitigating air pollution by airports? *International Journal of Hydrogen Energy, 35*, 2190-2202. http://dx.doi:10.1016/j.ijhydene.2009.12.022

Kistler, R., Collins, W., Saha, S., White, G., Woollen, J., Kalnay, E., ... & van den Dool, H. (2001). The NCEP-NCAR 50-year reanalysis: Monthly means CD-ROM and documentation. *Bulletin of the American Meteorological society, 82*(2), 247-267. Retrieved from http://journals.ametsoc.org/doi/pdf/10.1175/1520-0477%282001%29082%3C0247%3ATNNYRM%3E2.3.CO%3B2

Koo, J., Wang, Q., Henze, D. K., Waitz, I. A., & Barrett, S. R. H. (2013). Spatial sensitivities of human health risk to intercontinental and high-altitude pollution. *Atmospheric Environment, 71*, 140-147. http://dx.doi.org/10.1016/j.atmosenv.2013.01.025

Lee, D. S., Pitari, G., Grewe, V., Gierens, K., Penner, J. E., Petzold, A., … Sausen, R. (2010). Transport impacts on atmosphere and climate: Aviation. *Atmospheric Environment, 44*, 4678–4734. http://dx.doi.org/10.1016/j.atmosenv.2009.06.005

Luhar, A. K., & Hurley, P. J. (2004). Application of a prognostic model TAPM to sea-breeze flows, surface concentrations, and fumigating plumes. *Environmental Modelling & Software, 19*, 591–601. http://dx.doi.org/10.1016/j.envsoft.2003.08.011

MCT, (2010). Ministry of Science and Technology. 2st National Communication to the United Nations Framework Convention on Climate Change. Retrieved from http://www.google.com.br/search?q=Ministry+of+Science+and+Technology.+1st+National+Communication+to+the+United+Nations+Framework+Convention+on+Climate+Change.&ie=utf-8&oe=utf-8&client=firefox-b&gfe_rd=cr&ei=ISSrV72XNfGp8weRhoXwAQ

Miller, R. C. (1972). Notes on analysis and severe-storm forecasting procedures of the Air Force Global Weather Central, AWS Tech. Rpt. 200(rev), Air Weather Service, Scott AFB, IL, 1972.

Moussiopoulos, N., Sahm, P., Karatzas, K., Papalexiou, S., & Karagiannidis, A. (1997). Assessing the impact of the new Athens airport to urban air quality with contemporary air pollution models. *Atmospheric Environment, 31*, 1497-1511. http://dx.doi.org/10.1016/S1352-2310(96)00283-X

Oliveira, J. L. F. de. (2004). Spatial Analysis and Atmospheric Modeling Contributions Air Quality Manegement of Airshed III of Metropolitan Region of Rio de Janeiro (Ph.D. Thesis). Federal University of Rio de Janeiro, Rio de Janeiro. Retrieved from http://www.coc.ufrj.br/index.php/teses-de-doutorado/148-2004/978-jorge-luiz-fernandes-de-oliveira

Oliveira, M. M. F. de, Ebecken, N. F. F., Oliveira, J. L. F., & Aires, M. (2016). Numerical Modeling of Air Pollutants Emitted by Waterway Transportation. *Journal of Geoscience and Environment Protection, 4*, 123-136. http://dx.doi.org/10.4236/gep.2016.46011

Petterssen, S. (1956) *Weather Analysis and Forecasting*. New York, NY: McGraw-Hill.

Pison, I., & Menut, L. (2004). Quantification of the impact of aircraft traffic emissions on tropospheric ozone over Paris area. *Atmospheric Environment, 38*, 971-983. http://dx.doi.org/10.1016/j.atmosenv.2003.10.056

Sausen, R. (2010). Transport impacts on atmosphere and climate. *Atmospheric Environment, 44*, 46-47. http://dx.doi.org/10.1016/j.atmosenv.2010.02.033

Sioutas, M., Szilagyi, W., & Keul, A. (2013). Waterspout outbreaks over areas of Europe and North America: Environment and predictability. *Atmospheric Research, 123,* 167-179. http://dx.doi.org/10.1016/j.atmosres.2012.09.013

Wood, H. J., Long, G. R., & Morehouse, D. F. (2003). Long-term world oil supply scenarios: the future is neither as bleak or rosy as some asserts. Washington DC, USA: EIA- Energy Information Administration. Retrieved from http://www.eia.doe.gov

Yu, K. N., Cheung, Y. P., & Henry, R. C. (2004). Identifying the impact of large urban airports on local air quality by nonparametric regression. *Atmospheric Environment, 38*, 4501-4507. http://dx.doi.org/10.1016/j.atmosenv.2004.05.034

5

Forest Resource Management and Utilisation through a Gendered Lens in Namibia

Immaculate Mogotsi[1], Selma Lendelvo[2], Margaret Angula[3], & Jesaya Nakanyala[4]

[1] Gender Research Programme, Multidisciplinary Research Centre, University of Namibia, Namibia

[2] Life Science Division, Multidisciplinary Research Centre, University of Namibia, Namibia

[3] Department of Geography, History and Environmental Studies, Faculty of Humanities and Social Sciences, University of Namibia, Namibia

[4] Department of Wildlife Management, Faculty of Agriculture and Natural Resources, University of Namibia, Namibia

Correspondence: Immaculate Mogotsi, Gender Research Programme, Multidisciplinary Research Centre, University of Namibia, Namibia. E-mail: imogotsi@unam.na

Abstract

The shift in forestry policy towards resource management and access rights from state control to local community control has been a welcome step towards sustainable forest management in Namibia. The policy acknowledges the direct dependence on natural environmental resources by the proportional majority of the population that live in the rural areas of Namibia. This study was aimed at performing gender analysis by identifying relationships of various groups to natural resources. The study further assessed the influence these relationships have on control, access and use of forest resources, as well as on natural resource management and the implications thereof on various forest management efforts in the country. Data were collected from seven community forest institutions in Namibia and analysed using the Harvard Gender Analytical Framework. The findings show a gendered differentiated knowledge, control and access to forest resources and unequal participation in leadership and governance. Furthermore, the results suggest that unequal power relations among minority and vulnerable groups affect access to and control of forest resources. This study proposes participation of both men and women in the management, protection, access and utilisation of forest resources, as this will contribute to sustainable forest management and economic development of all members of society.

Keywords: Gender analysis, community forestry, access right, governance

1. Introduction

Degradation of natural forests and loss of biological diversity have been subjects of debate (Richardson, Sinclair, Reed, & Parkins, 2011; Nuggehalli & Prokopy, 2009) that prompted the need to assess the state of forest management (Nuggehalli & Prokopy, 2009; Pokharel, Neupane, Tiwarj, & Kohl, 2015). Sustainable forest management approaches have been accepted widely as the best framework for managing forest and forest resources. Such approaches are seen as the promising instruments that involve local people and integrate the environmental, social and economic values of people in the management of forest resources (Pokharel et al., 2015; Nuggehalli & Prokopy, 2009). However, developing countries, such as Namibia, have experienced challenges in applying the forest management approaches because local people rely heavily on forest resources for their livelihood (Pokharel et al., 2015). Namibia has also experienced issues of degradation resulting from high dependency on forest resources for the majority of rural people and this has led to challenges in forest management efforts. This has prompted a shift in the forestry legal framework towards resource management and access rights from the state to the local community. This has been a welcome step towards sustainable forest management in Namibia. The new trend in forest resource management acknowledges the direct dependence on natural environmental resources by the proportional majority of the population that live in rural areas in Namibia. This implies the importance of the involvement of local communities in the management and protection of forest resources, thereby improving their benefits. Both government and local communities recognise the importance of

Community Based Natural Resource Management (CBNRM), which is supported by the Namibian Association of CBNRM Support Organization (NACSO).

1.1 Conceptualising Forest Management and Gender

Sustainable forest management has its origin in the concept of Sustainable Development, which gained momentum after the Brundtland publication of *Our Common Future* in April 1987 (Pokharel et al., 2015). Further developments regarding the concepts of sustainable development were instrumental in the development of the non-legally binding statements presented at the 1992 Rio De Janeiro summit of the United Nations Conference on Environment and Development. These statements are known as "Forest Principles", in particular Principle 2b, which states that "Forest resources and forest lands should be sustainably managed to meet the social, economic, ecological, cultural and spiritual *needs of present and future generations"* (Pokharel et al., 2015). It was this development that led to sustainable forest management to be elevated to a globally acceptable goal (Nuggehalli & Prokopy, 2009).

The concept of sustainable forest management provides the essential framework for integrating governance, utilisation, conservation (precautionary approach), community-based management and participation considerations (Pokharel et al., 2015). Therefore, with the CBNRM system becoming popular a few years later (since mid-1990s), the community forestry programme has become a vehicle for applying sustainable forest management at local level, through devolution of management power to rural poor communities that largely rely on the forest resources (Republic of Namibia, 2005). Namibia followed Nepal's (pioneer for the implementation of sustainable forest management) foot-steps in shifting the National forestry policy from the top-down, state-dominated management approach towards a more integrative sustainable forest management in Namibia (Chhetri, Johnsen, Konoshima & Yoshimoto, 2013).

The CBNRM approach advocates for the co-management of forests and forest resources as well as the participation of users in decision-making, management and use of the resources. People or stakeholder participation in natural resource management has a long history dating back to the 1960s, starting with awareness raising (Reed, 2008), and it has become almost mandatory in planning development projects (Agarwal, 2001). Still, efforts to involve users and beneficiaries of natural resources are on-going (Reed, 2008; Lendelvo, Munyebvu and Suich, 2012). Over the past two decades, there have been debates concerning different forms of participation. These discussions concluded that at its narrowest, participation in a group is defined in terms of nominal membership and at its broadest, it is defined as a dynamic interactive process in which the disadvantaged have a voice and influence in decision-making (Agarwal, 2001). In addition, Lendelvo et al. (2012) argue that the degree of stakeholder participation could be shifting from nominal levels to interactive or empowerment levels of participation, involving minority communities and women. Furthermore, Agarwal (2009) argues that involving women in executive committees as well as in the decision-making processes extends the pool of citizens committed to forest conservation. Gender mainstreaming provides an opportunity for the disadvantaged and marginalised to participate in governing and managing CBNRM activities.

1.2 Gender and Forest Management in Namibia

1.2.1 Gender Relations in Namibia

Namibia, being a multicultural nation, has diverse ethnic groups with different cultures that define its gender roles. The different roles of men and women are based on whether the ethnic group or tribe is of matrilineal descent, bilateral descent or bifurcate descent (Jauch, Edwards & Cupido, 2009). These descent systems are also founded in cultural attitudes that tend to vary from relative equality to rigid inequality (UNAM and SARDC-WIDSAA, 2005). As in most African cultural communities, duties and responsibilities are divided among husband, wife and children based on stereotypes of what men and women should do and how they should behave, and not necessarily based on skills or ability (Iipinge, Phiri & Njabiri, 2000; UNAM and SARDC-WIDSAA, 2005). The majority of Namibian women are responsible for most of the household chores and have limited decision-making power within households and the community (Ruppel, 2008). The gender division of labour based on cultural views has shaped varied gendered relationships to the natural environment and natural resources in Namibia. The different levels of access and control regarding utilisation and management of forest products are therefore affected by these views (Iipinge, Phiri & Njabiri, 2000; UNAM and SARDC-WIDSAA, 2005).

Since Namibia's independence in 1990, the Constitution of Namibia has laid the foundation by stipulating in Article 10 that all people are equal and free from discrimination. Subsequently, several legal frameworks and programmes addressing gender relations and elimination of discriminatory practices against women were initiated and implemented (GRN, 1990). There is a notable change in most areas of Namibia with regard to discriminatory practices and gender representation at varying levels of institutions. However, attitudes and societal stereotypes towards gender are very slow to change (Angula, 2010) and by extension this can be seen in the management and decision making of natural resources including forest resources.

1.3 Forest Resource Management in Namibia

According to Mendelsohn, Jarvis, Roberts & Robertson (2002), forests in Namibia are found mainly in the north-central and north-eastern regions of the country where rainfall ranges from 400-700 mm per annum. Before independence, there were no major efforts invested in forest management in Namibia. The Nature Conservation Ordinance No.4 of 1975, which was used to govern forest resources in national reserves, was discriminatory as it did not make provision for communities to have rights to manage and benefit directly from forest resources. Forest management was confined to a few commercial species. A renewed focus on forest management came soon after independence in the 1990s, when a dedicated Directorate of Forestry was created with two divisions for management and research, supported by regional and field offices. This was followed by the first-ever promulgated Forest Policy of 1992, the Namibia Forestry Strategic Plan of 1996, the Forestry Policy and the Forest Act of 2001 (Republic of Namibia, 1996; 2001).

The current programmes of the forestry sector are shaped by the Namibia Forestry Strategic Plan of 1996, which brought the integrated approach to the forestry sector. The Strategic Plan included Community Level Management of Natural Forests, which led to the establishment of the Community Forestry programme in the country. The recognition of community participation has led to the rapid proclamation of community forest reserves and was strengthened by supportive legislation and local community training.

In 2008, a new strategic focus for forestry was developed as part of the overall Strategic Plan for the Ministry of Agriculture, Water and Forestry (MAWF) (2008–2012). One of the Ministry's strategies is Sustainable Natural Resources Management. The major result areas are stated as:

- Improved and integrated natural resources management;
- Coping mechanisms in place for addressing adverse effects of climate change on production (including global warming, floods, droughts and 'veld' fires);
- Reduced negative human and animal impact on environmental health (including pollution and waste management).

The above policy documents and strategies, strengthened by the Forest Act of 2001, have reinforced the need for production, protection and participation. This is consistent with the Namibian Constitution, Article 95(l) that promotes the maintenance of ecosystems, essential ecological processes and biodiversity as well as sustainable utilisation of natural resources for the benefit of all Namibians.

The community forestry programme has adopted a participatory approach to facilitate the participation of all people. The involvement of all local people, as the main implementers, is crucial for sustainable utilisation of the forest resources, leading to sustained benefits from such resources. Since 2008, the Community Forestry programme in Namibia has contributed considerably to the improvement of rural livelihoods. The establishment of community-managed forest areas in nine different regions of the country has proven to be an essential pillar of Namibia's CBNRM programme. In Kavango Region, for example, the four gazetted community forests (CFs) generated a total of N$98,100 (US$ 8,000-00) in 2009 (NACSO, 2010). Income varies between CFs, depending on the availability of resources, the efficiency of management procedures and the commitment of stakeholders (NACSO, 2010). Benefits are shared at regular intervals between traditional authorities, management bodies, communities' development funds and investments according to a benefit-sharing plan. CF management bodies and communities are trained to focus on sustainable forest and habitat management practices to prevent forest degradation and soil erosion, stop depletion of natural resources and protect wildlife habitats (NACSO, 2010). There is limited literature in Namibia that assesses the gendered dimension of community forest management, equitable sharing of resources and benefit sharing among rural communities.

2. Objectives of the Study

Given the intended objective of the Namibian community forestry programme to encourage inclusive participation in forest resource management and benefit sharing plan, this study is precisely aimed at investigating the equitable participation in harvesting and benefit of forest resources across gender and marginalized groups. Gender has become an important component for development in natural resources and forest management, in particular. This is because males and females have specific roles and responsibilities in the use of forest products, and therefore their opportunities to benefit from these products are not the same. In traditional societies, women are often disinclined to participate in activities that are seen to go against traditionally defined roles, which can present obstacles to participation in conservation efforts. Furthermore, in cases where participation is characterised by unequal power relationships, this will influence available livelihood options, decision-making abilities and development outcomes. When women are unable to express their voices openly because of a dominance of male counterparts, this may affect the outcomes of natural resource management.

3. Methodology and Approach

3.1 Study Design

Namibia has a total of 34 community forests (NACSO, 2014) and 25 that are considered 'emerging' as they have submitted the intention to register (B.K. Nathanael, MAWF, Personal Communication, 18 August 2015). The samples for this study were taken from both registered and emerging CFs and a total of seven (7) community forests (CFs) were targeted. The sampled community forests were selected from the thirteen community forests that were identified by the Ministry of Agriculture, Water and Forestry (MAWF) as target sites of a project called Namibia Forested Lands (NAFOLA). According to the Forest Act No. 12 of 2001, Clause 15(3), community forests are required to register and be gazetted through MAWF. Amongst the thirteen CFs that were identified by MAWF, only two were gazetted; one was randomly selected, i.e., Uukolonkadhi CF. In the case of the emerging CFs, six were randomly selected out of the eleven. The six selected emerging CFs are Ehirovipuka CF, Oshaampula CF, Otshiku-tshiitholonde CF, Okondjatu CF, Talismanus CF and Otjinene CF.

The criteria for sampling the seven research sites took into consideration the status and period of existence, national distribution, land use types, language and cultural variations of the CFs. Refer to Table 1 for sampling criteria.

Table 1. Description of the selected Community Forests (Cfs)

Community Forest/ hotspot	Status	Cultural/ ethnic groups	Land use type	Region
1. Ehirovipuka	Not registered but activities have been ongoing since 2008	Ovahimba and Ovaherero	Livestock farming, no crop cultivation, wildlife conservation (conservancy)	Kunene
2. Uukolonkadhi	Registered in 2006	Aakolonkadhi, Ovahimba, Aandongona, Ovazemba	Combination of crop cultivation and livestock farming, wildlife conservation (conservancy)	Omusati
3. Oshaampula	Not registered but has been in operation since 2000	Aandonga	Combination of crop cultivation and livestock farming	Oshikoto
4. Otshiku-Tshiithilonde	Not registered and concept introduced in 1998.	Aakwambi	Combination of crop cultivation and livestock farming	Oshana
5. Talismanus	Not registered and concept introduced in 1998.	Ovaherero and Ju|'hoansi	Livestock farming, wildlife conservation (conservancy)	Omaheke
6. Otjinene	Not registered and concept introduced in 1998.	Ovaherero and Ju|'hoansi	Livestock farming	Omaheke
7. Okondjatu	Not registered and concept introduced in 1998.	Ovaherero and Ju|'hoansi	Livestock farming and wildlife conservation (conservancy)	Otjozondjupa

3.2 Data Collection Procedure

The researchers used a qualitative research approach and the specific research methods used were key-informant interviews (KIIs) and focus group discussions (FGDs). Question guides were developed and engendered using the Harvard Gender Analytical Framework to ensure that gender dynamics were incorporated adequately. The Gender Analytical Framework assessed control and access to forest resources, gender, language and cultural representation in positions of leadership and decision making in forest resource management. The FGDs and KIIs solicited the respondents' perspectives on the management of the forest resources, the utilisation, access and control of forest resources and the support needed to ensure equitable control of forest resources by adults, youth and marginalised communities. Information on best practices in marketing forest resources and their contribution to livelihoods was also collected.

Key informant respondents were leaders in the communities, mainly managers of conservancies, community forests and water point committees, traditional leaders, officials from MAWF and the Ministry of Environment and Tourism (MET) who work closely in forest management activities. Focus Group Discussions (FGDs) were held with various community members from the seven research sites. Each FGD had a minimum of four (4) and a maximum of eleven (11) people. At each site FGDs were comprised of categories of adults, youth and people from different language groups and were further separated by gender. The study also paid particular attention to the utilisation of forest resources by marginalised and minority members and factors hindering their utilisation. During the interviews, special consideration was paid to any hierarchical and other cultural norms and taboos in the community. Finally, where necessary, translation was used during the interviews.

3.3 Data Analysis

Data analysis was done by reviewing the study's major objectives and questions in order to assist in the identification of thematic areas emerging from the interviews. Thereafter, the identified themes were categorised from topics derived from the FGDs and KIIs. The use of multiple analysts was a strategy to avoid bias, whereby all four researchers first independently identified the themes from the transcripts. The researchers then met to compare notes, identify commonalities including examples from participants' quotes and other key points. Major similarities and differences in the individual analyses were discussed until an agreement was reached among all researchers.

Among the themes that emerged from the interviews were gender representation in community forest leadership structures; capacity building in community forests; protection and monitoring of forest resources; awareness, communication and inclusion; harvesting forest resources; access and control of forest resources, marketing forest resources and contribution of forest products to livelihoods.

4. Results and Discussions

4.1 Gender Representation on the CF Management Committee

The study shows that the leadership structures of the different research sites were not the same across the board. Some CFs had independent committees responsible for forest management only, whilst others had integrated committees that were responsible for different natural resources, including forestry. Two of the CFs did not have a functional leadership committee because they were at a very early stage of development. This study has shown that for most of the emerging CFs, there is a trend towards adopting the integrated natural resource management approach, and thus avoiding the practice of managing natural resources within communities as isolated entities. Single-use management of forests should be avoided and be replaced with multiple-use management approaches that are aimed at integrating different resources found in forests for better conservation of biological diversity (Scialabba, 1998).

The sizes and gender representation of the CF leadership structures are presented as follow..

1) **Ehirovipuka CF** – This area uses an integrated natural resources management approach, whereby the conservancy committee governs all natural resources found in the area including forestry. This integrated conservancy committee is a 14-member structure with a gender proportion of 86% males and 14% females.

2) **Uukolonkadhi CF** - This area consists of different leadership structures for the Conservancy and Community Forestry. The community forestry has a two-layered leadership structure consisting of, firstly, a larger Forestry Management Committee (FMC) with a wider constituent representation; and, secondly, a smaller Executive Committee that manages the day-to-day activities. The FMC consists of 45 signed-up members with a gender proportion of 38% males and 62% females. About 5% of the FMC members are from the minority and vulnerable groups (Ovahimba and Ovazemba) with no females from these groups. The Executive Committee consists of 11 elected members with a gender representation of 82% males and 18% females.

3) **Oshaampula CF** - This area consists of a community forestry, which is a fully-functional body. The FMC has about 25 signed-up members, with a few members having designated tasks. The FMC for this area has an equal male and female representation, but more males are active.

4) **Otshiku-tshiitholonde CF** - This area also consists of the community forestry, with an FMC comprised of 18 elected members, having a gender proportion of 78% males and 22% females.

5) **Okondjatu CF** – This area has adopted an integrated natural resources management approach whereby the fully-functional committee governs all natural resources found in the area, including forestry. The Conservancy Committee consists of 12 elected members with a gender representation of 58% males and 42% females.

6) **Talismanus CF** - No functional management body.

7) **Otjinene CF** - No functional management body.

Representation of women and minority groups in leadership is important as it impacts on decision making (Mitchell, 2013).

4.2 Participation on CF Leadership

The participation of women in the leadership structure is important as they are the primary users of forests and this means that their contribution is valuable for forest protection (Mitchell, 2013; Reddy and Padmaja, 2014). Their involvement also permits an inclusive decision making process that is beneficial to all members of the community. The extent to which leadership structures are all-encompassing determines how community resources are accessed and controlled.

For most of the CFs with leadership committees in place, women and minority group representations on these structures were low. In addition, the chairpersons of most of these committees were predominantly men, with women mainly occupying positions such as treasurer, vice-treasurer, secretary or vice-secretary. At the Oshaampula CF, prominent committee positions such as chairperson and vice chairperson were mainly occupied by males. Whilst there was a significant number of women who were also members of the committee but did not hold any positions, most of these women were mostly delegated to attend workshops and training programmes as well as receiving visitors because of their fair educational backgrounds.

There are other crucial influences that contributed to the male dominance on these CFs or Conservancy committees. There is a compulsory representation of local Traditional Authority (TA) members on every CF committee. Thus, women felt that their participation in leadership structures was hindered by TAs, who preferred to nominate males for the CF committees. Furthermore, most CFs had a low representation of the youth and minority groups in their leadership structures. This was particularly the case in Omaheke and Otjozondjupa regions, where minority groups were excluded from leadership structures and thus could not make significant contribution.

The historically low representation of women in leadership structures is a result of discrimination and stereotyping (Stelter, 2002). During a mixed focus group discussion of both young and elderly women in Otjinene, a sentiment was raised by the discussants: *"We are discriminated against in terms of being given equal opportunities to participate in leadership structures as well as to be fully informed of the different activities and initiatives of CFs"* (Elderly female FGD participant). A similar view was also aired by women from the Ehirovipuka CF where they expressed that a separate females-only committee was needed because their male counterparts were not respecting their views or giving them opportunities to sufficiently contribute in meetings. In Uukolonkandhi CF, women and minority groups attributed their lack of participation in leadership to long distances that they had to travel to the forests.

Another phenomenon that also hinders women participation is lack of confidence to take up leadership positions in their areas. The researchers also observed that the interest of women to serve in leadership structures varied among the research sites. Among CFs in the north, women showed a lot of interest to serve in leadership structures, whilst in the eastern regions few women expressed the need to serve in leadership structures.

According to Varghese and Reed (2012), inclusiveness across a range of social groups was identified to assist in achieving social sustainability of forests. Therefore, there is a need to pay attention to how women and other marginalised groups might be excluded from sustainable forest management and/or how they can effectively participate in order to enhance inclusiveness.

Importantly, TAs are the custodians of communal land on which CFs are located. Their mandate as stipulated in the ACT involves the promotion of affirmative action among its members, in particular gender equality with regard to positions of leadership and ensuring the inclusion of marginalised communities within their areas of jurisdiction. Moreover, it is the role of the TA to provide leadership, support and consent for community forestry establishments. This research further shows that Traditional Authorities (TA) have emerged as one of the most important local institutions in the management of forest resources in CFs as they are represented on all Forest Management Committees (FMCs).

4.3 Capacity Building

There are numerous initiatives to improve the capacity of local communities, for better forest resource management. The initiatives entail, among others, empowering communities towards sustainable access and extraction of forest resources. The study revealed that both males and females in five of the seven CFs had participated in different capacity building initiatives. Some of the training initiatives were attended more by males than females. These are, especially, skills-based training on the harvesting methods of timber, veld-fire management, leadership and financial management. Women received training mainly on establishing and implementing activities that improve value addition, such as production of crafts, honey, skin oil, beads and crafts. The majority of those who attended training were young male and female members of the CFs. There were, however, training opportunities earmarked specifically for those in leadership and, at times, only for elderly women. Very limited training opportunities were targeted at benefiting marginalised communities such as Ju|'hoansi in Talismanus, Okondjatu and Otjinene.

Patterns of harvesting the *Devil's Claw* are changing as it is becoming an endangered species and it is increasingly found far from the settlements. Improved or modern harvesting methods for *Devil's Claw* have been introduced in most Southern African countries to cater for the high demand of this medicinal plant due to commercialisation (Grote, 2003). Historically, the *Devil's Claw* was harvested by women, and it was used mainly for household and medicinal purposes (Grote, 2003). However, over the years, the interest in harvesting it for commercial purposes has tremendously increased, thus raising concerns that it could be over-harvested. In an attempt to preserve it, a permit is needed to harvest the *Devil's Claw*. Furthermore, a harvesting permit for this species requires a proof of training attendance, a requirement which might disadvantage women to an extent. The fact that more males were involved than females, meant that women who possessed indigenous ecological knowledge of this species were no longer in control of this resource.

The main aim of CBNRM programme is to enhance sustainable utilization of forests and natural resources. Since harvest of the Devil's Claw started to attract commercial gains for community members it had led to unsustainable methods and rates of harvest that could lead to local extinction of this valuable resource. The social benefit which women used to accrue for their families and households to treat medicinal ailments had been threatened as a result. Despite training offered to both male and female, women still lag behind in the harvesting and commercial benefits of Devil's Claw plant in their respective CFs. Their participation could also be hampered by additional household chores, children rearing and other community activities that take up their times to harvest Devil's Claw in larger quantities for commercial sales.

Similarly, mopane worms were traditionally harvested by women for household food consumption however, this research noticed a shift in men showing interest in this forest resource. The involvement of men harvesting mopane worms creates competition with women over this resource. Both men and women have realised the commercial value of this forest product. The commercialisation resulted in the need to control the sustainable harvesting of this resource, through issuing of harvesting permit.

4.4 Control of Forest Resources

Control of forest resources mainly refers to protection and monitoring, which has been vested in the local communities as per the Forestry Policy of Namibia. Protection of forest resources varies, depending on their classification. Among them are those that are under strict protection as well as, monitoring and sustainable harvesting mechanisms for resources such as *Acacia erioloba* and *Devil's Claw*. This study established a number of issues related to the rights of use, implementing regulations, issuing of permits and fining of perpetrators. Most of these regulations are enforced by the Government, but in collaboration with the CF committees. Permits are issued to users through collaboration between MAWF and local CF committees as a form of control. Traditional authorities issue letters of consent if resources are harvested in their areas of jurisdiction.

Currently, forest resources in CFs are accessed by both men and women, either directly without requiring permission from any authority, or through acquisition of permits or consent from local leadership. Households within CFs are said to have been granted rights to forest resources through their community forest committees for harvesting on a limited scale for domestic use and, under such conditions, no harvesting permits are required. However, when harvesting on a large scale for commercial purposes, a permit is required. Harvesters from outside the CF are also required to acquire a permit for the quantity they are harvesting.

CFs that are not yet registered have difficulties in enforcing laws or regulations to control the harvesting of resources. Despite these difficulties, the Directorate of Forestry (DoF) within MAWF, together with CFs, ensures that permits are issued and well-monitored, especially for the resources harvested on a commercial basis or if they exceed a certain quantity. For instance, Oshaampula CF (not yet registered) has been in operation for more than 14 years and works together with the DoF in issuing permits to users. The Oshaampula CF and DoF has also dealt with illegal harvesters. It emerged from the study that most of the forest products that men harvest such as timber, high quantity firewood, are controlled through permits, as they are harvested for commercial purposes. On the other hand, most of the non-timber products such as thatching grass and edible and medicinal plants that women harvest for household use, do not require any permits or letter of consent from the TA leaders.

Contrarily, women also venture into traditionally male controlled forest resources. In terms of construction poles, this resource used to be harvested by men only, but currently women also apply for permits to harvest poles. During 2012, 40% of the permits in Uukolonkadhi for harvesting construction poles were issued to women. However, the study revealed that despite women accessing construction poles through permits, their direct engagement in harvesting the actual resources is still not common.

Control of resources also includes aspects of fire management and patrolling of these resources. Most functioning CFs indicated that they engage in fire management by putting up fire breaks to prevent its spread. Since this activity involves people having to travel and spend some days in the field making these cut lines, the activity is dominated by males. This study showed that women are subjected to gender-based violence when undertaking forest fire fighting duties. The TA in Uukolonkandhi narrated an incident where a young man was withdrawn from the fire-fighting exercise because of sexual harassment claims.

Patrolling of forests was done by appointed forest guards, of which the majority were males. The CF committees indicated that the functions of the forest guards included the patrol of the protected forest and reporting cases of over-harvesting, veld-fire outbreaks, illegal harvesting or any activities defeating sustainable forest management. It emerged from the interviews that active participation of women in the protection of forest resources could be hindered by issues such as the nature of the activities, long distances, odd times clashing with home responsibilities, abilities and strength required to carry out the tasks, fear of wild animals and sexual harassment from male colleagues. A young man in Uukolonkadhi referred to an old Oshiwambo saying, "*omulumentu okuti*", which equates men to a forest, to justify why he thought only men were appointed to patrol. Forest guards are remunerated either through DOF or CFs.

Participation of women in CF initiatives at Ehirovipuka is low because of how men and women are traditionally socialized. Similarly, limited participation of women in CF activities leads to women relying on their partners for most community-level information. The reliance of women on their male counterparts for information flow contributed to women having very little knowledge on the actual activities of the CFs. Low level of education and lack of information were identified as some factors that hindered active participation of women in both leadership and other forest activities in Nepal (Mitchell, 2014). Women in CFs such as Ehirovipuka showed an interest to attend and participate in community meetings. However, male dominance during community meetings seems to discourage the women from actively participating.

4.5 Accessing Forest Resources

Forest policy directives enable equitable and fair access to resources in CFs by emphasising that no one should be excluded from accessing these resources. Women and men access different forest resources. Wood resources emerged as an important forest product with the highest value to both men and women in all CFs. Although most female respondents indicated that they did not harvest wood resources, they still prioritised wood highly for its uses. Wood resources are used as construction poles, firewood, droppers and wooden crafts. Firewood is harvested by both men and women, but women focus mainly on collecting firewood within their vicinity for household use. This finding is in line with Reddy and Padmaja (2014) study that women do not venture far from their homestead when collecting firewood. In Otshiku-Tshiithilonde, Uukolonkadhi and Oshaampula the protected areas are more than 5km from the homestead, which is the vicinity where most women are fetching their firewood.

The male respondents indicated that they used donkey-carts and vehicles to collect firewood further afield, where the resources are abundant, especially for commercialisation. In general, however, distances to most resources are an inhibiting factor that could affect the participation of youth, marginalised groups and women in harvesting. Furthermore, access to forest resources is hindered by, competition over forest resources between locals and non-locals. Women in the Ehirovipuka CF lamented, *"We see more non-residents coming with their cars to camp in our forest and harvest mopane worms. We are concerned that this influx could lead to over-harvesting if not regulated or controlled; this also reduces our food supply and income generating opportunities"*.

The other inhibiting factor that this study came across was the limited access to forest resources by marginalised and poor women. They raised concerns of being unable to adequately access forest resources due to illegal fencing in their community, financial constraints to afford permits and lack of vehicles to harvest on large scale.

In certain CFs, there is still rigid gender division of labour and the fear of being stereotyped. For example, in Okondjatu, Talismanus and Otjinene, among the Otjiherero women, there was a reluctance to directly harvest construction poles and droppers. However, the study established that when the women wanted to access these resources, they would hire labour. Furthermore, in Otshiku-Tshiithilonde, which has a salt pan, it appeared that traditionally only men harvested this resource as the male group revealed that *"it is a taboo for women to collect salt"*. On the other hand, the women's group indicated that there is nothing wrong with women harvesting salt. Both men and women indicated the economic importance of this resource.

The study revealed that women mainly participate in less commercial non-timber forest resources, which include medicinal, edible plants, agro-forestry resources such as *marula* and palm tree products that contribute to household health and food security. Men stated that they also harvested these resources, but on a small scale.

A case of conflict over accessing forest resources outside the Uukolonkadhi CF surfaced during the study. The red ochre has been traditionally utilised by Ovahimba women as a cosmetic product is being commercialised by Ovazemba people. Ovahimba respondents were grieved by the fact that their traditional resource is being commercialised mainly by Ovazemba men. Whilst the demand for red ochre is increasing, Ovahimba people are resisting to change their traditional way of harvesting, thus losing the opportunity to commercialise it. The commercialisation of red ochre among the Ovazemba people is spearheaded by young men, as it is labour intensive and requires strength. The area where the ochre is mined falls outside the jurisdiction of the Forestry Committee, making it difficult to resolve the conflict.

4.6 Income Generation and Use of Harvested Resources

Studies in some African countries have reported the economic importance of forest products in rural households (Arnold and Townson, 1998; Mulenga, Richardson, Mapemba and Tembo, 2011). Mulenga et al. (2011) particularly found that non-timber forest products contributed 34% to household income in 2008. Income generated from forest resources contributes significantly to the livelihoods of rural communities. Various mechanisms are put in place to ensure access to forest resources. Out of the seven CF research sites, only two are generating income from the sale of forest resources. To access the wood resources from CFs, the DoF issues transport permits to interested harvesters, while the CF committee charges for the quantity of forest resources harvested. The Oshaampula CF case is illustrated in box.1 to reflect the income generating strategies at institutional and individual levels:

Box 1: Harvesting of forest resources in Oshaampula CF

In Oshaampula, the study revealed that those who acquire permits do not do the harvesting themselves, but the CF rather designates people from the community, mostly young males, to cut the required quantity. This process disadvantages women as it mainly involves men to cut the poles, and they receive direct compensation from the CF committee. The women in Oshaampula reported the increasing demand from outsiders for their resources, yet they do not have any direct benefits through employment and wages, except through the community benefit sharing process. Similarly, it emerged from the Oshaampula leadership that most of the CF funds are used for community development projects. A member of the Oshaampula leadership narrated thus: '*We recently built a kindergarten and a local clinic for antenatal care and immunisation*'. Although the income generating activities of this CF are male-dominated, the benefits are used for the greater good of the entire community.

The study revealed that community members can also generate income from forest resources by trading their labour or selling the harvested forest products. It was common in most of the CFs for young people or members of the marginalised groups to harvest forest resources for others in exchange for cash income. People who are relatively well-off seem to hire labour for harvesting forest resources. They either hire local Ju|'hoansi men or unemployed men from other regions. The Ju|'hoansi people, especially from the Talismanus, Otjinene and Okondjatu CFs, do not generate substantial income from selling forest resources compared to trading labour to harvest forest resources for others. The youth in these regions also showed little interest in harvesting forest resources, resulting in job opportunities going to non-locals. One Ovaherero man in Talismanus said, *"To harvest poles and droppers you need to recruit a strong man as these activities are labour intensive. We therefore hire strong men from other regions to harvest labour intensive activities such as cutting poles and droppers"*. Although women from the Ovaherero do not directly harvest poles and droppers, they are able to hire labour. These women sell some of the poles and droppers harvested through hired labour. In particular, OvaHerero women in Omaheke and Otjozundjupa regions said, *"We do not go to the forest to harvest, but if we want forest resources, we hire labourers and a car. People that we hire go to the forest and bring the resources to us."*

This study has highlighted the involvement of women and men in harvesting different forest resources for commercial purposes. Resources such as *Devil's Claw*, *mopane* worms, construction poles, droppers, firewood, *marama* beans and some edible plants are harvested commercially. Some of these resources have organised markets that provide opportunities to both males and females to sell high valued products such as *Devil's Claw.*

5. Conclusion

The gendered analysis of participation in the harvesting, benefits and management of forest resources in CFs revealed distinct inequalities in access, control, monitoring and use of forest resources. These inequalities are differentiated by gender, level of marginalisation and ethnicity. The institutional arrangements allow equal gender representation in management structures of all CF reserves. However, this study revealed that the representation of women in CFs committees is low. There are cases where the representation is equal or is characterised by a higher number of women in CF committees that govern the utilisation and benefit sharing of forest resources at local level. In such cases, men predominantly occupy strategic and prominent positions of chairperson and vice-chair person. This ultimately acts as a barrier for marginalised groups, minority ethnic groups and women's participation in decision-making. Women, being the carriers of indigenous knowledge of most forest products, will not be able to share this knowledge if they are not adequately involved in the decision-making processes. It has been proven that both scientific and indigenous knowledge are critical to contribute to resource sustainability.

The study also revealed that when forest resources that were traditionally harvested by women become commercialised, men also start to harvest these resources. This trend not only negatively affects the access of women to those resources, but also disrupts the social relations and learning process that women traditionally engaged in when harvesting these resources. The increasing participation of men in harvesting forest resources for commercial purposes coupled with women continuing to harvest for household consumption could lead to rapid depletion of some of the forest resources. Furthermore, as resources are getting scares near homestead, it becomes increasingly difficult for women to access them due to the physical distances women have to travel to access such resources. Consequently, less access for women and marginalised groups implies that income generated and supplies for household consumption from these forest resources is reduced, increasing the vulnerability of these households.

Factors that restrict women's participation are similar in other countries and regions of the world. Social norms define household chores and childcare as women's roles, thereby affecting effective participation in the harvest and monitoring of forest resources. Activities that require women to travel distances and spend more time away from their household chores inhibit the level of participation in the harvesting of commercial products by women. Women from low-income families and marginalised groups are facing challenges that inhibit their participation the most.

This study concludes that access to and control of forest resources have improved in Namibia if one looks at the number of harvesting permits issued to female applicants. However, control of and decision making regarding harvesting and benefits of forest resources may reveal a different pattern. Although the issuing of harvesting permit system has proven to be trusted control mechanism, caution needs to be put when harvesting for commercialization. This will require continuous and consistent monitoring of the resources and for use patterns to ensure sustainable equitable access to these forest resources.

This study therefore, recommends further research to interrogate the impact of benefits accrued from forest resources as well as control and decision-making power over the use and benefits of CF products.

References

Agarwal, B. (2001). Participatory exclusions, community forestry and gender: An analysis for South Asia and a conceptual framework. *World Development, 29*(10), 1623-1648. http://doi.org/:10.1016/S0305-750X(01)00066-3

Agarwal, B. (2009). Gender and forest conservation: The impact of women's participation in community forest governance. *Ecological Economics, 68*, 2785-2799. http://doi.org/:10.1016/j.ecolecon.2009.04.025

Angula, M. (2010). *Gender and climate change: Namibia case study*. Cape Town: Heinrich Böll Stiftung - Southern Africa.

Arnold, M., & Townson, I. (1998). Assessing the potential of forest product activities to contribute to rural incomes in Africa. ODI Natural Resource Perspectives, 37, 4p.

Chhetri, B. B. K., Johnsen, F. H., Konoshima, M., & Yoshimoto, A. (2013). Community forestry in the hills of Nepal: Determinants of user participation in forest management. *Forestry Policy and Economics, 30*, 6-13. http://dx.doi.org/10.1016/j.forpol.2013.01.010

Grote, K. (2003). The increased harvest and trade of Devil's Claw (harpagophytum procumbens) and its impacts on the peoples and environment of Namibia, Botswana and South Africa. Maccarese, Italy: Global Facilitation Unit for Underutilized Species.

Iipinge, E. M., Phiri, F. A., & Njabili, A. F. (2000). The national gender study. (vol.1). Windhoek: University of Namibia.

Jauch, H., Edwards, L., & Cupido, B. (2009). A rich country with poor people: Inequality in Namibia. Windhoek: LaRRI.

Lendelvo, S., Munyebvu, F., & Suich, H. (2012). Linking women's participation and benefits within the Namibian Community Based Natural Resource Management Program. *Journal of Sustainable Development, 5*(12), 27-39.http://dx.doi.org/10.5539/jsd.v5n12p27

Mendelsohn, J., Jarvis, A., Roberts, C., & Robertson, T. (2002). Atlas of Namibia: A portrait of the land and its people. Cape Town: David.

Mitchell, K.V. (2013). Identifying Barriers to Dalit and Janajati Women's Successful Leadership in Community Based Forest Management in Nepal, Hariyo Ban Report: 029, Nepal

Mulenga, B.P., Richardson, R.B., Mapemba, L., & Tembo, G. (2011). The contribution of non-timber forest products to rural household income in Zambia. (Working Paper No. 54, Food Security Research Project, Lusaka, Zambia, June 2011). Retrieved from http://www.aec.msu.edu/fs2/zambia/index.htm

NACSO. (2014). The state of community conservation in Namibia: A review of communal conservancies, community forests and other CBNRM initiatives (2013 Annual report, NACSO, Windhoek.

Namibia Association of CBNRM Support Organisation (NACSO). (2010). Namibia's communal conservancies: A review of progress and challenges in 2009. NACSO, Windhoek.

Nuggehalli, R. K., & Prokopy, L. S. (2009). Motivating factors and facilitating conditions explaining women's participation in co-management of Sri Lankan forests. *Journal of Forest Policy and Economics, 11*, 288-293. http://dx.doi.org/10.1016/j.forpol.2009.05.002.

Pokharel, R. K., Neupane, P. R., Tiwari, K. R., & Kohl, M. (2015). Assessing the sustainability in community based forestry: A case from Nepal. *Forestry Policy and Economics, 58*, 75-84. http://dx.doi.org/10.1016/j.forpol.2014.11.006

Reddy, T. B., & Padmaja, M. (2014). Income generation of women through the utilization of forest resources in tirumala foothill villages. *International Journal of Multidisciplinary Research and Development, 1*(6), 135-138. http://dx.doi.org/10.21746/ijbio.2015.11.007

Reed, M. (2008). Stakeholder participation for environmental management: A literature review. *Journal of Biological Conservation, 141*, 2417-2431. http://dx.doi.org/10.1016/j.biocon.2008.07.01

Republic of Namibia. (1990). Article 10 of the Namibia Constitution. Windhoek: Government of Namibia.

Republic of Namibia. (1996). Namibia Forestry Strategic Plan. Windhoek: Ministry of Agriculture Water and Forestry.

Republic of Namibia. (2001). Development Forestry Policy for Namibia. Windhoek: Ministry of Agriculture Water and Forestry.

Republic of Namibia. (2005). Community Forestry Guidelines. Windhoek: Ministry of Agriculture Water and Forestry.

Richardson, K., Sinclair, A.J., Reed, M. G., & Parkins, J. R. (2011). Constraints to participation in Canadian forestry advisory committees: A gendered perspective. *Canadian Journal of Forest Research, 41*(3), 524-532. http://dx.doi.org/10.1016/j.biocon.2008.07.01

Ruppel, O. (Ed.). (2008). Women and customs in Namibia: Cultural practice versus gender equality? Windhoek: Macmillan Education.

Scialabba, N. (ed.) (1998). Integrated coastal area management and agriculture, forestry and fisheries, FAO guidelines. Environmental and Natural Resources Services, FAO, Rome, 256p.

Stelter, N.Z., (2002). Gender differences in leadership: Current social issues and future organizational implications. *Journal of Leadership Studies, 8*(4), 88-99.http://dx.doi.org/10.1177/107179190200800408

UNAM and SARDC-WIDSAA. (2005). Beyond inequalities 2005: Women in Namibia. Windhoek and Harare: University of Namibia and Southern Africa Research and Documentation Centre.

Varghese, J., & Reed, M. G. (2012). Theorizing the implications of gender order for sustainable forest management. *International Journal of Forestry Research, 2012*, 11. http://dx.doi.org/10.1155/2012/257280

6

Ecological Landscape Planning and Design Strategies for Mangrove Communities (Hara Forests) in South-Pars Special Economic Energy Zone, Asalouyeh- Iran

Mohammad R. Masnavi[1], Neda Amani[2] & Ali Ahmadzadeh[3]

[1] Faculty of Environment, Department of Environmental Design, University of Tehran, Tehran, Iran

[2] Faculty of Environment, Department of Environmental Design, University of Tehran, Tehran, Iran

[3] Pars Special Economic Energy Zone Org, Asalouyeh, Iran

Correspondence: Mohammad R. Masnavi, Faculty of Environment, Department of Environmental Design, University of Tehran, Tehran, Iran. E-mail: masnavim@ut.ac.ir

Abstract

Along with the strategic role of Asalouyeh region as an industrial zone in recent decades, there have been some growing problems regarding the ecosystems of the region due to the heavy activities such as petrochemical, industrial, transportation and related development in the area. The major region of Mangrove communities in the Persian Gulf that contains some unique ecotones has established a dynamic sustainable ecosystem integrating the relationship between sea and plateau ecosystems. It is also considered as a rich habitat for the creatures in flood and Ebb conditions. Therefore, taking in to account the increasing process of ecological destruction in the existing specific protected regions- especially the Mangrove communities, it is essential to study the structural relationships between landscape elements and patches in order to balance the ecological, social relationships as well as the environmental remediation based on the preservation of ecological structures. This study tries to examine the ecological role of Mangrove communities and their habitat to introduce the threatening factors in the "South Pars Special Economic Energy Zone" and then it suggests some strategies for ecological planning and design for protecting Mangrove communities in the area i.e. storm water management, protection and development of degraded habitats and phytoremediation, to create a framework for eco-tourism and Sustainable Development in the region.

Keywords: ecosystem, ecological landscape design, ecological disturbance, south Pars special economic energy zone Asalouyeh, mangrove communities, Hara forests ecosystem

1. Introduction

1.1 Threatening Marginal Ecosystems by Active Industrial Areas

Active industrial areas, by utilizing technology and natural resources, have had lots of economic benefits. But parallel to that, they have created human and environmental issues, which have led some disorders to the natural ecosystems and urban societies' health. The pollutions of upstream industries, petrochemical, oil and gas, widely around exploration, refining facilities and locally on the transportation lines of these materials in oil-rich provinces, are an inevitable outcome of rapid population growth and industrialization process and marine ecosystems threat.

" Pars Special Economic Energy Zone " (PSEEZ) is one of the most important subsidiary companies of the south oil-rich area that extends an area of 46000 ha from Bushehr province to the north of Khuzestan province (Figure 1). Persian Gulf pollutions caused by oil waste disposal, sedimentation, habitat destruction, wastewater and agricultural fertilizers discharges, excavation and increased heavy metals are the most important threatening factors of Persian Gulf ecosystems (Mickwitz et al., 2006).

1.2 The Destruction of Mangrove Communities

Mangroves communities are coastal ecosystems of Persian Gulf that has been emphasized as the most important foci of biodiversity, throughout the world. While are these forests now threatened with various human activities (Price et al., 1994). Location of the South Pars (Pars 1) along the Nayband Bay in Asalouyeh has caused destruction and removal of forests areas and has converted them to other uses; and the most important it has leaded to ecological fragmentation (Yavari et al., 2006) and has produced bidirectional contaminations in land and water of these sensitive vegetation habitats (Tomlinson, 1986).

1.3 Rehabilitation of Ecosystems by landscape Ecology Principles

According to the rise of industrialization and the destruction of the ecological structures, guidance and rehabilitation of ecosystems is urgent based on landscape ecology principles by considering human perception in design that results aesthetic reactions to the ecological quality of the landscape. And can provide possibility to coordinate structure and function of organizations in industrial context with ecological conditions and the natural systems of the environment by integrating of sustainable development and perceptual visual concepts.

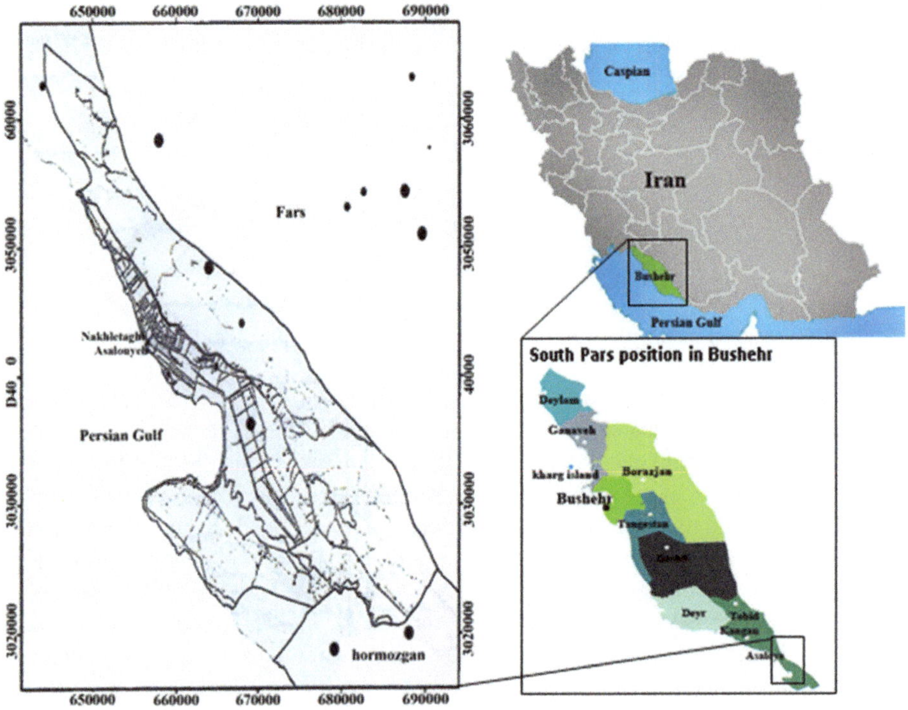

Figure1. Location of "Pars Special Economic Energy Zone 1" in Kangan city of Bushehr re-drawn based on 1982Geological maps and updated by GIS (Authors, 2014)

2. Materials and Method

This research conducted in three stages. Firstly the review of literature led to the identification and classification of major parameters influencing Mangrove communities' ecosystems and their life cycle; and their environmental planning and management. They are categorized as threats and opportunities to the Mangrove communities. In the next stage, the field observation was conducted in Asalouyeh- southern – Iran, to collect and rectify data about the Nayband bay Mangrove communities which is called Hara forests. The documentation of Nayband Hara forest has been done through preparing and mapping information. Several layers of information were prepared through GIS software such as vegetation covers, their location, their connectivity, distribution and fragmentation. Satellite images and GIS information then were used to prepare an overlaying method for analyzing data throughout the past five decades up to date. Finally the analyses of the prepared layers and transformation of Nayband Hara forest along with the landscape quality evaluation were used to develop ecological planning and design framework for the Nayband bay Hara forest.

2.1 Managing Perspective to the Ecosystem

Although in ecosystem management the maintaining of the all biodiversity and protected wide lands may be considered, according to Lackey, but have to remind that all these decisions benefits are for human. Managing approach has a revolutionary nature, and has proposed the ecological justice. From this approach, modern, linear, engineering and human oriented views are incorrect. He believes that the proponents of simple change of ecosystem management theory have two choices: 1) to prevent the ecosystem managing and declare that it is illegal, which is not practical or effective. 2) To accept the ecosystem management term within their framework (Lackey, 1995).

2.1.1 Human Ecology Approach in Environmental Management

Human ecology, is studying the relations between human beings and nature through a multidisciplinary approach, which has been considered from 1920 to examine the human factors in ecosystems. Human ecology approach has usage at different levels, from local to global, and support holistic studies. According to Baro the most ideal approach is the one that has a high adjusting capability and high clarity and its results must be tangible and visible for the public (Baro, 2002).

2.1.2 Bio Region theory

Bio Region is a distinguishable geographic area, which encompasses contiguous and self-maintained living systems with criteria such as watershed, ridge etc. It is considered for identification as the main areas of social harmony with nature, in the light of the ecology. According to this perspective, the human is a part of an ecosystem that is consisting: a) physical environment (climate, atmosphere, etc.), b) Soil as a separate substance in this physical environment, c) water, d) The vegetation consists of plants collections and, e) animals, which includes animals and community of human being. According to Don Flores, understanding the environmental changes over time in a region should begin by cognition of geology and climate history. However second base for life history of a region is beyond the environmental parameters, and includes the variety of the human cultures throughout the space and time (Yazdi & Ebrahimi, 1998) Therefore, in this approach, "society" is the fundamental pillar, solidarity and continuous relation between human and the environment that leads to establish a balance for having social and natural potential. Sustainable development is the result of these events in which there is no conflict between individual and common benefits of human and nature (Sarefi, 2000; Yazdi & Ebrahimi, 1998). This perspective therefore, within which the development of the local regional ecosystems combined with the potential of the area with emphasis on the rational utilization of resources, defines the use of technology in a way that the nature and human to be considered (Daneh kaar & Jalali, 1996).

2.2 Conceptual Common Ground between Visual and Ecological Landscape

The definition of landscape ecology given by the International Association for Landscape Ecology shows its integrated approach: "Landscape ecology is the study of the interactions between the temporal and spatial aspects of a landscape and its flora, fauna and cultural components" (IALE, 2008; Gobster et al., 2007). Although ecology and aesthetics share many common aspects they have also kept their own research worlds with ecology relying mainly on the natural science traditions while aesthetics has held closely to humanistic research traditions. This remains one of the great challenges in developing indicators for both interests (Palang & Fry,2003). Many of the approaches to visual landscapes, especially those relating spatial aspects of landscapes to human preferences, have been paralleled by conceptual developments of space use and resource exploitation in landscape ecology, e.g. the concepts of habitat complementation and supplementation (Taylor et al., 1993) (Figure 2). Therefore, to recognize and understand the relationship between ecology and aesthetics on a conceptual level are of great importance in landscape planning and management (Fry et al., 2009).

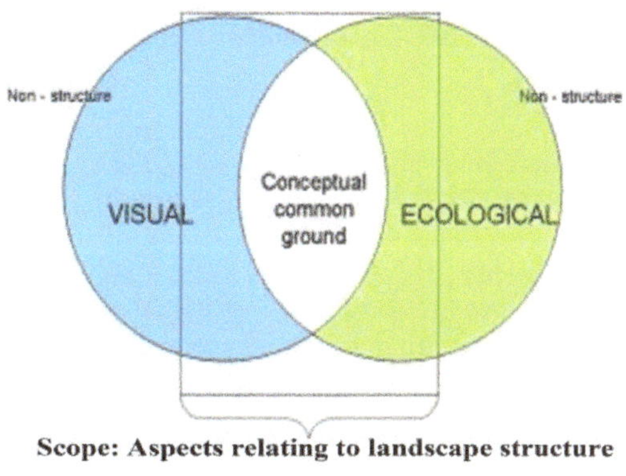

Figure 2: the conceptual common ground between visual and ecological landscape character, related to landscape structure (Gobster et al., 2007)

Visual character, these can be broadly divided into evolutionary and cultural preference theories (Appleton, 1975; Kaplan & Kaplan, 1989). Evolutionary theories explain visual landscape preferences as shaped by our common evolutionary history, claiming that we respond positively to features that enhance survival and well-being. On the other hand preferences are explained as shaped by cultural and personal experiences, with beauty essentially being in the eye of the beholder (Meinig et al., 1976).This can make important contributions to landscape planning and management, and is our basis for the development of visual indicators from a conceptual approach (Fry et al., 2009).

Ecological function, according to Forman, this is related to three fundamental components of landscape: structure, function and change. Identifying the main structural elements in the landscape, and their relation to ecological processes are essential for our understanding of how landscape change will affect species and communities (Forman, 1995).

Understanding the relationships between structure and function also enables us to predict the ecological consequences of proposed spatial solution(s). A common classification of landscape structural elements is the Patch–Corridor–Matrix approach. The Patch–Corridor– Matrix model (PCM) is fundamental to landscape ecology theory, as it defines the spatial arrangement of elements in a landscape. This in turn directs the way we represent and measure landscape to assess how changes in landscape content and configuration affect ecological processes (Young, 2000). Although this model may be an oversimplification of landscapes and poorly represent mosaic landscapes, it has been validated through many studies in many contexts linking aspects of biodiversity to landscape pattern .Landscape metrics enable us to quantify the spatial processes of attrition, fragmentation, etc., thereby capturing numerical descriptors of change. These metrics have commonly been grouped as describing patch area, edges, shape, diversity and configuration (Fry et al., 2009).

2.3 Landscape Ecological Aesthetics

The scope of landscape ecology has always included man as a central actor in landscape change and its impact. Hence landscape ecology has always included a cultural perspective and taken account of what is important for people (Ibid).

The phenomenon that is included in human perception results in diversifying aesthetic experiment of landscape that in turn creates conceptual processes and their effective responses , in which presents perceptible patterns of landscape. But the measurement in which human being are as creatures, can percept landscapes , is a scope that is so called " perceptible realm ", since in this measurement human changes landscapes deliberately , and these changes affect the environmental processes (Gobster et al., 2007).

Definition of landscape therefore, has not bounded alone in lexical definitions that focused on restricted zone or the scenes of the land surface. It is although a vast array of biological and environmental processes. Hence ecological approach of landscape studies biological and geological to extent which results in creation of perceptional properties on the land surface and reflection qualities of landscapes that are subdivisions of environmental main processes. At the other hand evaluation of landscape aesthetic quality, is a science that includes both perceptional (mental factors) and experimental (visual factors) processes that leads in to aesthetic quality before human eyes (Terry, 2001)

So it could be deducted that visual aesthetic quality of landscape is derived from practical features which can be seen in a landscape in relation to mental processes (perceptual, cognitive and emotional) in a human observer.

From this point of view, aesthetic experiments term, Sceneries with intact environments included in inaccessible lands and far from human activity, have quite natural landscape patterns and benefited from high ecological quality. Patterns which present sceneries that depend on ecological aesthetic terms are affected from integral human care. It is reaction of sustainable elements and shows aesthetic experiments.

Hence to ecological evaluation of landscape it is needed to have a complementary relation between aesthetic pleasure and environmental health in landscape. So it is recommended to make use of strategies, which equal human values and ecological goals. There are points to start operation on making the effective action practical on landscape changes, and in relation to them, which are answers to aesthetic behavior (Gobster et al., 2007).

2.4 Ecological Design: Comprehensive Application of the Art and Science of Environmental Design

"Ecological Design" is the intersection between theory of the design and the environment. Ecological design is sensitive to the local conditions with precise cognition of the place, facilities and site constraints, and considers the topological and chorological flows of the region ecosystem. Thus this knowledge prepares the sense of the place in current design by understanding the changes over time and place. The intersection of this approach is similar to the sustainable development basic framework and is based on four principles "reuse, recycle, renovate and restoration" (Makhzoumi & Pungetti,1999). Advocacy of ecological restoration and naturalistic landscapes has, at times, led to substantial controversy (Gobster & Hull, 2000), though not exclusively because of visual appearance (Kaplan, 2007).

Thus, ecological design based on the patterns and rules of the nature, aims to reduce the pollution, to promote the principles of landscaping in order to prevent destructive changes in biophysical capacity of an ecosystem and to restore it as its primary ecology and native conditions of the area.

In other word, it is a combination of natural reconstruction and rehabilitation that aims to reduce negative environmental impacts such as soil amendment, refining polluted water and prepare a condition for improvement and boosting the environment.

2.5 Mangroves Forest's Ecology

Mangrove communities grow in tropical and subtropical regions, and grow only in tidal areas along the coasts, deltas, river estuary, coves, and sandy clay islands. Mangrove trees have respiratory roots that appear in different ways. Mangrove trees' seed spends a part of its growth stages on mother root, and therefore they are viviparo us or pseudo viviparous (Tomlinson, 1986). These communities have an environment with ecotonic condition, which are in crossings of land and water and have been influenced by both environments. Thus they take advantage of the richness of the both environments while they are very sensitive because are threatened by both land and sea ecosystems (Tomlinson, 1986; Safyari, 2001; Majnoonian & Miraab Zadeh, 2003). Upper and lower bound of Mangrove expansion are determined by temperature. Temperatures around 5 ° C and frostbite limit the mangrove distribution (Tomlinson, 1986). Tides are also effective in moderating temperature, as they cause the movement of the water bed by their continuous movement and in total water and sometimes toxic substances, which are derived from the decomposition of aquatics, are transferred to the water surface and through this, the temperature reduces (Hogarth, 2007). Although Mangrove forests have pure mass and little plants diversity, but they have a lot of animal diversity. Invertebrates, crustaceans, fish and fish-eating birds are highly significant in this habitat (Ebrahimi & Riyahi, 2012).

Naybands' mangrove communities are only included tree species that named Avicennia marina and Mucronata Rhizophora (Figure 3 and Figure 4). Therefore in Iran mangrove forests are called Harraa. Herbaceous species are not counted as a part of the main species of mangrove (Daneh kaar & Jalali, 1996).

Figure 3. View of mangroves and some fish-eating birds (Authors, 2014)

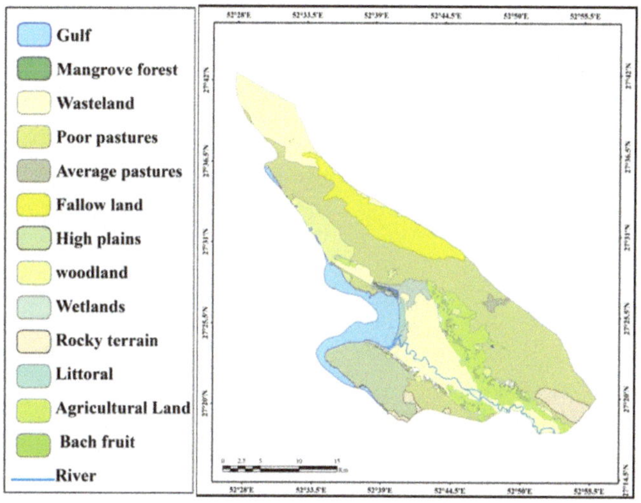

Figure 4. Classification of vegetation types and land use Pars Special Economic Energy Zone 1(Authors, 2014, based on Geological maps (1982) updated through GIS)

2.5.1 Effective Factors in Mangrove Forests' Life

Many factors are involved in the distribution and diversity of mangroves ,such as flooding cycle, temperature, salinity, marine currents, wind, waves, storms, beach slope, substrate, soil and unsustainable substrate, deposition processes, fresh water entrance (Tomlinson, 1986). It should be noted that these cases indicates the necessity of preserving ecological processes (upstream and downstream) for mangrove communities' survival.

2.5.2 The ecological importance of mangrove communities in protecting marine ecosystems are namely as followings:

• Tropical coastal ecosystems are important factors in preserving carbon balance (Sandilyan et al., 2012; Huang et al., 2003; Vane et al., 2009).

• Contributing to soil formation and protecting coastal communities against storms and leading shoreline to stability (FAO, 2007).

• Through the use and secretion salts and other solutes through the leaves are considered as filters for upland waters by turning seawater into freshwater (Odum et al., 1990).

• because of ecotonical conditions of mangrove communities streams , being rich in organic matter and nutrients from plant particles and also from the dry lands, they are valuable for many species of fishes. Hence they have a unique role in marine food chain (Ghosh et al., 2015; FAO, 2007; Safyari, 2001).

• High rate of production (2 ha /tons) during the year is such that protects the food cycle of sea and land (pelagic and benthic) and also carbon production and coastal fishing (Manson, 2005; Feller, 2002).

• Conserving biodiversity, which is included some endangered mammals, reptiles, amphibians and birds, and also through consolidation of coral reef, seaweed beds and sedimentation of suspended materials in the system to protect their roots (Hendy et al., 2014).

2.5.3 The Economic Importance of Mangrove Communities

They can be summarized as followings:

• Fuel: the most important use of them is as cellulosic fuels, because Mangrove is the only specie in coastal lands that withstand and grow in extreme environmental conditions. Mangroves' wood heating value is estimated of 222 to 322 kcal per kg.

• Animal feeding: in the growing area of mangrove forests of Hormozgan province, branches pruning are common to feed animals, especially camels. Branches and leaves of mangrove trees in addition to being palatable for animals, they contain equivalent nutritional value like hay and barley(Amiri et al., 2001).

• Construction materials: the wood of mangrove trees is used as building materials because of strength and resistance to termites attack.

• Productive purposes: this habitat is important for Apiculture, producing pulp, providing tannins and charcoal, crafts and make suitable shelter for spawning and nursery for many marine economical species, such as some fish, shrimp and crab (Safyari, 2001). Nearly 80% of the economical aquatics animals of the Persian Gulf spend their spawning period in the mangrove forests (FAO, 2007).

• Sustainable tourism: the social aspects of the mangroves are important in terms of attractiveness of its unique aesthetic and recreational and educational (FAO, 2007). Of course before anything else, should be considered a program that is based on sustainable tourism and two principles of the protection of natural resources and cultural heritage.

2.5.4 Threats and Limiting Factors of Mangrove Communities

Mangrove forests have declined vastly since the expansion of industrial development and refining facilities in South Pars in the Persian Gulf area. So that currently 12000/5 to 13000 hectares of mangrove habitat, that 80% of it is located on the coast of Iran, has remained. While in 70s decade 8500 ha of it was exclusively on the coast of Iran(Amiri et al., 2001). Existing data show that the population of some conserved species in the study area (such as Pazanan, ram, ewe and Jebeer Gazelle) have significantly endangered and due to few numbers couldn't be revived in any way (Masnavi & Amani, 2014).

2.6 Disturbances and Perturbations of Mangrove Habitat Affected by Asalouyeh Basin.

Disturbance can cause to change structure or physical and simple transformation of this system. At last, its structure pattern changes. In this manner, ecosystem can embroil degradation which shows the sections destructed need restoration in which returns to before disturbance manner. We point to some of artificial (human made) factors are effective to ecology disconnection and reduce of natural power of this ecosystem (Figure 5).

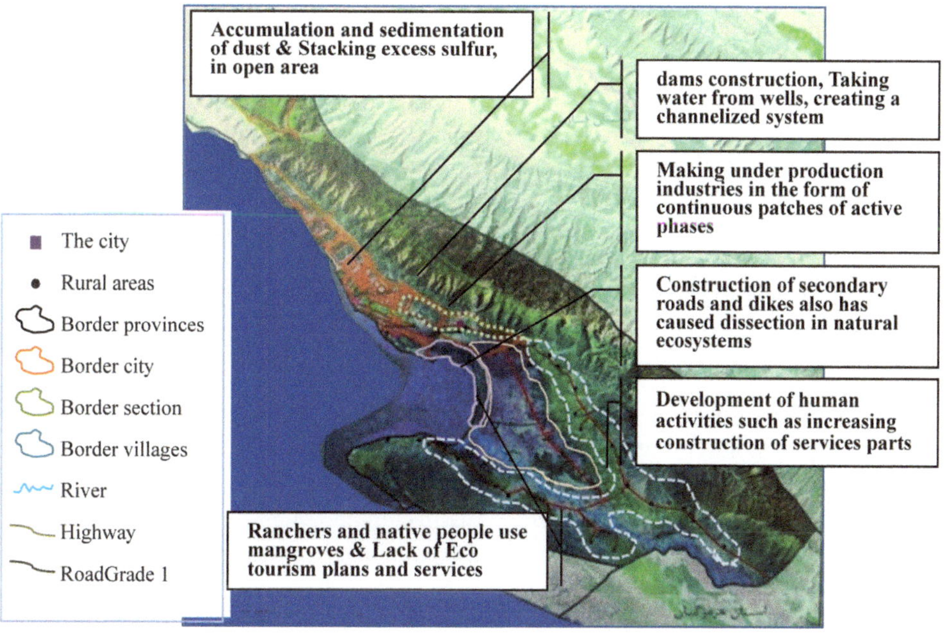

Figure 5. Disturbances and perturbations of mangrove habitat affected by Asalouyeh basin (own, 2014, based on Geological maps (1982) updated through GIS)

- Accumulation and sedimentation of dust remain like a cloud of the suspended particulate matter from the fuel gas industries, especially petrochemicals due to the mountains structure, petrochemical industry gathering in the west of the region, the effects of the moisture on temperature retention and dust heaviness, especially in the hot and humid months of the year (Amani, 2014).
- Stacking excess sulfur in open area that combines with particles from fuel gas flaring and other petrochemical facilities dusts. And after moisture absorption in the form of sulfuric acid, will cause pollution and destruction in the hydrologic system and native communities of the region.
- Groundwater contamination by heavy metals, cations and anions through various industrial processes as well as oil pollution from oil tankers traffic or marine accidents.
- Making under production industries in the form of continuous patches of active phases has led ecological connections, such as mountains, plains, and gulf (upstream and downstream) to fragmentation. In fact mangrove forests are in the form of an isolated patch between companies located in the region, and they have faced with reducing the ecological connections of (in the form of converting patch with connectivity body to the linking corridor) surrounding ecosystem (Amani, 2014).
- Construction of secondary roads and dikes also has caused dissection in natural ecosystems by cutting the ecological flow of the sea and mangrove communities. This dissection has led native patches to isolation and shrinkage and ultimately has caused reduction of the ecological capacity, as well as changing in process of the succession and balance in the native flora and fauna of the region, including: Disconnection of the tidal zone on both sides and the destruction of the mangrove habitats in the northern part (Torkianfar, 2007). Reduction of mangroves' root growth and increasing grass species of salted land that specify the water level during the tide. Increasing artificial wetlands that could be effective to change the balance between flora and fauna species in stagnant waters, particularly birds. Vehicles noise and air pollutions that are permitted to enter the area of the habitat without any limitation will result the water contamination, land degradation and ultimately will cause withering of these protected species.
- dams construction prevent the overflowing of the precipitation runoff from the specific flood of the desert areas flowing to the watershed and so this cause damage to the petrochemical Phases and companies and also cause the evaporation of such a large volumes of the fresh water resources basin.
- Taking water from wells, creating a channelized system and guiding surface water runoff despite providing a source of fresh water for the mangroves communities, result in decreasing of groundwater levels and thereby reducing the ecological capacity such as diversity in species associated with plateau and mountain in the region.

This water deficit will cause an increase in the concentration of the pollution and will interfere in sediment transportation. Therefore reducing the surface flow will effect on increasing penetration and subsidence of water and hydrological disturbances also impact on bioaccumulation changes (Yavari & Najmi Zadeh, 2006).

• Development of human activities such as increasing construction of services parts in the mangrove communities' habitats and use of seaside, which result in increasing of bacterial and chemical contamination because of the urban sewages (Torkianfar, 2007). Also this includes the development of agricultural, aquaculture, deformation of freshwater uses, disconnection of fresh water rivers, construction of shrimp ponds in the vicinity of mangrove forests and entering their wastewater in the sea, docks construction in timberlands (Guebas, 2000; Kairo, 2001; Huang, 2003).

• Ranchers and native people use mangroves, especially for grazing, are growing these trees annually about 30000 tones, According to the reports of Environment Agency, about 12000 tons of them are cropped for livestock (Amani et al., 2014).

• Lack of tourism plans and services: despite of travel and tourism attractions in this area, especially in the realm of protected habitats such as mangrove forests and their surrounding area.

• In destruction of mangrove forests due to natural factors, plant growth and development is threatened with series of aforementioned factors, along with environmental factors in region such as lack of nutrients, sediment in habitats that with changing in time and rate of daily accumulation. Increase of salinity that could be the result of droughts or changing in balance between precipitation and evaporation. And seasonal storms that uproot and destroy trees and plants, and finally inappropriate land morphology, have set Harra forests in an extremely sensitive and fragile conditions in the region (Odum et al., 1990), which in the lack of attention could cause an ecological crisis in the South Pars Special Zone.

2.7 Conclusion of the Potential and Limitations of the Region

Population growth, consideration the short-term benefits, lack of systematic management strategies and efficient techniques to restore and re-establish this ecosystem, unbalanced and inappropriate industrial units in the region are the most important threats to the region. There are deep shores with protected species, tourism and large financial resources in order to environmental care projects are the opportunities that should be considered to achieve an appropriate model for environmental management in the region. Of course mangroves' reaction due to defensive indicators against environmental changes such as the tension conditions is so that initially the plant growing will disturb and in severe level, the mangroves' seeding and flowers will impair.

2.8 Environmental Management Aspect, Appropriate for Asalouyeh the Special Region

Environmental management should seek the appropriate integration and balance between conservation and utilization of the biodiversity, that will be possible by estimating the national and regional needs, understanding the environmental impacts of gas and petrochemical industry's' functional process(Masnavi & Amani 2014). Some functional aspects of environmental management, which should be accessible in the form of short term, medium term and long term program, include:

• The regulation of the type, amount and distribution of land uses in the region due to the principles of the Land use planning.

• Caring and refining the ecological capability of resources and improving the damaged areas.

• Maintaining and enhancing biodiversity and preventing the coastal ecosystems, coral reefs, Nayband Bay and ecological functions sets from pollution and degradation.

• Preventing water, soil, air and sound pollution, with new and advanced technology.

• Providing a basis for active participation by promoting the environmental culture in special complex zone and their subsidiaries, that is responsible for developing and managing the surrounding environment.

• Definition, explanation, measurement and assessment of indexes and criteria in order that to be able to determine environmental conditions of the region and act as required data for a system feedback.

2.9 Asalouyeh Special Zone Guiding Strategies

• Development of the environment by strengthening and planting plant species in the region and expanding green space (Incursive strategy and developing).

• Decentralization and involving non-governmental sector in the planning and management according to the environmental and man-made threats which are in the region. (Defensive strategy)

• Environmental Management of Nayband region as the most important biological wealth of the region, by some methods including zoning and defining sensitive zones to prevent from more destruction and damage (Accommodating strategy).

• Coordinating affective elements on the regions' environmental management system in order to overcome internal weaknesses and minimizing internal weaknesses and minimizing weaknesses to overcome threats (Contingency strategy).

3. Ecological Design Strategies Nayband Bays' Mangrove Habitat Areas

Ecological restoration is to restore ecosystem even closer to the functions and conditions of the pre-disturbance. However, because of the dynamic nature of natural ecosystems, it is very difficult to determine the situation of the pre-disturbance and the aim of the complete restoration of the landscape cannot be entirely realistic (Masnavi & Amani 2014). In landscape ecology, the relationships between structures and components are more important than the individual components, these relations can be either vertical (into a spatial unit) or horizontal (between spatial units) (Figure 6).

Figure 6. Schematic relation of height (upstream) to the area of mangrove communities in the Bay area epiglottis (downstream) (Authors, 2014)

Most coastal development planning involves mangrove restoration and management, and rehabilitation or restoration of mangrove ecosystems has been reviewed (Hai Ren et al., 2010). Here, according to the domain of human perception are presented strategies for planning and ecological design of mangrove communities and habitats of Nayband Bay area (and Table 1; Figure 7).

Table 1. ecological design strategies of mangroves habitat in the Nayband bay (Authors, 2014)

Elevation	Strategies
Regional flood regime management	Maintain the moisture balance in warm and humid areas is an important principle of specific climate protection in deserts (Yavari & Najmi zadeh, 2006) Restoring the canalized system to conduct surface runoff from the mountains to provide required water and restoring biodiversity habitat that cause strengthening of ecological connectivity. B) Creating a permanent flow of fresh water for the mangrove communities.
Controlling urban and industrial development	A) restoring connecting corridors under construction areas with surrounded native ecosystems in order to create an urban ecological network B) controlling the development of servicing sectors according to the domain of influence of the protected area C) native design and use of renewable energy, specifically solar energy to supply service units resources(Amani et al. 2014)
Creating industries and uses associated with Harra	A) establish a comprehensive research based on preserve, restore, reform, development and utilization of forests Research base by creating an advanced data system allow recognition of forest structure and analysis of ecological complex relations (Abdulkader & Loughland, 2013), play a role in development and encouraging scientific and practical researches and creating guidelines for the use of mangrove forests' marginal industries for providing some parts of the research funding. B) Creating a collaborative cultivation and educating the ways of planting ,protecting and harvesting of mangrove trees (Soemodihardjo et al., 1996) C) Creating jobs for marginalized people in these forests by lopping of identified species
Planning and designing of ecotourism	Mangrove forests of aesthetic value, as an integrated and homogeneous ecological perspective, based on the close relationship between man and nature (Calatagan Park, 2011). Planning and designing tourism development based on the sustainable tourism principles and compatible with the sensitivity nature of the areas, so that provide alternative sources of income for local communities such as creating suitable environment station for survival, tourists tripping and making accommodation units and services in defined distance within the domain of influence of the mangrove habitat(Amani, 2014).
Conservation and development of degraded habitat	A) Natural regeneration as the best approach for the mangrove forests development 1) Rehabilitation and restoration of fragmented and isolated native patches and providing managed reproductive conditions 2) cleaning the invasive species on degraded lands 3) Selecting some areas for cultivation alongside the freshwater regime 4) silviculture compatible with mangrove forests at the interface between forest communities 5) creating mangroves' artificial habitat, along the shorelines and areas with growing capability 6) controlling the weeds, pruning and caring of drainage and irrigation (Safyari, 2001). The major advantages of remote sensing data is monitoring of change periodically (Satapathy et al., 2007). We can to apply a multi-temporal and multi-scale approach based on historical maps and remote sensing data to detect changes in mangrove cover (Ghosh et al., 2015). B) Use of resistant plants to pollutants, especially to heavy metals 1) Development of green belt around pollution producing industries to separate contaminants and to trap atmospheric pollutants (Ebrahimi, 2012), 2) use of superabsorbent plants that have a high capacity of absorption and accumulation of heavy metals in their body by use of their resistance mechanisms and can be used for cleaning the contaminated soils. Also, pollutants accumulated, detoxicated and decomposed in some kinds of plants, depending on the level of planting and clean air without any pollutants emitted (ibid).
Directing the tiding flow of the sea water Bay	In the mechanism of this ecosystem, sea water enter evenly, freely from the bay mouth in all directions during the flood tide and cover the entire mangrove forest but by ancillary roads construction, water, just have possibility of pass and flow through the embedded vent that retreated to the last point before the ebb tide. Possibility of widespread penetration of sea water in computed location of under roads and built dikes with creating channels which can be effective in reducing the effects of landscape fragmentation and also protecting patches continuity of native mangroves

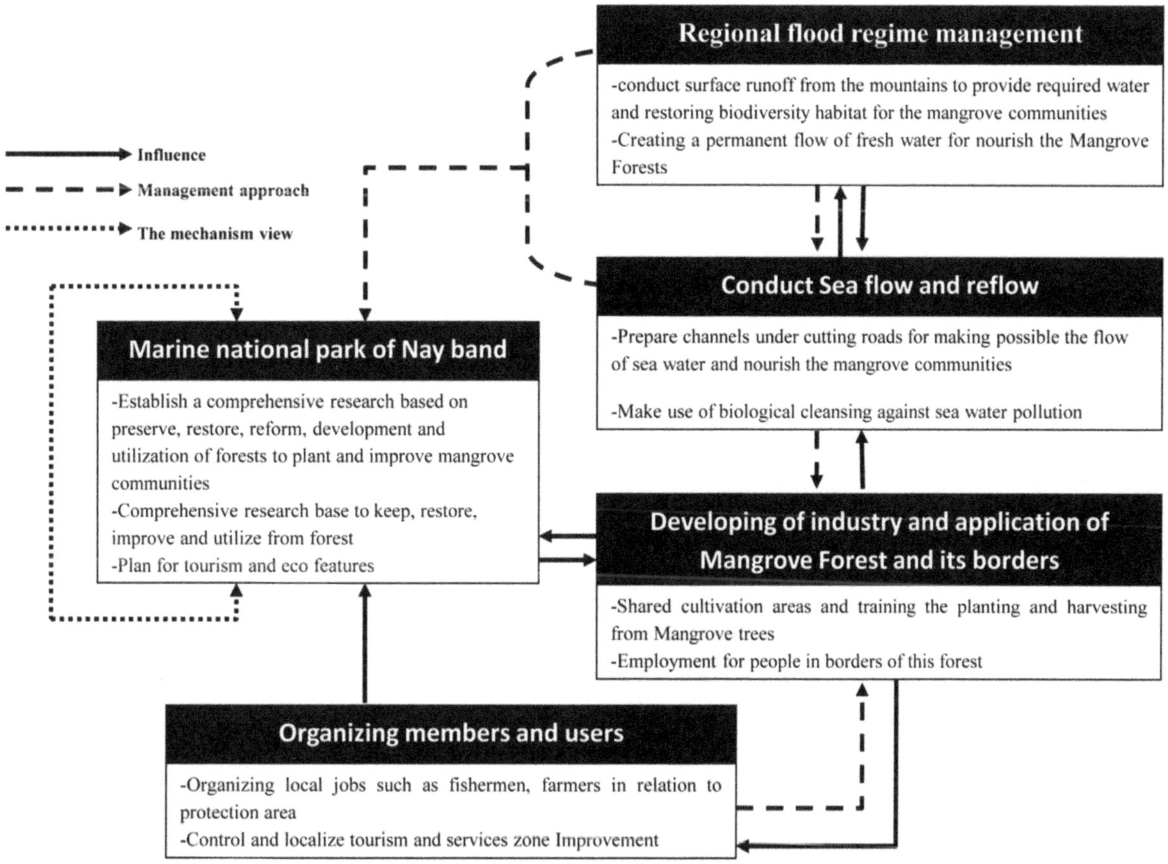

Figure 7. Communities surrounding the mangrove ecosystem conservation relations (Masnavi & Amani, 2014)

Note. dependence on Nay band mangrove communities' protective strategies according to the level of impact on the surrounding ecosystem, the influence of larger scale cycle management approach and the importance of having a single point of view in relation to the surrounding ecosystem

4. Conclusion

It should be noted that the presence of associated systems is important, both in terms of vertical bank connections and also in time and location dimensions of resources' access (at different elevations during warm and cold seasons) as well as the connecting nature between these components by the weather. This is also an important issue in planning that is not considered yet. Water flow, causes sediment and vegetation growth in habitats, habitat and health reduction (Impaired migration, life cycle and migration), and disease increasing will be the result of each disturbance and increased pollution. Currently coastal areas around the world are at risk by the irregular exploitation of their natural resources. Pollution and conversion of coastal mangroves have grown so much that they has been faced with an uncertain future, so that many Mangrove communities have been wiped out and demolished during the development process. The conservation and sustainable management of the coastal zone therefore, is an important and fundamental objective to overcome the mangroves destruction in many countries. Overall, ecological capability has been declined in mangrove habitats area, which is related to the upstream and downstream connections (the upper mountain to the bay), located in Asalouyeh in the Nayband Bay. According to nested hierarchy utilization, restoration and conservation of the resources, regarding the spatial distribution and structural features of landscape are found with great importance. These have to be combined with vertical connection between them in terms of structure and function with the aim of the restoration of the natural landscape that has been contaminated, such as the use of engineering techniques and rehabilitation and revitalization according to the new standards.

Acknowledgments

This paper is based on findings of the Masters Dissertation of the second author, and has been supported by the Research and Technology Management Section: N.I.O.C Pars Special Economic Energy Zone Organization (PSEEZ)., Asalouyeh, Iran.

References

Abdulkaderm, K., & Loughland, R. (2013). *Mangrove EcoPark* (pp. 30-31) Envir News, issue No. 21, *EED.* Retrieved from http://www.saudiaramco.com/en/home.html

Amani, N., Masnavi, M. R., & Nabi, G. H. R. (2014). *Environmental Design of Pars Special Economic Energy Zone - Asalouyeh (Pars 1).* Unpublished Master's Thesis: Department of Environmental Design, University of Tehran

Amiri, S., & Sadoogh, V. (2001). Appliance of plants branches from satellite data IRS to determine Nayband Mangrove forest surface in the vicinity to Oil facilities of Southern Pars. *National conference of Iran's environment effects* (pp. 38-43).

Baro, J. C. (2002). *Principles and methods of environmental management.* Kongreh publication, Tehran.

Calatagan Mangrove Forest Conservation Park. (2011). *Marine Protected Area.* Barangay Quilitisan. Calatagan, Batangas.

Dahdouh-Guebas, F., Mathenge, C., Kairo, J. G., & Koedam, N. (2000). Utilization of mangrove wood products around Mida Creek (Kenya) amongst subsistence and commercial users. *Economic Botany, 54*, 513–527.

Daneh kaar, A., & Jalali, S. (1996). *Checking the Mangrove forests construction in Khamir and Gheshm(Qeshm) Zone (Hormozgan State) using transect, 35*, 9-32

Ebrahimi. Z., & Riyahi. A. (2012). Petroleum pollution in mangrove forests sediments from Qeshm Island and Khamir Port Persian Gulf, Iran.

FAO. (2007). The World's mangroves 1980-2005.

Feller, I. C. (2002). The role of herbivory by wood-boring insects in mangrove ecosystems in Belize. *Oikos, 97*, 1657-176.

Forman, R. T. T. (1995). *Land Mosaics: The Ecology of Landscapes and Regions.* Cambridge University Press, Cambridge

Fry, G., Tveit, M. S., Ode, A, & Velarde, M. D. (2009). The ecology of visual landscapes: Exploring the conceptual common ground of visual and ecological landscape indicators. *Ecological Indicators, 9*, 933 – 947

Gobster P. H., Nassauer, J. I., Daniel, T. C., & Fry, G. (2007). The shared landscape: what does aesthetics have to do with ecology. *Landscape Ecol., 22*, 959-972.

Gobster, P. H., & Hull, R. B. (Eds.) (2000). *Restoring Nature: Perspectives from the Social Sciences and Humanities.* Island Press, Washington, DC.

Ghosh, A., Schmidt, S., Fickert, T., & Nüsser, M. (2015). The Indian Sundarban mangrove forests: history, utilization, conservation strategies and local perception. *Diversity, 7*(2), 149-169.

Haines-Young, R. (2000). Sustainable development and sustainable landscapes: defining a new paradigm for landscape ecology. *Fennia, 178*(1), 7-14.

Hai Ren., Xiaoming Wu., Tianzhu Ning., Gu Huang., Jun Wang., Shuguang Jian., Hongfang Lu., (2010). Wetland changes and mangrove restoration planning in Shenzhen Bay, Southern China, International Consortium of Landscape and Ecological Engineering and Springer

Hendy, I. W., Michie, L., & Taylor, B. W. (2014). Habitat creation and biodiversity maintenance in mangrove forests: teredinid bivalves as ecosystem engineers. *PeerJ, 2*, e591.

Hogarth, P. J. (2007). The biology of Mangroves and Sea grass (2nd ed.). Oxford University Press. New York.

Huang, L., Tan, Y., Song, X., Huang, X., Wang, H., Zhang, S., Dong, J., & Chen, R. (2003). The status of the ecological environment and a proposed protection strategy in Sanya Bay, Hainan Island, China. *Marine Pollution Bulletin., 47*, 180–186.

Kairo, J. G., Dahdouh-Guebas, F., Bosire, J., & Koedam, N. (2001). Restoration and management of mangrove systems – a lesson for and from the East African region. *South African Journal of Botany, 67*, 383–389.

Kaplan, R. (2007). Employees' reactions to nearby nature at their workplace: The wild and the tame, Elsevier. *Landscape and Urban Planning, 82*(2007), 17–24.

Lackey R. T. (1995). Ecosystem management: implications for fisheries management. *Renewable Resources Journal, 13*(4), 11-13.

Majnoonian, H., & Miraab, Z. (2003). *Mangrove forest, protected coast zone*. Environmental Protection Agency publication

Makhzoumi, J., & Pungetti, G. (1999). *Ecological landscape design and planning*. E and FN spon, London, UK

Manson, F. J., Loneragan, N. R., Harch, B. D., Skilleter, G. A., & Williams, L. (2005). A broad-scale analysis of links between coastal fisheries production and mangrove extent: a case-study for northeastern Australia. *Fisheries Research, 74*, 69–85.

Masnavi, M. R., & Amani, N. (2014). Ecological Landscape Planning And Design Strategies For Mangrove Communities (Hara Forests) In South- Pars Special Economic Energy Zone, Asalouyeh - Iran. *International Conference on Environmental Planning And Management (ICPM)*.

Masnavi, M. R., & Amani, N. (2014). Industrial and Urban Development of South Pars Special Economic Energy Zone, Asalouyeh and strategies to reduce the environmental impact on marine protected area Nayband. First National Conference on Architecture, Urban Development and the Environment, University of martyr Mofateh. Hamedan.

Meinig, D. W. (1976). The beholding eye. Ten versions of the same scene. *Landscape Architecture (January)*, 47–54.

Mickwitz, P., Melanen, M., Rosenström, U., & Seppälä, J. (2006). Regional eco-efficiency indicators–a participatory approach. *Journal of Cleaner Production, 14*(18), 1603-1611.

Odum, W. E., & McIvor, C. C. (1990). Mangroves. In R. L. Myers, & J. J. Ewel (Eds.), *Ecosystems of Florida* (pp. 517-548). University of Central Florida Press.

Palang, H., & Fry, G. (Eds.) (2003). *Landscape Interfaces: Cultural Heritage in Changing Landscapes* (vol. 1). Kluwer Academic Publishers, Dordrecht.

Price, A. R. G., Downing, N., Fowler, S. W., Hardy, J. T., Le Tissier, M., Mathews, C. P., Wrathall, T. J. (1994). *The 1991 Gulf War: Environmental Assessments of IUCN and Collaborators*. IUCN, Gland, Switzerland in collaboration with WWF, IAEA and IOC.

Ren, H., Wu, X., Ning, T., Huang, G., Wang, J., Jian, S., & Lu, H. (2011). Wetland changes and mangrove restoration planning in Shenzhen Bay, Southern China. *Landscape and Ecological Engineering, 7*(2), 241-250.

Safyari, S. (2002). *Mangrove forests in World*. Research Institute of Forests and Rangelands. Iran

Sandilyan. S., & Kathiresan, K. (2012). Mangrove conservation: a global perspective. Springer Science+Business Media Dordrecht. *Biodivers Conserv, 21*, 3523–3542.

Sarefi, M. (2000). Principles of Regional Planning. PBO Publication, Tehran.

Satapathy, D. R., Krupadam, R. J., Kumar, L. P., & Wate, S. R. (2007). The application of satellite data for the quantification of mangrove loss and coastal management in the Godavari estuary, East Coast of India. *Environmental monitoring and assessment, 134*(1-3), 453-469.

Soemodihardjo, S., Waroatmodgo, P., Mulia, F., & Harahap, M. K. (1996). Restoration of Mangroves in Indonesia: A case study of Tembilaham, Sumatra. In C. D. Field, (Ed.), *Restoration of Mangrove Eco systems*. International society for mangrove ecosystem, Okinawa, Japan. pp 97-110

Taylor, P. D., Fahrig, L., Henein, K., & Merriam, G. (1993). Connectivity is a vital element of landscape structure. Oikos. *Springer Science+Business Media Dordrecht, 68*, 571–573.

Terry C. D. (2001). Whither scenic beauty? Visual landscape quality assessment in the 21 st century. *Springer Science+Business Media Dordrecht, 54*, 267-281

Tomlinson, P. B. (1986). *The botany of mangroves Cambridge*. UK, Cambridge University Press.

Torkianfar F. (2007). *Evaluation of environmental effects of Southern Pars Construction on Persian Gulf Shore*. Masters dissertation, University of Tehran.

Vane, C. H., Harrison, I., Kim, A. W., Moss-Haysa, V., Vickers, B. P., & Hong K. (2009). Organic and metal contamination in surface mangrove sediments of South China. *Marine Pollution Bulletin, 58*, 134-144.

Yavari, A., & Najmi Zadeh, S. (2006). Evaluation of environment outcomes. *National conference of Iran's environment effects, 6*, 66-51.

Yazdi, M., & Ebrahimi, M. (1998). *Rural improvement theories*. Samt publication, Tehran.

7

Influence of Temperature and Humidity on the Physiological Indices of Stress in the Obudu Mountain Landscape Environment, Nigeria

Henry Ojobo[1], Bukar Usman Wakawa[2] & Aminu Umar[2]

[1] Department of Architecture, Faculty of Environmental Sciences, Kaduna State University, Nigeria

[2] Department of Architecture, School of Environmental Technology, Abubakar Tafawa Balewa University, Nigeria

Correspondence: Henry Ojobo, Department of Architecture, Faculty of Environmental Sciences, Kaduna State University, Nigeria. Email: ojheny@gmail.com

Abstract

Stress can be viewed in terms of the contribution of urbanization, lifestyle changes and the ameliorating potential of nature related environments. This study explored the influence of temperature and humidity of the mountain landscape environment on the physiological indices of individuals. 38 respondents formed a single within-group study sample. Measures of Physiological indices including blood pressure, pulse rate and respiratory rate as well as ambient environment conditions were carried out both at the urban and mountain landscape environments. Findings imply that temperature and humidity are aspects of the mountain landscape environment conditions that combine to influence human physiological wellbeing. Therefore, Individuals confronted with many sources of stress from daily engagements in urban environments can obtain short term relief in the mountain landscape environment.

Keywords: temperature, humidity, physiological indices, stress, mountain landscape, Obudu

1. Introduction

In healthcare management, stress is vital in considering the etiology of a diverse range of common health challenges, including cardiovascular diseases, anxiety disorders, obesity and depression (Lee and Oh, 2010; Probst, 2013). In landscape studies, stress is viewed in terms of the contribution of urbanization, lifestyle changes and the ameliorating potential of nature-related environments (Hartig et al., 2014). Researchers have variously tried to conceptualize the negative attributes of stress (e.g. forgetfulness, distractions, mistakes and illness) and also critically assess the positive aspects of restoration linked to nature (e.g. feeling relaxed, effectiveness, productivity and wellbeing) (Bergdahl and Bergdahl, 2002; Kaplan, 2001; Kaplan and Kaplan, 2011; Tsunetsugu et al., 2013). The assessment of the physiological stress response requires a stressor that is critical enough to produce a palpable response and also sufficient to show variations between individuals (Kajantie and Phillips, 2006). The physiological responses that are the outcome of stressful experiences (Campbell and Ehlert, 2012) manifest in a variety of ways (Hughes et al., 2014). They include changes in respiration, pulse rate, blood pressure, electro-dermal activity and salivary cortisol (Nater et al., 2005; Pan et al., 1993).

Consequently, several researchers have suggested an association between ambient environmental conditions and physiological responses (Chen et al., 2013; Hozawa et al., 2011; Modesti et al., 2006; Murakami et al., 2011). For example, Kampmann and Bröde (2009) suggested that the combined effect of temperature and humidity generally influences stress. Abdulla and Taka (1988) claimed that seasonal climatic variations in temperature and humidity have a small overall effect on blood pressure. Also, Chifamba et al. (1998) asserted that low temperatures heighten sympathetic nervous system activity and increase catecholamine secretion, which in turn influence heart rate and blood pressure increases. However, quite a number of studies pertaining to variations in physiological responses of individuals have been carried out in natural environments, especially forests, compared with urban areas (Horiuchi et al., 2014; Park et al., 2007; Tsunetsugu, et al., 2013; van den Berg et al., 2010).

Particularly, Yamaguchi et al. (2006) measured salivary amylase activity before and after walking in urban and forest environments. It was found that salivary amylase activity was a good indicator of variations in sympathetic nervous activity. In addition, findings from the study revealed that people experience less stress during contact with forest environments. Another study that provided evidence that contact with natural environments, especially forests, evokes physiological responses is that of Lee et al. (2009). In their study, they compared the responses of participants who visited both forest and urban environments. It was revealed that variations in responses in terms of salivary cortisol concentration, diastolic blood pressure and pulse rate in a forest landscape environment indicate stress mitigation, enhancement of autonomic nervous system relaxation and increased positive emotion. On the whole, natural environments stimulate physiological responses which could be partly due to psychological reactions and the restorative attributes inherent in nature (Brown et al., 2013). On the other hand, urban environments stimulate physiological responses as a result of their stress-inducing attributes (Lin and Lai, 1995; Ulrich et al., 1991; Yang et al., 2012). Whereas studies concerning the influence of environmental conditions on wellbeing have utilized forests as study contexts, no study was found to have used mountains. Mountains are high altitude environments: therefore, ambient conditions such as temperature and humidity are expected to be significantly different from those in urban environments.

This study examined variations in blood pressure, pulse rate and respiratory rate of individuals between contact with the urban and mountain landscape environments. It set out to establish whether there is sufficient evidence to conclude that there is an adequate difference between the ambient conditions of the urban and mountain environment to influence variations in physiological responses. This is in keeping with the assertion of Berto (2014) that environmental conditions are precursory factors related to stress mechanisms which mediate the relationship between the environment and human wellbeing. It also illuminates the relationship between ambient environment conditions, physiological responses and stress mitigation. According to Cairns Jr et al. (1993), changes in human physiological responses are highly sensitive early alerts indicating stress. Therefore, this study explores the influence of ambient conditions of the mountain landscape environment, specifically temperature and humidity, on the physiological indices of individuals in a manner that would mitigate stress and enhance wellbeing. To achieve this objective, the study specifically tests the following hypotheses:

1) Ambient mountain landscape environment conditions have significant influence on systolic blood pressure.
2) Ambient mountain landscape environment conditions have significant influence on diastolic blood pressure.
3) Ambient mountain landscape environment conditions have significant influence on pulse rate.
4) Ambient mountain landscape environment conditions have significant influence on respiratory rate.

2. Methodology

2.1 Study Population

The main units of assessment in this study were urban dwellers. A non-probability convenience sampling method was deployed in determining the sample. Because of the experimental nature of the study and following the idea of Creswell (2012), the sample was made up of only individuals who volunteered and agreed to be studied. A few of the volunteers who were known acquaintances at the Benue state university, Nigeria were initially co-opted by one of the authors. These initial volunteers then co-opted others to make up the sample. Also, individuals' behaviour was a determinant in the selection of volunteers who eventually constituted the sample (Weathington et al., 2010). Only individuals who were non-smokers (Lee et al., 2013), not on any form of cardiovascular related drug and not at the time of study suffering from any acute illnesses were selected (Abdulla and Taka, 1988). This process was conducted through verbal questioning (Okada and Kakehashi, 2014). Forty respondents, comprising lecturers and students between the ages of 20 and 40 years, were recruited from the urban environment of Makurdi, Benue State, Nigeria. A day before the commencement of the experiment, two of the respondents opted out: one was unable to participate due to time factors and the other admitted not being psychologically stable for the experiments. Thirty-eight respondents, including 16 males and 22 females, gave informed written consent to participate in the study at no fee. All the thirty-eight respondents formed a single within-group experimental study sample.

2.2 Study Locations

This study involved comparison between an urban and a mountain landscape environment. Makurdi, the urban environment, is the capital city of Benue State, Nigeria. It was used as the pre-test environment. It has a tropical savanna climate and lies on an altitude of about 104m above sea level. The intervention environment was the Obudu mountains in Obanliku local government area of Cross River State, Nigeria. It has a semi-temperate mountain climate and an altitude of between 1700m and 1765m above sea level. It is characterized by diverse landscape attributes such as waterfalls, grottos, river, forest reserve, a 70m long canopy walkway, bird watching platforms and a cable car ride. The cable car ride affords a moving view of prominent undulating mountain formations covered by near dense but fascinating green vegetation. Hence, the environment offers serene ambience suitable for stress mitigation.

2.3 Materials and Procedure

The physiological indices data were collected through observations. The observations involved physical measurements and recording of the respondents' physiological indices and ambient environment conditions. The experimentation was carried out between 28th January and 3rd February 2014.The experimental protocol spanned seven days, as shown in Figure 1. Pre-test measures were carried out at the urban environment on 28th, 29th and 30th January, which represented the first three days of the study. Intervention measures were carried out in the mountain landscape environment during the following three days, 1st, 2nd and 3rd February. Three qualified medics assisted in carrying out measurements in the urban environment while two assisted in the mountain landscape environment. The respondents were verbally briefed at the beginning of the study on the measures to be taken, which included systolic blood pressure, diastolic blood pressure, pulse rate and respiratory rate measurements. This briefing was done in order to alleviate any fear and anxiety that the respondents might be feeling concerning the procedure (O'Brien et al., 2003).

The pre-test centre for the measurements was set at the Benue State University Makurdi medical school private dining hall, which was within a 10 km radius and about fifteen minutes' drive from the location of each respondent. Daily activities engaged in by respondents in the pre-test environment included walking from one lecture venue to another as well as receiving lectures in non-air conditioned crowded halls. Measurements were carried out between 6 pm and 8 pm during the three-day period in the urban environment. Each respondent was allowed to rest in a seated position on arrival at the pre-test centre for at least five minutes before their blood pressure, pulse rate and respiratory rate were taken. Respondents were asked to switch off their phones to avoid distraction and disturbance. Systolic blood pressure (BP), diastolic blood pressure (BP), pulse rate and respiratory rate of each respondent were recorded for the three consecutive days.

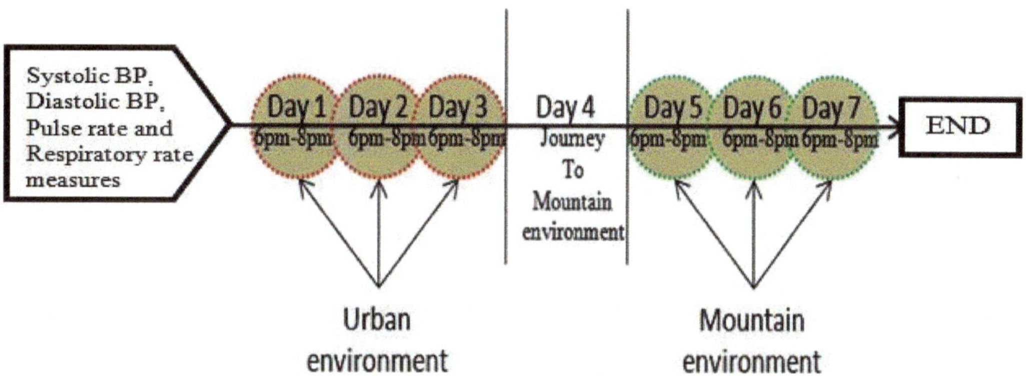

Figure 1. The seven-day experimental procedure showing measurement time

Respondents were transported by road on a journey that lasted four hours to the Obudu mountain landscape environment on the 31st of January 2014. They were allowed to rest and interact freely on arrival but measurements were not taken that day in order to check the effect of 'travel fatigue' (Waterhouse et al., 2004). The activities individuals engaged in during the three days post-test intervention study at the Obudu mountain landscape environment were both active and passive. The active activities included forest walks and swimming in the river while the passive activities involved meditative moments in the forest, mountain vantage point and river areas. Systolic BP, diastolic BP, pulse rate and respiratory rates were carried out every day throughout the three-day stay in the mountain landscape environment using the same equipment and process as obtained in the urban environment. Similar to the procedure in the urban environment, measures were carried out between 6 pm and 8 pm each day. The intervention measures were also carried out within the same time frame (6pm to 8pm) in order to check the influence of circadian effects (Biaggioni, 2008; Ochiai et al., 2015). Blood pressure measurement was performed by the medics using the standard mercury sphygmomanometer and the auscultatory technique. The pulse rate and respiratory rate measurements immediately followed the blood pressure measurement in a consecutive pattern.

To provide real-time measurement of temperature and humidity, an Onset Hobo U12 model data logger was used. Its real-time collection and presentation of data with sensors that are able to respond to parameters beyond the normal range makes the data logger an invaluable instrument for experimental data collection and analysis (Waghmare and Chatur, 2012). The temperature and humidity data extraction took place at the same time of day as the physiological indices of respondents were being measured in both the urban and mountain landscape environments. The device was set up and launched in the urban environment a day before the commencement of measures of physiological indices. This was to ensure that the correct range of data was recorded. It recorded series of real-time temperature and humidity data. On the day of the journey to the mountain landscape environment, the device was removed from its position and recording stopped. On arrival at the mountain landscape environment on the 31st of January 2014, the device was re-launched. It was fixed outdoors following the same method employed in the urban environment. The device recording was stopped on conclusion of physiological measures at the mountain landscape environment.

This set of data captured the three-day experimental period in the urban environment and similarly at the mountain environment, with one-hour intervals between measurements (Jamaludin et al., 2013). The temperature and humidity data were eventually downloaded from the device into a personal computer for data analysis. Data which did not fall within the experimental days were excluded.

3. Data Analysis

The data collected were analysed using the Statistical Package for the Social Sciences (SPSS version 19) software. T-tests were used to ascertain the differences in temperature and humidity between the urban and mountain landscape environments. Multiple linear regression analysis was employed to determine the contribution and association of temperature and humidity to changes in systolic and diastolic BP, pulse rate and respiratory rate.

4. Results

The values acquired from temperature and humidity data were analysed and the results of the t-test obtained are shown in Table 1. From the table, highly significant difference exist in temperature ($t = 21.31$, $P < 0.05$) and humidity ($t = -11.815$, $P < 0.05$) between the urban and the mountain environment. The mean values of temperature (urban: 30.55°C ± 4.22, mountain: 21.86°C ± 2.26) and humidity (urban: 45% ± 20.10, mountain: 75% ± 10.47) indicate reduced temperature and increased humidity in the mountain environment. The mean difference in temperature is 8.7°C, while the mean difference in humidity is 30%.

Table 1. Difference in ambient conditions between the urban and mountain environment

Ambient condition	Environment	Mean	Std. Deviation	t	Sig. (2-tailed)
Temperature	Urban	30.5536	4.22536	21.313	.000
	Mountain	21.8662	2.26017		
Humidity	Urban	44.9367	20.10641	-11.815	.000
	Mountain	75.0163	10.47910		

To ascertain the contribution and association of temperature and humidity to the mean change in systolic and diastolic BP, multiple linear regression analysis was employed. Table 2 shows a competing regression model between the urban and the mountain environment relative to the association of temperature and humidity on systolic BP. Based on the coefficients of determination (urban: R^2 0.28, $F = 8.60$, $P < 0.05$; mountain: R^2 0.47, $F = 7.20$, $P < 0.05$), it could be concluded that the model is a good one, given that other latent factors may exist which were not included in the regression model. However, the model indicates that the independent variables account for about 28% and 47% of the variance in systolic BP in the urban and mountain landscape environments, respectively. The model also shows that temperature and humidity are significantly related to systolic BP ($P < 0.05$) in both the urban and the mountain landscape environment.

Having shown the overall amount of variance in the dependent variable on account of the independent variables, the contribution of temperature and humidity in predicting systolic BP is examined. Temperature (urban: $\beta = -0.69$; mountain: $\beta = -0.72$) and humidity (urban: $\beta = -0.68$; mountain: $\beta = -0.88$) are inversely related to systolic BP in the urban and mountain environments. This suggests that a unit increase in temperature and humidity in both environments would lead to a decrease in systolic BP. This signifies that when humidity is held constant in the model, a unit increase in temperature accounts for about a 0.69 mmHg and 0.72 mmHg decrease in systolic

BP in the urban and mountain landscape environments respectively. On the other hand, when temperature is held constant, a unit increase in humidity accounts for a 0.68 mmHg and 0.88 mmHg decrease in systolic BP in the urban and mountain landscape environment respectively.

Table 2. Multiple regression analysis of temperature and humidity on systolic blood pressure

	Unstandardized Coefficients		Standardized Coefficients	t	Sig.
	B	Std. Error	Beta		
(Constant)	208.471	21.680		9.616	**.000****
Temperature	-2.197	.576	-.694	-3.813	**.001****
Humidity	-.453	.121	-.681	-3.739	**.001****
F = 8.605, P< 0.05, R = .569^b, R^2 = .286					
Model	Unstandardized Coefficients		Standardized Coefficients	t	Sig.
(mountain environment)	B	Std. Error	Beta		
(Constant)	427.230	55.087		7.756	**.000****
Temperature	-7.936	1.525	-0.722	-5.203	**.000****
Humidity	-1.708	.326	-0.881	-5.232	**.000****

F = 7.209, P< 0.05, R = .683^b, R^2 = .466

a. Dependent Variable: Systolic blood pressure ** Significant at 0.05% (Significant),

b. Independent Variables: Temperature, Humidity

Turning now to the regression model of temperature and humidity on diastolic BP, Table 3 reveals a good model based on the values of the coefficients of determination (urban: R^2 0.52, F = 19.40, P < 0.05; mountain: R^2 0.48, F = 16.43, P < 0.05), excluding other factors not considered in the model. This model suggests that temperature and humidity account for about 52% and 48% of the variance in diastolic BP in the urban and mountain environments, respectively. In addition, the model indicates that temperature and humidity are significantly related to diastolic BP (P < 0.05) in both the urban and the mountain landscape environment. The contribution of temperature and humidity in predicting diastolic BP is also examined in this case. Similarly, the result shows that temperature (urban: β = -0.89; mountain: β = -0.87) and humidity (urban: β = -0.85; mountain: β = -0.82) are also inversely associated with diastolic BP in the urban and the mountain landscape environment. This shows that a unit increase in each of the independent variables (temperature, humidity) in the model would lead to a decrease in diastolic BP in both environments. As revealed by the standardized beta coefficient (β), when humidity is held constant, a unit increase in temperature accounts for a decrease of more than 0.80 mmHg in diastolic BP in both the urban and the mountain landscape environment, and vice versa.

Table 3. Multiple regression analysis of temperature and humidity on diastolic blood pressure

Model	Unstandardized Coefficients		Standardized Coefficients	t	Sig.
(urban environment)	B	Std. Error	Beta		
(Constant)	178.684	16.256		10.992	**.000****
Temperature	-2.506	.432	-.890	-5.799	**.000****
Humidity	-.502	.091	-.849	-5.528	**.000****
F = 19.403,P< 0.05, R = .720^b, R^2 = .519					
Model	Unstandardized Coefficients		Standardized Coefficients	t	Sig.
(mountain environment)	B	Std. Error	Beta		
(Constant)	302.265	39.148		7.721	**.000****
Temperature	-5.859	1.084	-0.872	-5.405	**.000****
Humidity	-1.274	.232	-0.824	-5.492	**.000****

F = 16.437,P< 0.05, R = .691^b, R^2 = .477

a. Dependent Variable: Diastolic blood pressure ** Significant at 0.05% (Significant),

b. Independent Variables: Temperature, Humidity

The results in Table 4 reveal a model similar to the one in Tables 2 and 3. In this case, the coefficients of determination (urban: R^2 0.34, F = 9.28, P <0.05; mountain: R^2 0.30, F = 7.92, P < 0.05) also indicate that the model is reliable, barring other factors that are not included. This means that temperature and humidity explain about 34% and 30% of the variation in pulse rate in both the urban and mountain environments. It is apparent from the results that the predictive strengths of temperature (urban: β = -0.67; mountain: β = -0.93) and humidity (urban: β = -0.74; mountain: β = -0.71) are inversely related to pulse rate. Holding humidity constant in the model, a unit increase in temperature accounts for a variation in pulse rate of more than 0.67 beats/min in the urban environment and more than 0.93 beats/min in the mountain landscape environment. On the other hand, humidity accounts for a difference in pulse rate of more than 0.70 beats/min in both environments when temperature is held constant.

Table 4. Multiple regression analysis of the effect of temperature and humidity on pulse rate

Model	Unstandardized Coefficients		Standardized Coefficients	t	Sig.
(urban environment)	B	Std. Error	Beta		
(Constant)	117.163	9.624		12.174	**.000****
Temperature	-.949	.256	-.666	-3.708	**.001****
Humidity	-.220	.054	-.736	-4.094	**.000****
F = 9.288,P< 0.05, R = .583ᵇ, R^2 = .340					
Model	Unstandardized Coefficients		Standardized Coefficients	t	Sig.
(mountain environment)	B	Std. Error	Beta		
(Constant)	157.539	21.923		7.186	**.000****
Temperature	-2.410	.607	-.934	-3.971	**.000****
Humidity	-.394	.130	-.713	-3.033	**.004****

F = 7.927,P< 0.05, R = .553[b], R^2 = .306

a. Dependent Variable: Pulse rate ** Significant at 0.05% (Significant),

b. Independent Variables: Temperature, Humidity

The results that emerged from the regression model of temperature and humidity on respiratory rate reveal a divergent pattern from the models pertaining to systolic BP, diastolic BP and pulse rate. As shown in Table 5, the coefficients of determination (urban: R^2 0.06, F = 1.19, P > 0.05; mountain: R^2 0.11, F = 2.38, P > 0.05) indicate that the model is poor. This means that temperature and humidity account for very little variation (6%, 11%) in respiratory rate in both environments. Thus, it could be said that no significant relationship exists between temperature, humidity and respiratory rate (P > 0.05) in the two environments.

Table 5. Multiple regression analysis of temperature and humidity on respiratory rate

Model	Unstandardized Coefficients		Standardized Coefficients	t	Sig.
(urban environment)	B	Std. Error	Beta		
(Constant)	18.631	6.067		3.071	**.004**
Temperature	.102	.161	-.135	-.632	**.531***
Humidity	-.022	.034	-.139	-.647	**.522***
F = 1.195,P> 0.05, R = .250ᵇ, R^2 = .062					
Model	Unstandardized Coefficients		Standardized Coefficients	t	Sig.
(mountain environment)	B	Std. Error	Beta		
(Constant)	29.475	7.888		3.737	**.001**
Temperature	-.350	.218	-.425	-1.602	**.118***
Humidity	-.019	.047	-.110	-.415	**.681***

F = 2.387,P> 0.05, R = .342[b], R^2 = .117

a. Dependent Variable: Respiratory rate, Significant at 0.05% (*Not Significant),

b. Independent Variables: Temperature, Humidity

5. Discussion

On the whole, the findings of this study suggest that there is a statistically significant difference between the urban and the mountain landscape environment in terms of ambient conditions. They demonstrate that ambient environmental conditions influence the stress response mechanism of the individual. The foregoing builds on findings from Madsen and Nafstad (2006), who suggested that the effect of exposure to environmental conditions like temperature on the cardiovascular system is evident in blood pressure increases. Their study also revealed that a 10ºC decrease in temperature is associated with an increase in blood pressure for both men and women. Specifically, the outcome of this study resembles that of Park et al. (2009), which revealed that diastolic BP and pulse rate decreased significantly after walking in the forest environment compared to the city, indicating enhanced relaxation.

These findings are also congruent with the work of Jansen et al. (2001), who revealed that ambient temperature has a moderate but significant influence on blood pressure. Additionally, they corroborate the study by Kunutsor and Powles (2010), which revealed a significant inverse relationship between ambient temperature and both systolic and diastolic BP. (Kunutsor and Powles (2010) found that systolic BP was lowered by 5 mmHg per 10°C increase in temperature. This is also evident in the findings of the present study, which indicate that a unit increase in temperature and humidity accounts for a decrease in systolic BP and diastolic BP in the urban and mountain landscape environment respectively. These findings are, however, not in alignment with those of Charach et al. (2013), who suggested that variations in humidity are completely unrelated to blood pressure levels.

Similar to the findings pertaining to blood pressure, the inverse association of temperature and humidity with pulse rate implies that increases in temperature and humidity influence a reduction in pulse rate. While there have been several studies on the pulse rate response of individuals based on the comparison between urban and forest environments (Lee et al., 2012; Lee et al., 2014; Park, et al., 2009), none was found to consider the association between ambient conditions and pulse rate responses in a mountain environment. However, the significant relationship between ambient conditions and pulse rate indicates that blood pressure as well as pulse rate is considerably associated with temperature and humidity. Whereas systolic BP, diastolic BP and pulse rate showed significant relationships with temperature and humidity, respiratory rate did not show a significant relationship. The predictive strength of temperature and humidity on respiratory rate was weak, indicating that the relationship, though inverse, lacks much influence. In sum, this suggests that the relationship between ambient environment conditions and respiratory rate is not significant: thus, its influence on physiological response is minimal. Hence, the findings in this study suggest that a significant increase in temperature and relative humidity influenced reduced systolic BP, diastolic BP and pulse rate, whilst no change in respiratory rate was noted.

However, limitations based on variations in the physiological outcomes of individuals open more scope for further studies. One of these limitations stems from the measures of blood pressure, which was taken only once on each experimentation day due to cost constraints. It is known that blood pressure varies throughout the day depending on the psychological state of the individual, which greatly affects physiological outcomes. Another limitation of this study pertains to the continual fluctuations of ambient environment conditions, particularly temperature and humidity. Hence, ambulatory measurements of physiological indices should be utilized in further studies to check against variations influenced by fluctuations in blood pressure and ambient environment conditions.

6. Conclusion

Temperature and humidity are aspects of the mountain landscape environment that combine to influence human physiological wellbeing. Our findings indicate significant changes in blood pressure and pulse rate when individuals interact with varied ambient temperatures and humidity conditions. Further, comparison of these conditions between the urban and the mountain landscape environment revealed the significance of their effect on wellbeing. Therefore, essential empirical data is provided for use and implementation by landscape practitioners and policy makers in harnessing and managing mountain landscape environments. Hence, individuals confronted with many sources of stress from daily engagements in urban environments can obtain short term relief in the mountain landscape environment.

References

Abdulla, K., & Taka, M. (1988). Climatic effects on blood pressure in normotensive and hypertensive subjects. *Postgraduate Medical Journal, 64*(747), 23-26. http://dx.doi.org/10.1136/pgmj.64.747.23

Bergdahl, J., & Bergdahl, M. (2002). Perceived stress in adults: prevalence and association of depression, anxiety and medication in a Swedish population. *Stress and Health, 18*(5), 235-241. http://dx.doi.org/10.1002/smi.946

Berto, R. (2014). The role of nature in coping with psycho-physiological stress: a literature review on restorativeness. *Behavioral Sciences, 4*(4), 394-409. http://dx.doi.org/10.3390/bs4040394

Biaggioni, I. (2008). Circadian clocks, autonomic rhythms, and blood pressure dipping. *Hypertension, 52*(5), 797-798. http://dx.doi.org/10.1161/HYPERTENSIONAHA.108.117234

Brown, D. K., Barton, J. L., & Gladwell, V. F. (2013). Viewing nature scenes positively affects recovery of autonomic function following acute-mental stress. *Environmental science & technology, 47*(11), 5562-5569. http://dx.doi.org/10.1021/es305019p

Cairns Jr, J., McCormick, P. V., & Niederlehner, B. (1993). A proposed framework for developing indicators of ecosystem health. *Hydrobiologia, 263*(1), 1-44. http://dx.doi.org/10.1007/BF00006084

Campbell, J., & Ehlert, U. (2012). Acute psychosocial stress: does the emotional stress response correspond with physiological responses? *Psychoneuroendocrinology, 37*(8), 1111-1134. http://dx.doi.org/10.1016/j.psyneuen.2011.12.010

Charach, G., Shochat, M., Argov, O., Weintraub, M., Charach, L., Rabinovich, A., ... George, J. (2013). Seasonal changes in blood pressure: cardiac and cerebrovascular morbidity and mortality. *World J Hypertens, 3*(1), 1-8. http://dx.doi.org/10.5494/wjh.v3.i1.1

Chen, Q., Wang, J., Tian, J., Tang, X., Yu, C., Marshall, R. J., ... Lv, J. (2013). Association between ambient temperature and blood pressure and blood pressure regulators: 1831 Hypertensive patients followed up for three years. *PloS one, 8*(12), e84522. http://dx.doi.org/10.1371/journal.pone.0084522

Chifamba, J., Mufunda, J., Sigola, L., Matenga, J., & Sparks, H. (1998). Effect of variation in environmental temperature on blood pressure: is it important? *The Central African journal of medicine, 44*(2), 37-40.

Creswell, J. W. (2012). *Educational research: Planning, conducting, and evaluating quantitative* (4th ed.). Prentice Hall.

Hartig, T., Mitchell, R., De Vries, S., & Frumkin, H. (2014). Nature and health. *Annual Review of Public Health, 35*, 207-228. http://dx.doi.org/10.1146/annurev-publhealth-032013-182443

Horiuchi, M., Endo, J., Takayama, N., Murase, K., Nishiyama, N., Saito, H., & Fujiwara, A. (2014). Impact of Viewing vs. Not Viewing a Real Forest on Physiological and Psychological Responses in the Same Setting. *International Journal of Environmental Research and Public Health, 11*(10), 10883-10901. http://dx.doi.org/10.3390/ijerph111010883

Hozawa, A., Kuriyama, S., Shimazu, T., Ohmori-Matsuda, K., & Tsuji, I. (2011). Seasonal variation in home blood pressure measurements and relation to outside temperature in Japan. *Clinical and experimental hypertension, 33*(3), 153-158. http://dx.doi.org/10.3109/10641963.2010.531841

Hughes, A. M., Volante, W. G., Stowers, K., Leyva, K., Oglesby, J. M., Bisbey, T., ... Vidulich, M. A. (2014). Cognition and Physiological Response Towards a Model of Validated Physiological Measurement. *Proceedings of the 2014 Proceedings of the Human Factors and Ergonomics Society Annual Meeting: Sage Publications,* 1009-1013. http://dx.doi.org/10.1177/1541931214581211

Jamaludin, A. A., Hussein, H., Daud, N. K., Ariffin, M., & Rosemary, A. (2013). Living behaviour assessment at residential college building with bioclimatic design strategies.

Jansen, P., Leineweber, M., & Thien, T. (2001). The effect of a change in ambient temperature on blood pressure in normotensives. *Journal of human hypertension, 15*(2), 113-117. http://dx.doi.org/10.1038/sj.jhh.1001134

Kajantie, E., & Phillips, D. I. (2006). The effects of sex and hormonal status on the physiological response to acute psychosocial stress. *Psychoneuroendocrinology, 31*(2), 151-178. http://dx.doi.org/10.1016/j.psyneuen.2005.07.002

Kampmann, B., & Bröde, P. (2009). Physiological Responses to Temperature and Humidity compared with Predictions of PHS and WBGT. *Environmental Ergonomics XIII, University of Wollongong, Wollongong,* 54-58.

Kaplan, R. (2001). The nature of the view from home psychological benefits. *Environment and behavior, 33*(4), 507-542. http://dx.doi.org/10.1177/00139160121973115

Kaplan, R., & Kaplan, S. (2011). Well-being, Reasonableness, and the Natural Environment. *Applied Psychology: Health and Well-Being, 3*(3), 304-321. http://dx.doi.org/10.1111/j.1758-0854.2011.01055.x

Kunutsor, S. K., & Powles, J. W. (2010). The effect of ambient temperature on blood pressure in a rural West African adult population: a cross-sectional study: cardiovascular topics. *Cardiovascular journal of Africa, 21*(1), 17-20.

Lee, H.-Y., & Oh, B.-H. (2010). Aging and arterial stiffness. *Circulation Journal, 74*(11), 2257-2262. http://dx.doi.org/10.1253/circj.CJ-10-0910

Lee, J., Park, B.-J., Tsunetsugu, Y., Kagawa, T., & Miyazaki, Y. (2009). Restorative effects of viewing real forest landscapes, based on a comparison with urban landscapes. *Scandinavian Journal of Forest Research, 24*(3), 227-234. http://dx.doi.org/10.1080/02827580902903341

Lee, J., Park, B.-J., Tyrväinen, L., Li, Q., Kagawa, T., Miyazaki, Y., & Tsunetsugu, Y. (2012). *Nature therapy and preventive medicine.* INTECH Open Access Publisher. http://dx.doi.org/10.5772/37701

Lee, J., Tsunetsugu, Y., Takayama, N., Park, B.-J., Li, Q., Song, C., ... Kagawa, T. (2014). Influence of forest therapy on cardiovascular relaxation in young adults. *Evidence-Based Complementary and Alternative Medicine*, 2014. http://dx.doi.org/10.1155/2014/834360

Lee, M.-S., Park, B.-J., Lee, J., Park, K.-T., Ku, J.-H., Lee, J.-W., ... Miyazaki, Y. (2013). Physiological relaxation induced by horticultural activity: transplanting work using flowering plants. *Journal of Physiological Anthropology, 32*(1), 15. http://dx.doi.org/10.1186/1880-6805-32-15

Lin, N., & Lai, G. (1995). Urban stress in China. *Social Science & Medicine, 41*(8), 1131-1145. http://dx.doi.org/10.1016/0277-9536(94)00426-T

Madsen, C., & Nafstad, P. (2006). Associations between environmental exposure and blood pressure among participants in the Oslo Health Study (HUBRO). *European Journal of Epidemiology, 21*(7), 485-491. http://dx.doi.org/10.1007/s10654-006-9025-x

Modesti, P. A., Morabito, M., Bertolozzi, I., Massetti, L., Panci, G., Lumachi, C., ... Lonati, L. (2006). Weather-related changes in 24-hour blood pressure profile effects of age and implications for hypertension management. *Hypertension, 47*(2), 155-161. http://dx.doi.org/10.1161/01.HYP.0000199192.17126.d4

Murakami, S., Otsuka, K., Kono, T., Soyama, A., Umeda, T., Yamamoto, N., ... Kitaura, Y. (2011). Impact of outdoor temperature on prewaking morning surge and nocturnal decline in blood pressure in a Japanese population. *Hypertension Research, 34*(1), 70-73. http://dx.doi.org/10.1038/hr.2010.176

Nater, U. M., Rohleder, N., Gaab, J., Berger, S., Jud, A., Kirschbaum, C., & Ehlert, U. (2005). Human salivary alpha-amylase reactivity in a psychosocial stress paradigm. *International Journal of Psychophysiology, 55*(3), 333-342. http://dx.doi.org/10.1016/j.ijpsycho.2004.09.009

O'Brien, E., Asmar, R., Beilin, L., Imai, Y., Mallion, J.-M., Mancia, G., ... Palatini, P. (2003). European Society of Hypertension recommendations for conventional, ambulatory and home blood pressure measurement. *Journal of Hypertension, 21*(5), 821-848. http://dx.doi.org/10.1097/00004872-200305000-00001

Ochiai, H., Ikei, H., Song, C., Kobayashi, M., Takamatsu, A., Miura, T., ... Imai, M. (2015). Physiological and Psychological Effects of Forest Therapy on Middle-Aged Males with High-Normal Blood Pressure. *International Journal of Environmental Research and Public Health, 12*(3), 2532-2542. http://dx.doi.org/10.3390/ijerph120302532

Okada, M., & Kakehashi, M. (2014). Effects of outdoor temperature on changes in physiological variables before and after lunch in healthy women. *International Journal of Biometeorology, 58*(9), 1973-1981. http://dx.doi.org/10.1007/s00484-014-0800-1

Pan, R.-C., Chen, B., & Li, J.-J. (1993). Noninvasive monitoring of autonomic cardiovascular control during stress. *Proceedings of the 1993 Bioengineering Conference, 1993., Proceedings of the 1993 IEEE Nineteenth Annual Northeast*: IEEE, 187-188. http://dx.doi.org/10.1109/nebc.1993.404361

Park, B.-J., Tsunetsugu, Y., Kasetani, T., Hirano, H., Kagawa, T., Sato, M., & Miyazaki, Y. (2007). Physiological effects of Shinrin-yoku (taking in the atmosphere of the forest)-using salivary cortisol and cerebral activity as indicators. *Journal of Physiological Anthropology, 26*(2), 123-128. http://dx.doi.org/10.2114/jpa2.26.123

Park, B.-J., Tsunetsugu, Y., Kasetani, T., Morikawa, T., Kagawa, T., & Miyazaki, Y. (2009). Physiological effects of forest recreation in a young conifer forest in Hinokage town, Japan. *Silva Fennica, 43*(2), 291-301. http://dx.doi.org/10.14214/sf.213

Probst, T. M. (2013). Conducting Effective Stress Intervention Research: Strategies for Achieving an Elusive Goal. *Stress and Health, 29*(1), 1-4. http://dx.doi.org/10.1002/smi.2482

Tsunetsugu, Y., Lee, J., Park, B.-J., Tyrväinen, L., Kagawa, T., & Miyazaki, Y. (2013). Physiological and psychological effects of viewing urban forest landscapes assessed by multiple measurements. *Landscape and Urban Planning, 113*, 90-93. http://dx.doi.org/10.1016/j.landurbplan.2013.01.014

Ulrich, R. S., Simons, R. F., Losito, B. D., Fiorito, E., Miles, M. A., & Zelson, M. (1991). Stress recovery during exposure to natural and urban environments. *Journal of environmental psychology, 11*(3), 201-230. http://dx.doi.org/10.1016/S0272-4944(05)80184-7

van den Berg, A. E., Maas, J., Verheij, R. A., & Groenewegen, P. P. (2010). Green space as a buffer between stressful life events and health. *Social Science & Medicine, 70*(8), 1203-1210. http://dx.doi.org/10.1016/j.socscimed.2010.01.002

Waghmare, M., & Chatur, P. (2012). Temperature and Humidity Analysis using Data Logger of Data Acquisition System: An Approach. *International Journal of Emerging Technology and Advanced Engineering, 2*(1), 102-106.

Waterhouse, J., Reilly, T., & Edwards, B. (2004). The stress of travel. *Journal of Sports Sciences, 22*(10), 946-966. http://dx.doi.org/10.1080/02640410400000264

Weathington, B. L., Cunningham, C. J., & Pittenger, D. J. (2010). *Research Methods for the Behavioral and Social Sciences*. John Wiley & Sons.

Yamaguchi, M., Deguchi, M., & Miyazaki, Y. (2006). The effects of exercise in forest and urban environments on sympathetic nervous activity of normal young adults. *Journal of International Medical Research, 34*(2), 152-159. http://dx.doi.org/10.1177/147323000603400204

Yang, T., Rockett, I. R., Lv, Q., & Cottrell, R. R. (2012). Stress status and related characteristics among urban residents: a six-province capital cities study in China. *PloS One, 7*(1), e30521. http://dx.doi.org/10.1371/journal.pone.0030521

8

Role of Environmental Pollutants in Skeletal Muscle Insulin Resistance and Mitochondrial Dysfunction

Lucia Chehade[1,2]*, Audrey Caron[1,2]* & Céline Aguer[1,2]

[1] Institut de recherche de l'Hôpital Montfort, Ottawa, ON K1K 0T2, Canada

[2] Faculty of Medicine, Biochemistry, Microbiology and Immunology Department, University of Ottawa, Ottawa, ON K1H 8L1, Canada

Correspondence: Céline Aguer, Institut de recherche de l'Hôpital Montfort, Ottawa, ON K1K 0T2, Canada. Email: celineaguer@montfort.on.ca

* These authors contributed equally to this work

Abstract

In the last decade, the incidence of diabetes in Canada has nearly doubled and is now estimated to affect one in three individuals. Type 2 diabetes (T2D) is a serious public health problem: governments and health care professional are working to control its propagation, offer better treatment alternatives and reduce its impact on patient quality of life. Insulin resistance is an early event in the development of T2D. Due to its mass and important role in the maintenance of glucose homeostasis, skeletal muscle is believed to play a central role in the development of insulin resistance. The development of this metabolic disorder is multifaceted with obesity, physical activity and diet receiving the most research interest. It is recognized that mitochondrial dysfunction, increased oxidative stress and inflammation are implicated in the development of insulin resistance in muscle. Recently, the environmental hypothesis has been advanced to explain the increased number of patients with T2D. Various persistent organic pollutants (POPs), such as polychlorinated biphenyls (PCBs), bisphenol A (BPA) and p, p-dichlorodiphenylchloroethane (DDT) are being investigated in their relation to T2D. However, despite the importance of skeletal muscle in the development of insulin resistance and T2D, very few studies have focused on the effect of POPs on skeletal muscle energy metabolism. This review will highlight the implication of POPs in the development of diabetes and present work being done to asses POPs' involvement in observed metabolic disarrangements, specifically at the level of skeletal muscle.

Keywords: Glucose metabolism, Inflammation, Insulin signaling, Mitochondrial function, Oxidative stress, Persistent organic pollutants, Skeletal muscle, Type 2 Diabetes

1. Introduction

Diabetes is a chronic disease affecting over 382 million people worldwide, now considered an epidemic level (Carvalho-Santos et al., 2015). Its prevalence is only expected to rise, projecting numbers to reach 592 million individuals affected by 2035 (Guariguata et al., 2014). It has been declared as a major threat to public health in the United States and worldwide (Centers for Disease Control and Prevention [CDC] 2014; World Health Organization [WHO], 2016). Characterized by insulin resistance (IR) in targeted tissues followed by a decrease in insulin production due to β-cell failure (Carvalho-Santos et al., 2015), type 2 diabetes (T2D) is responsible for 90% of all diabetes cases (Cameron et al., 2010). While it has been shown that lifestyle, exercise, and food intake are main causes behind these staggering statistics; some studies have considered the impact of environmental pollution as a contributor to metabolic disarrangements leading to the development and advancement of T2D. Today, accumulating evidence is indeed pointing to an association between environmental contaminants and the increased risk of T2D.

Persistent organic pollutants (POPs) are organic compounds that resist photolytic, biological and chemical degradation (Ngwa, Kengne, Tiedeu-Atogho, Mofo-Mato, & Sobngwi, 2015). The chemical structure of these pollutants allows them to be soluble in lipids leading to a facilitated bioaccumulation of these compounds in adipose tissue (Pavlikova, Smetana, Halada, & Kovar, 2015). Human exposure to POPs such as polychloro-dibenzopara dioxins (PCDD) and polychloro-dibenzo furans (PCDF), bisphenol A (BPA), polychlorinated biphenyls (PCBs) and organo-chlorine pesticides such as dicholordiphenyltrichloroethane (DDT) and 1,1-dichloro- 2,2-bis(p-chlorophenyl)ethylene (DDE) occurs primarily through the consumption of animal fat found in dairy products, fish and meats (Ngwa, Kengne, Tiedeu-Atogho, Mofo-Mato, & Sobngwi, 2015). In fact, POPs remain in fatty tissues for many years because they are highly resistant to metabolic degradation (Kiviranta, Vartiainen, & Tuomisto, 2002). Many of existing POPs are still used in agriculture around the world while the potential risk to human exposure is under-evaluated. While the usage of some POPs such as DDT and PCBs have been banned in North America since the late 1970s, traces are still found in animal and plant tissue in countries that have stopped the usage of the product. In Canada, due to its resistance and semi-volatile properties, DDT has managed to be introduced into the Arctic marine ecosystems, thus increasing local human exposure (Risebrough, Walker, Schmidt, De Lappe, & Connors, 1976). Based on the Canadian Quality Soil Guidelines for the Protection of Environmental and Human Health, Herbet et al. (1994) reported concentrations of DDT and DDE, a metabolite of DDT, that range from 1.7 to 131.7 mg.kg^{-1} and 4.0 to 342.6 mg.kg^{-1}, respectively in Ontario, Canada soil. Concentrations of DDT and DDE have also been detected in maternal blood plasma, umbilical cord plasma and human breast milk in the Canadian population (D'Amour, Lye, & Murray, 2011). Similarly, a report issued by Health Canada in 2000 showed that in Ottawa, 0.2 to 12 428.6 ppt per day of PCBs (a chemical that was used mainly in electrical equipment) were found in consumable foods (Health Canada, 2013). Likewise, the public is exposed to Bisphenol A (BPA), a chemical used in the manufacture of food and beverage containers such as water bottles, storage containers, and infant bottles, through migration of the chemical from food packaging and the exposure rates in Canada for the general population range from 0.06 to 4.30 μg/kg body weight/day (D'Amour, Lye, & Murray, 2011).

2. Link Between Type 2 Diabetes and Persistent Organics Pollutants Exposure: Epidemiological Studies

With the increase in T2D incidence at a globally alarming rate, researchers have sought to elucidate whether the exposure to POPs has a physiological impact contributing to the development of IR-associated metabolic diseases. Various POPs such as PCBs, BPA and DDT have been linked to the development of IR and T2D (Neel & Sargis, 2011; Everett et al., 2007; Sargis, 2014). Epidemiological studies have reported 62% higher levels of POPs in people living with T2D compared to the rest of the population (Fierens et al., 2003; Rignell-Hydbom, Rylander, & Hagmar, 2007; Wang & Wang, 2005). In a nested case-control study, low dose POPs predicted incident T2D and it was concluded that simultaneous exposure to various POPs such as organo-chlorine pesticides and PCBs may contribute to development of obesity, dyslipidemia and IR, which are common precursors of T2D (Lee et al., 2011). Within this same study, both p,p'-DDE and PCB178 showed positive significant associations with body mass index, triglycerides, and negative association with HDL-cholesterol (Lee et al., 2011). Furthermore, based on the Nurses' Health Study, it was found that hexachlorobenze and PCB concentrations were associated with diabetes (Wu et al., 2013).

High exposure to POPs is a result of the rapid agricultural and industrial development. For example, in India, a country with high DTT usage, prevalence of diabetes has increased from 5.8% in 2000 to 8.6% in 2013 (Jaacks & Staimez, 2015). Similarly, abnormal glucose regulation has been reported in Chinese factory workers exposed to pesticides compared to unexposed people (Jaacks & Staimez, 2015). A recent cross-sectional study conducted in South Korea also showed increased risk of diabetes with increasing DTT concentration in visceral adipose tissue (Kim et al., 2014). The U.S. National Toxicology Program Workshop in 2011 also concluded that evidence is sufficient to support a positive association between DDE exposure and T2D (Kuo, Moon, Thayer, & Navas-Acien, 2013). Furthermore, many other epidemiological studies have made an association between DDE, DDT or DDD (Dichlorodiphenyldichloroethane, another metabolite of DDT) and diabetes. An analysis of workers engaged in agricultural tasks in Michigan, Hawaii, and California showed that there was an apparent association between high serum organochlorine pesticides and the appearance of hypertension, arteriosclerotic cardiovascular disease and diabetes (Morgan, Lin, & Saikaly, 1980). Also, a study performed on a cohort of Great Lakes Sport Fish Consumers established in the early 1990s and followed through 2005, tested serum for DDE and assessed diabetes diagnosis, demographics, and fish consumption. This study noted consistent dose-related associations of DDE with incident diabetes and that DDE and PCBs were higher in participants who subsequently developed diabetes, thus confirming previous studies and suggesting the possibility of a casual relationship (Turyk, Anderson, Knobeloch, Imm, & Persky, 2009).

Similarly, several studies have shown an association between PCB exposure and the development of T2D. A cross-sectional study conducted on a Native American population residing in the provinces of Ontario and Quebec in Canada, concluded that there is a significant association between PCBs and hexachlorobenzene exposure with an increased prevalence of diabetes (Aminov et al., 2016). Another study conducted on a Japanese population exposed to these pollutants (PCBs, polychlorinated dibenzo-p-dioxins, polychlorinated dibenzofurans and dioxin-like-polychlorinated biphenyls), showed a significant association with metabolic syndrome, in particular, the highest quartiles of PCB126 and PCB105 (Uemura et al., 2008). Furthermore, clinical examinations of elderly residing on dozen islands in the Northern Atlantic between Norway and Iceland showed a high prevalence of T2D and those with T2D or impaired fasting glycaemia also had higher circulating PCB concentrations (Grandjean et al., 2011).

Based on a cross-sectional study, a strong association was also made between exposure to PCDDs and PCDFs and IR (Chang, Chen, & Lee, 2016). In a similar study, conducted in non-diabetic individuals, a significant association was found between high levels of PCDD/Fs and other metabolic syndromes (Chang et al., 2009). Finally, the National Health and Nutritional Examination Survey conducted between 2003 and 2008, indicated a positive association between BPA levels in urine and T2D (Shankar & Teppala, 2011; Lang et al., 2008).

In summary, based on the numerous epidemiological studies conducted, the link between exposure to POPs (such as PCBs, DDT, DDE, PCDDs, and BPA) and the development or increased risk of T2D is clear, suggesting that POP exposure is, at least in part, responsible for the increased prevalence of T2D development in the last decades.

3. Skeletal Muscle, Insulin Resistance and Type 2 Diabetes

3.1 Importance of Skeletal Muscle in the Development of Insulin Resistance and Type 2 Diabetes

Insulin resistance is a characteristic feature of T2D and is largely implicated in the pathogenesis of the disease. In the early stages, pancreatic cells adapt to IR by increasing mass and β-cell function. While β-cell failure is considered to be ultimately responsible for the development of T2D, skeletal muscle IR is likely the primary defect, evident before hyperglycaemia develops or β-cell failure (DeFronzo & Tripathy, 2009). Indeed, skeletal muscle plays a crucial role in the maintenance of glucose homeostasis (DeFronzo & Tripathy, 2009) since it accounts for 40% of total body weight in lean individuals. Furthermore, muscle cells are the primary site of glucose storage in the form of glycogen following physical exercise or in the postprandial period accounting for 80% of whole-body glucose disposal (Yang, 2014; Kersten, 2014).

Following the consumption of a meal, blood glucose levels rise and insulin production increases to allow for the storage of glucose in tissues. Insulin activates glucose transporter 4 (GLUT4) translocation to the plasma membrane thus increasing glucose uptake into muscle cells. When muscle cells are resistant to insulin, glucose uptake is disturbed leading to hyperglycaemia (Yoshida et al., 2008). Being the major organ for glucose disposal, skeletal muscle IR leads to compensatory systems within the body, including increased production of insulin by β-cells. However, if IR and resulting hyperglycaemia persists, β-cells can no longer cope and may die through apoptosis, causing a decrease in insulin production and hyperglycaemia, establishing T2D. Thus, impaired muscle glucose uptake is thought to play a central role in the development or initiation of T2D (Gaster, Petersen, Hojlund, Poulsen, & Beck-Nielsen, 2002). A great importance therefore lies in understanding the mechanisms responsible for the development of IR in this tissue. Most studies have looked at the consequences of the exposure to POPs in tissues such as the brain, liver and pancreas. However, while skeletal muscle is largely implicated in the development of IR and T2D, studies on the effect of POPs on skeletal muscle IR are lacking.

3.2 Insulin Signaling Pathway and Insulin Resistance in Skeletal Muscle

Insulin is responsible for maintaining blood sugar levels by activating glucose uptake into cells using specialized glucose transporters. Insulin secretion and action can be affected by various factors including hyperglycaemia, hyperlipidemia, genetic factors and exposure to POPs (Evans, Goldfine, Maddux, & Grodsky, 2003). At the cellular level, IR develops when there is a problem in the insulin signaling pathway. Mutations in genes coding for kinases or proteins, such as Akt (also known as protein kinase B (PKB)), PDK (phosphoinositide-dependant kinase 1), the insulin receptor, and imbalances in the number of phosphoinositide 3-kinase (PI3K) subunits, may hinder the insulin signaling pathway and lead to IR in muscle (Sargis, 2014; Saini, 2010).

A predisposing factor in the development of T2D is the development of muscle IR. Interestingly, many studies have demonstrated that offspring with normal glucose tolerance of two parents with T2D exhibit moderate to severe skeletal muscle IR (Vaag & Beck-Nielsen, 1992; Gulli, G., Ferrannini, Stern, Haffner, & DeFronzo, 1992; Tripathy et al., 2003). The implication of skeletal muscle IR in the development of T2D was further elucidated by a study demonstrating that genetically predisposed T2D individuals had increased IRS (Insulin Receptor Substrate)-1 serine phosphorylation and decreased PI3K activity in skeletal muscle (Morino et al., 2005). This observed increased Ser/Thr phosphorylation of IRS-1 was previously shown to impair insulin signaling. Indeed, in skeletal muscle, phosphorylation of IRS-1 on serine residues generates a negative feedback that inhibits the tyrosine phosphorylation of both insulin receptor and IRS-1 which in consequence inhibits the insulin signaling pathway and GLUT4 translocation (Morino et al., 2005).

3.3 Mitochondrial Dysfunction and Insulin Resistance in Muscle

Muscle IR is thought to develop as a result of mitochondrial dysfunction leading to disarrangement in the insulin signaling pathway. In a study done by Morino et al. (2005), it was observed that the rate of insulin-stimulated glucose uptake in muscle was 60% lower in IR offspring of parent with T2D than the control subjects associated with a 38% lower mitochondrial density (Morino et al., 2005). This study confirms the results of many others that have shown that mitochondrial function and/or number is altered in individuals diagnosed with T2D (Petersen et al., 2004; Befroy et al., 2007, Kelley et al., 2002, Ritov et al., 2005). Alteration of mitochondrial function results in intramyocellular accumulation of lipids, which has been associated with a decrease in insulin sensitivity in the rodent model and patients with T2D or obesity (Krssak, Falk Petersen, Dresner, DiPietro, Vogel, Rothman, et al., 1999; Pan et al., 1997; G. Perseghin et al., 1999; An et al., 2004). Inefficient fatty acid β-oxidation is also linked to increased muscle accumulation of lipid intermediaries such as ceramides, diacylglycerol, acyl-CoA and acylcarnitines (Zhang, Keung, Samokhvalov, Wang, & Lopaschuk, 2010). These metabolites participate in the development of muscle IR by interfering with the insulin signaling pathway (Coll et al., 2008; Weigert et al., 2004; Aguer et al., 2015). Inversely, an elevation of plasma free fatty acids induces mitochondrial dysfunction by decreasing oxidative phosphorylation and increasing IR (Brehm, Krssak, Schmid, et al., 2006).

3.4. Muscle Oxidative Stress and Insulin Resistance in Muscle

Oxidative stress is the accumulation of reactive oxygen species (ROS) deemed to be damaging. While ROS act as important signaling molecules, their effects are dose dependant and high levels could exert negative effects and alter gene expression affecting their ability to grow, differentiate, and adapt (Musarò, Fulle, & Fanò, 2010). Emerging evidences have assigned an important role to oxidative stress in muscle homeostasis and in the physiopathology of skeletal muscle (Musarò, Fulle, & Fanò, 2010). Mitochondria are a large producer of ROS under basal conditions, and when dysfunctional can produce even higher levels of ROS (Czajka et al., 2011). Skeletal muscle is a major source of oxidants in the body, since it contains a large number of mitochondria. Usually, there is a balance between the production of oxidants and antioxidants. However, if there is an imbalance in the ratio oxidants/antioxidants, ROS and other oxidants will accumulate and cause oxidative stress and cell damage. In fact, ROS can cause mutations in the DNA, lipid peroxidation and disrupt the function and expression of proteins, leading to dysfunctions in cellular processes (Matsuda & Shimomura, 2013). Generally, under conditions leading to metabolic disorders such as IR and T2D, there is an imbalance between ROS and antioxidant production by mitochondria, leading to harmful ROS accumulation. ROS can damage the components of mitochondria, directly leading to mitochondrial dysfunction. In fact, according to Bonnard et al. (2008), mitochondrial defects do not appear before IR development but is the result of oxidative stress that affects both the insulin signaling pathway and mitochondrial function. This claim is supported by data showing an increase in muscle ROS production in mice exposed to a high-fat / high-sucrose diet (Bonnard et al., 2008). Those effects were similarly observed *in vitro* when exposure of muscle cells to high glucose and high fatty acid concentrations induced ROS production and altered mitochondrial density and function (Bonnard et al., 2008). Interestingly, treatment with antioxidants decreased muscle ROS levels and restored mitochondrial integrity in both the mouse model as well as the *in vitro* muscle cell model (Bonnard et al., 2008).

Further, it has been noted that the establishment of IR in skeletal muscle causes a decrease in electron transport chain efficiency (lower ATP production rate) occurring with elevated mitochondrial ROS production (Zabielski et al., 2015). Furthermore, with decreasing muscle mitochondrial function and fatty acid β-oxidation, there is an increase in skeletal muscle accumulation of ceramides and acylcarnitines which contribute to an increase in ROS production, mitochondrial integrity alteration and IR (Zabielski et al., 2015; Aguer et al., 2015; Schmitz-Peiffer et al., 1999; Chavez & Summers, 2012; Bastie et al., 2004; Pickersgill et al., 2007).

Another study suggests that a large redox imbalance in muscles leads to the activation of proteolysis and massive oxidation of proteins, which become more susceptible to degradation (Pellegrino et al., 2011). NADPH oxidase and electron transport chain complexes I and III (found in mitochondrial inner membrane) are main cellular sources of ROS (Aguer & Harper, 2012). A cross talk between mitochondria and NADPH oxidases exists and may represent a feed-forward vicious cycle of ROS production (Dikalov, 2011). When nutrients such as glucose or lipids are present in excess or a metabolic disarrangement is established, ROS production is increased (Henriksen, Diamond-Stanic, & Marchionne, 2011). This production of ROS in skeletal muscle mitochondria was shown to inhibit the insulin signaling pathway by acting on glycogen synthase kinase-3β (Henriksen, Diamond-Stanic, & Marchionne, 2011). Oxidative stress is also inducing a number of stress-sensitive signaling pathways (NF-κβ (nuclear factor-kappa B, transcription factor used in mediating immune and inflammatory responses), JNK (Jun N terminal Kinase) / SAPK (Stress-activated protein kinases) (SAPK/ JNK are members of the MAP Kinase family and are activated by a variety of environmental stresses, inflammatory cytokines, and growth factors), and p38 MAPK) leading to IR by interfering with IRS-1 and/or Akt (Rains & Jain, 2011; Henriksen, Diamond-Stanic, &

Marchionne, 2011) (Figure 1). The activation of NF-κβ was also shown to regulate pro-inflammatory cytokines such as tumour necrosis factor (TNF-α), which directly affects insulin signaling pathways by inhibiting IRS-1 (Wang et al., 2015). Also other serine/threonine kinases such as PKC (protein kinase C) are activated by ROS, which disrupt the ability to recruit and activate molecules that contain SH-2 (Src Homology 2 domain) hence disrupting the interaction of IRS protein with the insulin receptor (Rains & Jain, 2011).

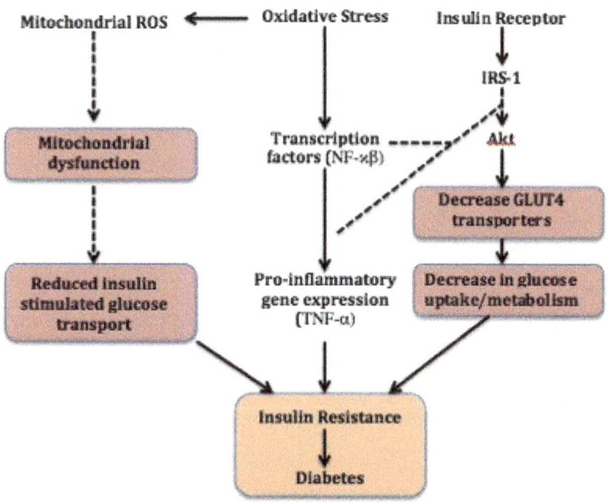

Figure 1. The effect of oxidative stress on the insulin signaling pathway and on mitochondrial dysfunction and its link to type 2 diabetes

4. Adipose Tissue and Skeletal Muscle Crosstalk in the Development of Insulin Resistance

Obesity is a risk factor in the etiology of T2D, demonstrating that there could be a link between the increase in body fat and IR development. Adipose tissue plays a role in storing excess energy as well as in the synthesis and secretion of hormones and cytokines/adipokines. The secretion of these adipokines by adipose tissue is altered by an excessive accumulation of fat, exposure to pollutants or in response to oxidative stress, consequentially affecting the function of other tissues including skeletal muscle. Moreover, cytokines and adipokines secreted by adipose tissue may have extrinsic effects on the development of IR in skeletal muscle (Kadowaki et al., 2006; Steinberg, 2007; Gadupudi et al., 2015; Quinn et al., 2005). Dysfunctional adipose tissue generally increases the secretion of pro-inflammatory cytokines and reduces that of anti-inflammatory cytokines. These mediators have various effects on the insulin signaling pathway and insulin sensitivity, as well as the expression of other cytokines.

4.1 Implication of Adipokines in the Development of Insulin Resistance

Among the adipokines secreted by adipose tissue, adiponectin is a hormone that plays a role in energy homeostasis, glucose and lipid metabolism (Yamauchi et al., 2002; Imbeault, 2007). Adiponectin sensitizes the body to insulin, and as such could be a target to fight against the development of IR (Kadowaki et al., 2006; Saini, 2010). In muscle cells, adiponectin activates AMPK (AMP-activated protein kinase) and PPAR-α (peroxisome proliferator-activated receptor) (Preedy & Hunter, 2011; Yoon et al., 2006). Upon AMPK activation, mitochondrial fatty acid oxidation and glucose uptake are increased independently of insulin signaling. Similarly, PPAR-α, a transcription factor, promotes muscle fatty acid oxidation, via upregulation of genes involved in the catabolism of lipids (Burri, Thoresen, & Berge, 2010; Kersten, 2014). As a result, adiponectin stimulates muscle fatty acid oxidation and mitochondrial oxidative capacities (Civitarese et al., 2006; Iwabu et al. 2010). It was also shown that adiponectin corrects mitochondrial dysfunction in high-fat fed mice and restores glucose homeostasis (Liu et al., 2013). Adiponectin also has a beneficial effect on insulin activity via indirect reduction of inflammation and oxidative stress (Jorlay et al. 2012).

Leptin, the first adipokine to be discovered, mediates long-term regulation of nutrient intake and energy expenditure. The amount of leptin secreted is dependent on adipose tissue mass, with higher leptin secretion when adipose tissue mass increases. In skeletal muscle, leptin regulates fatty acid transport, fatty acid oxidation, and mitochondrial function through AMPK-dependent pathways (Steinberg, Parolin, Heigenhauser, & Dyck, 2002;

McClelland et al., January 2004; Henry, Andrews, Rao, & Clarke, 2011). In non-obese individuals, leptin may also increase insulin activity, inhibit hepatic glucose production and increase cellular glucose uptake (Preedy & Hunter, 2011). However, obesity increases adipose tissue mass and thus, the secretion of leptin, resulting in leptin resistance in peripheral tissues (Steinberg, Parolin, Heigenhauser, & Dyck, 2002). However, when cells present leptin resistance, the positive impact of leptin on muscle metabolism is decreased (Aguer & Harper, 2012; Steinberg, Parolin, Heigenhauser, & Dyck, 2002). For example, in obese rats, activation of AMPK by leptin was reduced which decreased muscle fatty acid metabolism and mitochondrial content (Steinberg et al., 2006). It was also shown that chronic administration of leptin reduces muscle fatty acid uptake, via decreased expression of the muscle fatty acid transporters fatty acid translocase (FAT)/CD36 and FABPpm (plasma membrane fatty acid-binding protein) (Steinberg et al., 2001).

Resistin is another adipokine that plays a role in the development of IR in the liver and adipose tissue, but also in skeletal muscle, due to its inhibitory action on AMPK (Satoh et al., 2004; McTernan, Kusminski, & Kumar, 2006; Saini, 2010), which decreases muscle glucose transport (Steinberg et al., 2007). Furthermore, resistin inhibits the insulin signaling pathway in skeletal muscle, possibly via the activation of toll-like receptor 4 (TLR-4), which further increases blood glucose levels (Steppan et al., 2001; Benomar et al., 2013). Other studies have shown that resistin induces the expression of SOCS-3 (suppressor of cytokine signaling-3) and PTP-1B (protein-tyrosine phosphatase 1B), both inhibitors of the insulin signaling pathway, by decreasing the phosphorylation and expression of the insulin receptor (Kusminski, Mcternan, & Kumar, 2005; Benomar et al., 2013). The role of resistin in IR development was further demonstrated by the removal of visceral fat that reduced the serum levels of resistin and led to reversal of IR in rats (Borst, Conover, & Bagby, 2005).

4.2 Secretion of Inflammatory Factors and the Development of Muscle Insulin Resistance

During obesity, increased inflammation in adipose tissue is related to the infiltration of macrophages within adipocytes. Those macrophages secrete the pro-inflammatory cytokines IL-6 (interleukin 6) and TNF-α (tumor necrosis factor α) (Wiseberg et al., 2003). TNF-α and IL-6 were shown to negatively impact the regulation of muscle glucose and lipid metabolism. For example, chronic exposure of muscle cells to TNF-α or IL-6 promotes IR (Plombard et al., 2005; Peraldi & Spiegelman, 1998; Bhatnagar et al., 2010; Wang, Xiaowen, & Du, 2010; Preedy & Hunter, 2011; Nieto-Vazquez, Fernandez-Veledo, de Alvaro, & Lorenzo, 2008). In muscle cells, it has been shown that TNF-α not only increases the phosphorylation of IRS-1 on serine residues which results in the inhibition of Akt phosphorylation and decreased GLUT4 translocation (Plomgaard et al., 2005; Steinberg, 2007; Saini, 2010; Bhatnagar et al., 2010), but also reduces the expression of key proteins of the insulin signaling pathway such as the insulin receptor, IRS-1, and GLUT4 (Cawthorn & Sethi, 2008). Similarly, chronic exposure to IL-6 impairs GLUT4 translocation due to an increased phosphorylation of IRS-1 on serine residues. Taken together these effects of TNF-α and IL-6 on muscle cells result in decreased muscle glucose uptake. This might be done through increased inflammation and oxidative stress (Bhatnagar et al., 2010; Nieto-Vazquez et al., 2008), which activate JNK and NFκB signaling pathways (Cawthorn & Sethi, 2008; Nieto-Vazquez, Fernandez-Veledo, de Alvaro, & Lorenzo, 2008). It has also been shown that IR might be caused by the effect of TNF-α on the transcriptional activity through downregulation of proteins in the insulin signaling pathway (Cawthorn & Sethi, 2008). In addition, TNF-α reduces the activity of AMPK by upregulating PP2C (protein phosphatase 2 C) gene expression (Steinberg, 2007; Cawthorn & Sethi, 2008), therefore reducing glucose uptake in muscle cells but also muscle fatty acid oxidation leading to fatty acid intermediaries' accumulation involved in the development of muscle IR. Furthermore, some targets of TNF-α decrease the expression of genes that regulate fatty acids and glucose uptake in adipose cells (Preedy & Hunter, 2011). TNF-α also induces inducible nitric oxide synthase (iNOS) and stimulates nitrite production by muscle cells causing a reduction in insulin action (Bédard, Marcotte, & Marette, 1997). Moreover, TNF-α, as well as interleukin-6 (IL-6) decrease adiponectin secretion by the adipose tissue, which reduces insulin sensitivity and mitochondrial biogenesis, and increases ROS production via mitochondrial dysfunction in muscle cells (Fasshauer et al., 2003; Nieto-Vazqquez et al., 2008).

During an inflammatory state, MCP-1 (monocyte chemoattractant protein 1) is another mediator that is secreted by the adipose tissue, which usually attracts immune cells to the physiologically wounded sites. It has been shown that macrophage recruitment plays a role in the development of T2D (Patsouris et al., 2014) and that mice lacking one of the MCP-1 receptor, CCR2 (C-C chemokine receptor type 2), are partially protected against the development of obesity-related IR (Shoelson, Lee, & Goldfine, 2006).

In conclusion, multiple studies have demonstrated a link between adipose tissue function, regulation of its production of cytokine/adipokines and the development of IR in skeletal muscle, showing that the communication between adipose tissue and skeletal muscle is central in the development of IR in patients with obesity.

5. Role of POPs in the Development of Insulin Resistance and Mitochondrial Dysfunction

5.1. Effects of POPs on Insulin Sensitivity

Animal studies have previously associated POPs with the development of IR and glucose intolerance. For example, it has been shown that certain mixtures of PCBs (i.e. Aroclor) diminish the action of insulin in adipose cells (Park et al., 2013). Similarly, a chronic exposure (20 wk) of 36 mg/kg/wk of Aroclor 1254, which contains several PCBs induced hyperinsulinemia and exacerbated IR in mice (Gray et al., 2013). Moreover, Ibrahim et al. (2011) demonstrated that mice on a high fat diet with salmon fillet intake (containing POPs) exhibited resistance to insulin more than mice fed a diet with reduced levels of POPs. In a similar study in rats, intake of purified salmon oil had low impact on the development of IR and obesity, whereas rats that consumed crude salmon oil containing POPs over a period of 10 weeks (including PCBs and DDT) showed exaggerated obesity and IR development (Midtbo et al., 2013).

Exposure to 1.7 mg/kg body weight to prenatal DDT from gestational day 11.5 to postnatal day 5 in rats impaired energy expenditure and metabolism in adults by increasing adiposity, impairing insulin secretion, decreasing glucose tolerance and elevating gluconeogenesis (La Merrill et al., 2014). While the analyses were performed on the liver, the results suggest that exposure to DDT during development increases the risk for developing metabolic syndrome (La Merrill et al., 2014).

BPA may have similar effects on insulin sensitivity. Long-term exposure to BPA leads to metabolic dysfunctions in adult male mice including glucose intolerance, decreased secretion of adiponectin and muscle IR without affecting β-cell function (Moon et al., 2015). Furthermore, exposure to BPA during pregnancy mimicked the effects of a high-fat diet (Garcia-Arevalo et al., 2014). These groups reported fasting hyperglycaemia, glucose intolerance and high levels of non-essential fatty acids in plasma of BPA-exposed mice compared to controls. Furthermore, glucose stimulated insulin release was also disrupted.

In humans, it was demonstrated that metabolically abnormal obese individuals had higher plasma concentrations of POPs (PCBs, trans-nonachlor) compared to metabolically healthy obese people, suggesting a link between circulating POP concentrations and the development of the metabolic syndrome (Gauthier et al., 2014). Also, insulin-stimulated glucose uptake was impaired in adipocytes (3T3L1) when treated with a mixture of POPs (Gauthier et al., 2014).

Taken together, these studies strongly suggest that exposure to POPs induce glucose intolerance and the development of IR in multiple models. The mechanisms by which POPs affect insulin action, including increased oxidative stress, inflammation and mitochondrial dysfunction will be discussed below.

5.2 Effects of POPs on Oxidative Stress

As discussed previously, POPs such as DDT, DDE, BPA and PCBs impair glucose homeostasis in animal models. These observed effects could be the result of increased ROS production induced by POP exposure. In *Hypostomus commersoni,* an important catfish to study both ecological impacts and a vehicle to human exposure, POPs were found at significant concentrations in the liver and muscle (Bussolaro et al., 2012). Results of that study associated banned pesticides such as aldrin, dieldren and DDT with various histopathological findings in the liver and gills (Bussolaro et al., 2012). Interestingly, a negative correlation between the concentration of several POPs and glutathione S-transferase and glucose-6-phosphatate dehydrogenase were also observed (Bussolaro et al., 2012). Glucose-6-phosphate dehydrogenase is an enzyme involved in the pentose phosphate pathway which produces NADPH and pentoses. NADPH maintains the levels of glutathione which helps protect cells against oxidative damages (Mazulis et al., 2015). Exposure to POPs may therefore increase oxidative damage through modulating glutathione levels.

POPs are present and metabolized in the liver of exposed organisms and are known to cause liver damage via oxidative stress. As expected, hepatocytes exposed to DTT showed elevated ROS content and in return ROS increased NF-kB activation in those cells (Jin et al., 2014). Following NF-kB activation, caspase-dependent cell death is triggered causing an increase in mitochondrial potential and apoptosis. This confirms that DDT intoxication contributes to the generation of free radicals in cells and tissues (Jin et al., 2014).

Similarly, exposure to PCB126 in endothelial cells, hepatocytes and chondrocytes increased ROS levels (Hennig et al., 2002; Lee & Yang, 2012), oxidative stress (Hennig et al., 2002; Ramadass et al., 2003), and caused mitochondrial defects (Park et al., 2013). Mitochondrial electron transport chain defects also result in an increase in the production of ROS (Park et al., 2013). ROS production and oxidative stress by coplanar PCBs for example happen through the activation of AhR, which in turn activates the transcription of genes of cytochrome P450 (CYP1A1/2 and CYP1B1) and enzyme ligands metabolism (Park et al., 2013; Baker et al., 2012). For instance, PCB77 induces oxidative stress by acting as an agonist to AhR, which subsequently activates JNK/SAPK and eventually leads to glucose and insulin impairment (Slim, Toborek, Robertson, Lehmler, & Hennig, 2000). *In vivo* and *in vitro* studies have shown that oxidative stress induces inflammation through the activation of NF-κβ (Hennig et al., 2002). Hennig's further studies demonstrated that PCB exposure induces oxidative stress, activating inflammatory pathways involved in a number of pathologies such as atherosclerosis and cardiovascular disease (Hennig et al., 2002; Hennig et al., 2005). The role of oxidative stress in endothelial cell damage mediated by PCBs was demonstrated by treating endothelial cells with antioxidants such as vitamin E, which rescued PCB-induced oxidative stress and endothelial cell dysfunction (Hennig, Hammock, Slim, Toborek, Saraswathi, & Robertson, 2002).

An oral chronic exposure in adult female rats to a representative mixture of POPs composed of endosulfan, chlorpyrifos, naphthalene and benzopyrane caused an increase in mitochondrial superoxide dismutase (SOD) (enzyme that alternately catalyzes the partitioning of the superoxide (O_2^-) radical into either ordinary molecular oxygen (O_2) or hydrogen peroxide (H_2O_2)) and other mitochondrial oxidative parameters suggesting the POP mixture induced oxidative stress particularly in mitochondria (Lahouel et al., 2016). In another study, a mixture of POPs containing perfluorinated (PFC), brominated mixture, PCBs, DDT, hexachlorobenzene, chlordanes, thexachlorocyclohexanes and dieldrin, not only induced the production of ROS but also induced apoptosis in a human hepatocarcinoma cell line (Wilson et al., 2016).

5.3 Effects of POPs on Inflammation

POP exposure is therefore linked to an increase in ROS production and oxidative stress, which in turn activates inflammatory pathways. Indeed, for example, PCB77 treatment in adipocytes and endothelial cells (5 μmol/L for 16 hours) activated inflammatory pathways, increased the production of pro-inflammatory adipokines (TNF-α, IL-6 and MCP-1) and reduced the expression of anti-inflammatory adipokines (adiponectin and leptin) (Wang, Xiaowen, & Du, 2010; Arsenescu et al., 2008). In addition, it has been demonstrated that PCB77 blocks Akt activation by insulin in endothelial cells, due to an increased secretion of TNF-α; while the increased secretion of IL-6 may be responsible for the phosphorylation of IRS-1 on serine residues (Wang, Xiaowen, & Du, 2010). Finally, it has been demonstrated *in vivo,* in mice, and *in vitro* in human multipotent adipose-derived stem cells, that TDCC, PCB126 and PCB153 increase the activity of the inflammatory response by binding to the AhR (aryl hydrocarbon receptor) (Kim et al., 2012). The same observations were made when adipocytes were exposed to POPs and the proposed mechanism involves the AhR receptor, inflammation and mitochondrial dysfunction (Imbeault et al., 2012). In conclusion, it is proposed that POPs increase proinflammatory messengers and decrease anti-inflammatory mediators (Imbeault et al., 2012), which in turn have an impact on the insulin signaling pathway and therefore lead to the development of IR.

Some PCBs also decrease adipocyte differentiation (Gadupudi et al., 2015; Arsenescu et al., 2008) responsible for adipose tissue dysfunction. For example, PCB126 inhibits adipogenesis of human preadipocytes (Gadupudi et al., 2015). However, PCB77 does not seem to have a similar effect (Chapados et al., 2012). Furthermore, treatment of mouse adipocytes (3T3L1) with low concentrations of PCB77, and of human preadipocytes with PCB126 increases the expression and production of pro-inflammatory adipokines such as TNF-α and MCP-1, and decreases the expression of adiponectin (Arsenescu et al., 2008; Hennig et al., 2002; Kim et al., 2012). These effects may be partly caused by an increased fat cell size and were not observed by PCB153 exposure which has a different structure than PCB126 or PCB77 (PCB126 and PCB77 are coplanar PCBs whereas PCB153 is non-coplanar) (Arsenescu et al., 2008). Given the role of TNF-α, MCP-1, and adiponectin in the development of IR as previously discussed, the effects of these pollutants on the secretion of these adipokines/cytokines by adipose tissue could thus result in the perturbation of glucose homeostasis at the whole-body level.

5.4 Effects of POPs on Mitochondrial Function

POPs have shown to have a negative effect on energy metabolism and the expression and/or functioning of mitochondrial proteins. For example, in isolated rat-liver mitochondria, DDT acts as a weak ATP-ase inhibitor, which can be explained by the disruption of the integrity of the inner membrane by DDT molecules (Nishihara & Utsumi, 1985). On the other hand, Byczkowski (1976) demonstrated that DDT and its metabolites are inhibitors of the electron transfer chain in mitochondria at the site between NADH (Nicotinamide adenine dinucleotide coenzyme) and Coenzyme Q/ubiquinone pointing to mitochondrial function disturbance following exposure. Furthermore, evidence showed that a decrease in oxidative phosphorylation in brain and liver mitochondria occurred in rats treated with 600 mg of DDT for 24h due to the uncoupling of oxidative phosphorylation (Byczkowski, 1976). Rat liver mitochondria were shown to be more sensitive to the uncoupling action of DDT than rat brain mitochondria, suggesting an alternate mode of action in the brain (Byczkowski & Tłuczkiewicz, 1978). A significant decrease in repolarization potential (after a phosphorylation cycle), state 3 respiration and an increase in uncoupled respiration were also shown in testicular mitochondria after exposure to DDE (Mota et al.,

2011). Further, death of Sertoli cells (cells of the testicles part of the seminiferous tubules) by apoptosis through the mitochondrial pathway, and an increase in ROS production dependent on DDE concentration were also observed (Xiong et al., 2006). Moreover, a decrease in SOD was present confirming that oxidative stress may be an important component in DDE-induced toxicity (Shi et al., 2010). In liver and testis mitochondria, DDE also decreased oxygen consumption stimulated by ADP (State 3) as well as the activity of succinate-cytochrome c reductase (Mota et al., 2011).

Similarly, PCB exposure also causes mitochondrial dysfunction. Indeed, a variety of PCBs (tetra-chlorinated biphenyls), hydroxylated PCBs or mix of PCBs (Aroclor 1254) were shown to act as inhibitors of the electron transport chain as well as uncoupling agents (Ebner & Braselton 1987; Hasegawa, Ogata, & Tomokuni, 1982; Narasimhan, Kim, & Safe, 1991; Nishihara & Utsumi, 1987). PCB exposure likely causes mitochondrial dysfunction through AhR-mediated inflammation. Indeed, 2,3,7,8-tetrachlorodibenzodioxin (TCDD), which has a similar action to co-planoar PCBs, inhibits the interaction between mitochondrial AhR and an ATP synthase domain, which reduces the effectiveness of mitochondrial ATP synthesis in endothelial cells (Park et al., 2013). Moreover, mitochondrial function is reduced when there is an activation of cytochrome P450 1A1, which is caused by the interaction of PCB126 with AhR in adipose tissue and liver (Gadupudi et al., 2015; Song & Freedman, 2005).

BPA also affects mitochondrial function. It was shown that BPA disturbs mitochondrial function by inducing an overload of Ca^{2+} (Hughes et al., 2000; Orrenius et al., 2003). It was then confirmed that these observed effects are the result of ROS production and oxidative stress induced by BPA exposure (Wang et al., 2016).

It has also been proposed that the effect of POPs (TCCD, PCB and DDT) could be caused by a direct accumulation of pollutants in the mitochondria of endothelial cells (Park et al., 2013). Moreover, it has been observed that exposure to those POPs reduced the expression of PGC1-α, which decreased mitochondrial biogenesis and function and fatty acid oxidation at the whole-body level (Ruzzin et al., 2010).

6. What is Known About the Effects of POPs on Skeletal Muscle Energy Metabolism?

The direct effects of POPs on muscle energy metabolism have not been adequately studied. Most studies, as discussed above, have focused on the effects of these products on the liver, endothelial cells and adipose tissue. Muscle metabolism may differ from other tissues in that oxidative phosphorylation enzymes are differently expressed, making the muscle more sensitive to respiratory-chain deficiency (Rossignol, Letellier, Malgat, Rocher, & Mazat, 2000). While GLUT4 insulin-dependent transporters influence glucose uptake within the muscle and adipose tissue, the other organs such as the liver, kidneys, and brain take up glucose using non-insulin glucose transporter isotypes GLUT2 and GLUT3, respectively (Scheepers, Joost, & Schurmann, 2004). GLUT4 transporter in skeletal muscle plays a central role in whole-body metabolism and its function is a key determinant of glucose homeostasis (Huang & Czech, 2007), as previously discussed in the present review.

However, knowing that skeletal muscle is the largest glucose consumer in the body and participates to glucose homeostasis, the effects of POPs such as DDT, PCBs and BPA on muscle glucose metabolism and oxidative stress must be understood in order to evaluate their implication in the development of IR and T2D. An interesting study has shown that POP exposure in mice through the consumption of farmed salmon decreased muscle insulin-stimulated glucose uptake in association with a decrease in insulin-stimulated Akt phosphorylation and an increase in intramyocellular lipid accumulation (Ibrahim et al., 2011). Furthermore, rats exposed to a mixture of PCBs showed a decrease in muscle glucose uptake in association with a decrease in GLUT4 translocation (Williams et al., 2013). These results are in accordance with recent data from our laboratory showing that exposure to PCB126 induced a 20% decrease in glucose uptake and glycolysis in L6 muscle cells (Mauger, Nadeau, Caron, Chapados, & Aguer, 2015). Similarly, an exposure to BPA for 8 days at 100 μg/kg in mice caused skeletal muscle insulin-stimulated tyrosine phosphorylation of the β subunit on the insulin receptor to be impaired associated with reduced Akt phosphorylation. Muscle and liver also showed upregulation of IRS-1 expression, while MAPK pathway was impaired in muscle (Batista et al., 2012).

Some data also suggest that exposure to some POPs results in muscle mitochondrial dysfunction and disrupts skeletal muscle energy metabolism. For example, it was demonstrated that an increase in circulating levels of pollutants following weight loss in humans was linked to a decrease in mitochondrial enzymatic activity in skeletal muscle (Imbeault, Tremblay, Simoneau, & Joanisse, 2002). Furthermore, preliminary results from our laboratory demonstrated that exposure to DDT in rats significantly decreased muscle mitochondrial function (Chapados, Tremblay, Haddad, & Aguer, 2014). Rats exposed to 40 mg/kg of DDT by a single intraperitoneal injection for a week have shown a significant reduction in oxygen consumption rate measured with complex I substrates (glutamate, malate and pyruvate) in permeabilized muscle fibers (Chapados, Tremblay, Haddad, & Aguer, 2014). Interestingly so, exposure of rat L6 muscle cells *in vitro* to DDE at 1 μM and 10 μM showed a shift from oxidative metabolism to glycolytic metabolism and both DDT and DDE have shown a general tendency to increase glycolysis and glucose uptake (unpublished results). Preliminary studies from our laboratory have also shown that rats exposed to 1.05 µg/kg PCB126 have decreased mitochondrial respiration as measured in permeabilized muscle fibers (Chapados, Tremblay, Haddad, & Aguer, 2014). However, exposition of L6 muscle cells to PCB126 *in vitro* did not result in major mitochondrial respiration defects (Mauger, Nadeau, Caron, Chapados, & Aguer, 2016), suggesting that the effect of PCB126 *in vivo* on muscle mitochondrial function does not involve a direct mechanism since it was not reproduced *in vitro*. This could be explained by extrinsic mechanisms, which are not observed in the absence of other tissues, such as adipose tissue. Since PCBs are stored in adipose tissue and PCB exposure alters adipokine secretion by adipose tissue (Arsenescu et al., 2008; Hennig et al., 2002; Kim et al., 2012), we propose that PCB126 exposure affects not only skeletal muscle mitochondrial function, but also muscle insulin sensitivity through increased inflammatory cytokines (such as TNFα) or altered adipokine secretion from adipose tissue (Figure 2). Further studies are thus needed to explore this hypothesis.

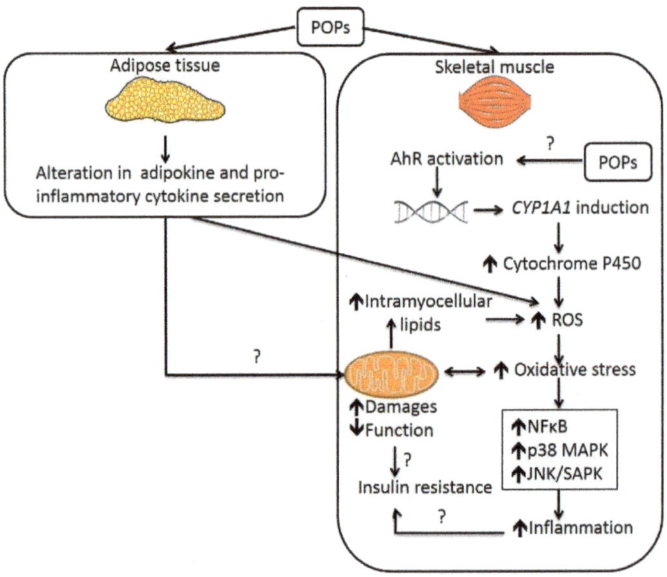

Figure 2. Schematic representation summarizing the direct effects of POPs on adipose tissue and skeletal muscle, with subsequent communication between adipose tissue and muscle in the development of insulin resistance. Question marks represent areas that are still under investigation. The hypothesis that exposure to POP alters adipose tissue cytokine/adipokine release leading to muscle mitochondrial dysfunction and IR is still under investigation

7. Conclusion

Due to its mass, skeletal muscle is largely implicated in the maintenance of glucose homeostasis and is thus a major player in the development of IR and T2D. Exposure to POPs is known to alter whole-body insulin sensitivity in rodents and to be associated with T2D in humans. This effect of POPs is probably through an increased oxidative stress and inflammation, which also alters mitochondrial function in different tissues and cells. Despite the important role of skeletal muscle in the pathogenesis of IR and T2D, the effect of POPs on muscle insulin sensitivity and mitochondrial function has been underappreciated. Mechanisms responsible for muscle mitochondrial dysfunction and glucose uptake disarrangement following exposure to DDT, DDE, PCB126, or BPA (and potentially other POPs) need to be studied in order to develop strategies (e.g. anti-oxidant or anti-inflammatory treatments) to improve muscle protection against POP aggressions and ultimately prevent and treat the development of IR and T2D due to POP exposure.

Acknowledgments

This project was supported by NSERC (Natural Sciences and Engineering Research Council of Canada) Discovery grants (grant numbers 2015–06263 to CA) and operating funding from Institut de recherche de l'Hôpital Montfort to CA. Master scholarships to LC were provided by the Institut de recherche de l'Hôpital Montfort and Canadian Institutes of Health Research (CIHR) and to AC by NSERC. We thank David Patten for critically reading the manuscript.

References

Aguer, C., & Harper, M.-E. (2012). Skeletal muscle mitochondrial energetics in obesity and type 2 diabetes mellitus: endocrine aspects. *Best Practice & Research. Clinical Endocrinology & Metabolism*, *26*(6), 805–819. http://doi.org/10.1016/j.beem.2012.06.001

Aguer, C., McCoin, C. S., Knotts, T. A., Thrush, A. B., Ono-Moore, K., McPherson, R., … Harper, M.-E. (2015). Acylcarnitines: potential implications for skeletal muscle insulin resistance. *The FASEB Journal*, *29*(1), 336–345. http://doi.org/10.1096/fj.14-255901

Aminov, Z., Haase, R., Rej, R., Schymura, M. J., Santiago-Rivera, A., Morse, G., … Carpenter, D. O. (2016). Diabetes Prevalence in Relation to Serum Concentrations of Polychlorinated Biphenyl (PCB) Congener Groups and Three Chlorinated Pesticides in a Native American Population. *Environmental Health Perspectives*, *124*(9), 1376–1383. http://doi.org/10.1289/ehp.1509902

An, J., Muoio, D. M., Shiota, M., Fujimoto, Y., Cline, G. W., Shulman, G. I., … Newgard, C. B. (2004). Hepatic expression of malonyl-CoA decarboxylase reverses muscle, liver and whole-animal insulin resistance. *Nature Medicine*, *10*(3), 268–274. http://doi.org/10.1038/nm995

Arsenescu, V., Arsenescu, R. I., King, V., Swanson, H., & Cassis, L. A. (2008). Polychlorinated biphenyl-77 induces adipocyte differentiation and proinflammatory adipokines and promotes obesity and atherosclerosis. *Environmental Health Perspectives*, *116*(6), 761–768. http://doi.org/10.1289/ehp.10554

Baker, N. A., Karounos, M., English, V., Fang, J., Wei, Y., Stromberg, A., … Cassis, L. A. (2013). Coplanar polychlorinated biphenyls impair glucose homeostasis in lean C57BL/6 mice and mitigate beneficial effects of weight loss on glucose homeostasis in obese mice. *Environmental Health Perspectives*, *121*(1), 105–110. http://doi.org/10.1289/ehp.1205421

Bastie, C. C., Hajri, T., Drover, V. A., Grimaldi, P. A., & Abumrad, N. A. (2004). CD36 in myocytes channels fatty acids to a lipase-accessible triglyceride pool that is related to cell lipid and insulin responsiveness. *Diabetes*, *53*(9), 2209–2216. http://dx.doi.org/10.2337/diabetes.53.9.2209

Batista, T. M., Alonso-Magdalena, P., Vieira, E., Amaral, M. E. C., Cederroth, C. R., Nef, S., … Nadal, A. (2012). Short-term treatment with bisphenol-A leads to metabolic abnormalities in adult male mice. *PloS One*, *7*(3), e33814. http://doi.org/10.1371/journal.pone.0033814

Bedard, S., Marcotte, B., & Marette, A. (1997). Cytokines modulate glucose transport in skeletal muscle by inducing the expression of inducible nitric oxide synthase. *The Biochemical Journal*, *325 (Pt 2)*, 487–493.

Befroy, D. E., Petersen, K. F., Dufour, S., Mason, G. F., de Graaf, R. A., Rothman, D. L., & Shulman, G. I. (2007). Impaired mitochondrial substrate oxidation in muscle of insulin-resistant offspring of type 2 diabetic patients. *Diabetes*, *56*(5), 1376–1381. http://doi.org/10.2337/db06-0783

Benomar, Y., Gertler, A., De Lacy, P., Crépin, D., Ould Hamouda, H., Riffault, L., & Taouis, M. (2013). Central Resistin Overexposure Induces Insulin Resistance Through Toll-Like Receptor 4. *Diabetes*, *62*(1), 102–114. http://doi.org/10.2337/db12-0237

Bhatnagar, S., Panguluri, S. K., Gupta, S. K., Dahiya, S., Lundy, R. F., & Kumar, A. (2010). Tumor necrosis factor-alpha regulates distinct molecular pathways and gene networks in cultured skeletal muscle cells. *PloS One*, *5*(10), e13262. http://doi.org/10.1371/journal.pone.0013262

Bloch-Damti, A., & Bashan, N. (2005). Proposed mechanisms for the induction of insulin resistance by oxidative stress. *Antioxidants & Redox Signaling*, *7*(11-12), 1553–1567. http://doi.org/10.1089/ars.2005.7.1553

Bonnard, C., Durand, A., Peyrol, S., Chanseaume, E., Chauvin, M.-A., Morio, B., … Rieusset, J. (2008). Mitochondrial dysfunction results from oxidative stress in the skeletal muscle of diet-induced insulin-resistant mice. *The Journal of Clinical Investigation*, *118*(2), 789–800. http://doi.org/10.1172/JCI32601

Borst, S. E., Conover, C. F., & Bagby, G. J. (2005). Association of resistin with visceral fat and muscle insulin resistance. *Cytokine*, *32*(1), 39–44. http://doi.org/10.1016/j.cyto.2005.07.008

Boucher, M.-P., Lefebvre, C., & Chapados, N. A. (2015). The effects of PCB126 on intra-hepatic mechanisms associated with non alcoholic fatty liver disease. *Journal of Diabetes and Metabolic Disorders*, *14*, 88. http://doi.org/10.1186/s40200-015-0218-2

Brehm A, Krssak M, Schmid AI, Nowotny P, Waldhausl W, et al. (2006) Increased lipid availability impairs insulin-stimulated ATP synthesis in human skeletal muscle. Diabetes 55: 136–140. http://dx.doi.org/10.2337/diabetes.55.01.06.db05-1286

Burri, L., Thoresen, G., & Berge, R. (2010). The Role of PPARαActivation in Liver and Muscle. *PPAR Research*, *2010*, 1-11. http://dx.doi.org/10.1155/2010/542359

Bussolaro, D., Filipak Neto, F., Glinski, A., Roche, H., Guiloski, I. C., Mela, M., ... Oliveira Ribeiro, C. A. (2012). Bioaccumulation and related effects of PCBs and organochlorinated pesticides in freshwater fish Hypostomus commersoni. *Journal of Environmental Monitoring: JEM*, *14*(8), 2154–2163. http://doi.org/10.1039/c2em10863a

Byczkowski, J. Z. (1976). The mode of action of p,p'-DDT on mammalian mitochondria. *Toxicology*, *6*(3), 309–314. http://dx.doi.org/10.1016/0300-483X(76)90034-2

Byczkowski, J. Z., & Tluczkiewicz, J. (1978). Comparative study of respiratory chain inhibition by DDT and DDE in mammalian and plant mitochondria. *Bulletin of Environmental Contamination and Toxicology*, *20*(4), 505–512. http://dx.doi.org/10.1007/BF01683556

Cameron, A. J., Sicree, R. A., Zimmet, P. Z., Alberti, K. G. M. M., Tonkin, A. M., Balkau, B., ... Shaw, J. E. (2010). Cut-points for waist circumference in Europids and South Asians. *Obesity (Silver Spring, Md.)*, *18*(10), 2039–2046. http://doi.org/10.1038/oby.2009.455

Carey, A. L., Steinberg, G. R., Macaulay, S. L., Thomas, W. G., Holmes, A. G., Ramm, G., ... Febbraio, M. A. (2006). Interleukin-6 increases insulin-stimulated glucose disposal in humans and glucose uptake and fatty acid oxidation in vitro via AMP-activated protein kinase. *Diabetes*, *55*(10), 2688–2697. http://doi.org/10.2337/db05-1404

Carvalho-Santos, A., Ribeiro-Alves, M., Cardoso-Weide, L. C., Nunes, J., Kuhnert, L. R. B., Xavier, A. R., ... Carvalho-Pinto, C. E. (2015). Decreased Circulating Levels of APRIL: Questioning Its Role in Diabetes. *PloS One*, *10*(10), e0140150. http://doi.org/10.1371/journal.pone.0140150

Cawthorn, W. P., & Sethi, J. K. (2008). TNF-α and adipocyte biology. *FEBS Letters*, *582*(1), 117–131. http://doi.org/10.1016/j.febslet.2007.11.051

CDC (Centers for Disease Control and Prevention) (2011). National Diabetes Fact Sheet: National Estimates on Diabetes. Available: http://www.cdc.gov/diabetes/pubs/estimat es11.htm [accessed 20 april 2016].

Chang-Chen, K. J., Mullur, R., & Bernal-Mizrachi, E. (2008). Beta-cell failure as a complication of diabetes. *Reviews in Endocrine & Metabolic Disorders*, *9*(4), 329–343. http://doi.org/10.1007/s11154-008-9101-5

Chang, J.-W., Chen, H.-L., Su, H.-J., & Lee, C.-C. (2016). Abdominal Obesity and Insulin Resistance in People Exposed to Moderate-to-High Levels of Dioxin. *PloS One*, *11*(1), e0145818. http://doi.org/10.1371/journal.pone.0145818

Chang, J., Chen, H., Su, H., Liao, P., Guo, H., & Lee, C. (2009). Association Between Insulin Resistance and Co-Exposure to Dioxins and Mercury in Taiwanese Living Near a Deserted Pentachlorophenol and Chloralkali Factory. *Epidemiology*, *20*, S15. http://dx.doi.org/10.1097/01.ede.0000362221.66727.79

Chapados, C. Tremblay Laganière, J. Haddad, C. Aguer. (2014). Une exposition à des polluants persistants organiques affecte le métabolisme énergétique du muscle squelettique in vivo et in vitro. 15ème réunion annuelle de la Société Québécoise de Lipodologie, de Nutrition et de Métabolisme, February 19-21, 2014, Québec, Canada. (poster, national conference, Seahorse Travel Award).DOI: N/A

Chapados, N. A., Casimiro, C., Robidoux, M. A., Haman, F., Batal, M., Blais, J. M., & Imbeault, P. (2012). Increased proliferative effect of organochlorine compounds on human preadipocytes. *Molecular and Cellular Biochemistry*, *365*(1-2), 275–278. http://doi.org/10.1007/s11010-012-1268-0

Chavez, J. A., & Summers, S. A. (2012). A ceramide-centric view of insulin resistance. *Cell Metabolism*, *15*(5), 585–594. http://doi.org/10.1016/j.cmet.2012.04.002

Civitarese, A., Ukropcova, B., Carling, S., Hulver, M., DeFronzo, R., & Mandarino, L. et al. (2006). Role of adiponectin in human skeletal muscle bioenergetics. *Cell Metabolism*, *4*(1), 75-87. http://dx.doi.org/10.1016/j.cmet.2006.05.002

Coll, T., Eyre, E., Rodríguez-Calvo, R., Palomer, X., Sánchez, R., & Merlos, M. et al. (2008). Oleate Reverses Palmitate-induced Insulin Resistance and Inflammation in Skeletal Muscle Cells. *Journal Of Biological Chemistry*, *283*(17), 11107-11116. http://dx.doi.org/10.1074/jbc.m708700200

Czajka, A., Ajaz, S., Gnudi, L., Parsade, C. K., Jones, P., Reid, F., & Malik, A. N. (2015). Altered Mitochondrial Function, Mitochondrial DNA and Reduced Metabolic Flexibility in Patients with Diabetic Nephropathy. *EBioMedicine*, *2*(6), 499–512. http://doi.org/10.1016/j.ebiom.2015.04.002

D'Amour, M., Lye, E., & Murray, J. (2011). Human biomonitoring of environmental chemicals in Canada—Overview of recent activities. *Toxicology Letters*, *205*, S74.

DeFronzo, R. A., & Tripathy, D. (2009). Skeletal muscle insulin resistance is the primary defect in type 2 diabetes. *Diabetes Care, 32 Suppl 2*, S157-163. http://doi.org/10.2337/dc09-S302

Dikalov, S. (2011). Cross talk between mitochondria and NADPH oxidases. *Free Radical Biology & Medicine*, *51*(7), 1289–1301. http://doi.org/10.1016/j.freeradbiomed.2011.06.033

Ebner, K.V, & Braselton, W.E., Jr. (1987). Structural and chemical requirements for hydroxychlorobiphenyls to uncouple rat liver mitochondria and potentiation of uncoupling with aroclor 1254. *Chem Biol Interact 63*, 139-155. http://doi.org/10.1016/0009-2797(87)90094-9

Evans, J. L., Goldfine, I. D., Maddux, B. A., & Grodsky, G. M. (2003). Are oxidative stress-activated signaling pathways mediators of insulin resistance and beta-cell dysfunction? *Diabetes*, *52*(1), 1–8. http://dx.doi.org/10.2337/diabetes.52.1.1

Everett, CJ. et al. (2007). Association of a polychlorinated dibenzo-*p*-dioxin, a polychlorinated biphenyl, and DDT with diabetes in the 1999-2002 National Health and Nutrition Examination Survey. *Environmental Research*, vol. 103, p.413-418. http://dx.doi.org/10.1016/j.envres.2006.11.002

Fasshauer et al. (2003). Adiponectin gene expression and secretion is inhibited by interleukin-6 in 3T3-L1 adipocytes. *Biochemical and Biophysical Research Communications*, 301(4), p.1045-1050. http://dx.doi.org/10.1016/S0006-291X(03)00090-1

Fierens, S., Mairesse, H., Heilier, J. F., De Burbure, C., Focant, J. F., Eppe, G., et al. (2003). Dioxin/polychlorinated biphenyl body burden, diabetes and endometriosis: Findings in a population-based study in belgium. *Biomarkers: Biochemical Indicators of Exposure, Response, and Susceptibility to Chemicals, 8*(6), 529-534. http://dx.doi.org/10.1016/S0006-291X(03)00090-1

Gadupudi, G. et al. (2015). PCB126 inhibits adipogenesis of human preadipocytes. *Toxicology in Vitro*, vol.29, p.132-141. http://doi.org/10.1016/j.tiv.2014.09.015

García-Arevalo, M., Alonso-Magdalena, P., Rebelo Dos Santos, J., Quesada, I., Carneiro, E., & Nadal, A. (2014). Exposure to Bisphenol-A during Pregnancy Partially Mimics the Effects of a High-Fat Diet Altering Glucose Homeostasis and Gene Expression in Adult Male Mice. *Plos ONE*, *9*(6), e100214. http://doi.org/10.1371/journal.pone.0100214

Gaster, M., Petersen, I., Hojlund, K., Poulsen, P., & Beck-Nielsen, H. (2002). The Diabetic Phenotype Is Conserved in Myotubes Established from Diabetic Subjects: Evidence for Primary Defects in Glucose Transport and Glycogen Synthase Activity. *Diabetes*, *51*(4), 921-927. http://dx.doi.org/10.2337/diabetes.51.4.921

Gauthier, M., Rabasa-Lhoret, R., Prud'homme, D., Karelis, A., Geng, D., van Bavel, B., & Ruzzin, J. (2014). The Metabolically Healthy But Obese Phenotype Is Associated With Lower Plasma Levels of Persistent Organic Pollutants as Compared to the Metabolically Abnormal Obese Phenotype. *The Journal Of Clinical Endocrinology & Metabolism*, *99*(6), E1061-E1066. http://doi.org.10.1210/jc.2013-3935

Grandjean, P., Henriksen, J., Choi, A., Petersen, M., Dalgård, C., Nielsen, F., & Weihe, P. (2011). Marine Food Pollutants as a Risk Factor for Hypoinsulinemia and Type 2 Diabetes. *Epidemiology*, *22*(3), 410-417. http://doi.org/10.1097/EDE.0b013e318212fab9

Gray, S. L., Shaw, A. C., Gagne, A. X., & Chan, H. M. (2013). Chronic exposure to PCBs (Aroclor 1254) exacerbates obesity-induced insulin resistance and hyperinsulinemia in mice. *Journal of Toxicology and Environmental Health. Part A*, *76*(12), 701–715. http://doi.org/10.1080/15287394.2013.796503

Guariguata, L., Whiting, D. R., Hambleton, I., Beagley, J., Linnenkamp, U., & Shaw, J. E. (2014). Global estimates of diabetes prevalence for 2013 and projections for 2035. *Diabetes Research and Clinical Practice, 103*(2), 137-149. http://doi.org/10.1016/j.diabres.2013.11.002

Gulli, G., Ferrannini, E., Stern, M., Haffner, S., & DeFronzo, R. A. (1992). The metabolic profile of NIDDM is fully established in glucose-tolerant offspring of two mexican-american NIDDM parents. *Diabetes, 41*(12), 1575-1586. http://dx.doi.org/10.2337/diab.41.12.1575

Hasegawa, T., Ogata, M., & Tomokuni, K. (1982). Inhibition of oxidative phosphorylation of rat liver mitochondria by PCB and ferric chloride: a preliminary report. *Ind Health, 20*, 273-275. http://doi.org/10.2486/indhealth.20.273

Health Canada. (2003). ARCHIVED - Concentrations (pg/g wet wt.) of total PCBs in fatty foods from Total Diet Study in Ottawa, 2000 Food code Description Concentration (ppt). *Food and Nutrition*. Retrieved March 15 2016, from http://www.hc-sc.gc.ca/fn-an/surveill/total-diet/concentration/pcb_conc_dpc_ottawa2000-fra.php

Hennig, B., Hammock, B., Slim, R., Toborek, M., Saraswathi, V., & Robertson, L. (2002). PCB-induced oxidative stress in endothelial cells: modulation by nutrients. *International Journal Of Hygiene And Environmental Health, 205*(1-2), 95-102. http://dx.doi.org/10.1078/1438-4639-00134

Hennig, B., Reiterer, G., Majkova, Z., Oesterling, E., Meerarani, P., & Toborek, M. (2005). Modification of Environmental Toxicity by Nutrients: Implications in Atherosclerosis. *Cardiovascular Toxicology*, *5*(2), 153-160. http://dx.doi.org/10.1385/ct:5:2:153

Henriksen, E., Diamond-Stanic, M., & Marchionne, E. (2011). Oxidative stress and the etiology of insulin resistance and type 2 diabetes. *Free Radical Biology and Medicine*, *51*(5), 993-999. http://doi.org/10.1016/j.freeradbiomed.2010.12.005

Henry, BA., Andrews, ZB., Rao, A. & Clarke, IJ. (July 2011). Central Leptin Activates Mitochondrial Function and Increases Heat Production in Skeletal Muscle. *Endocrinology*, 152(7), p.2609-2618. http://doi.org/10.1210/en.2011-0143

Huang, S. & Czech, M. (2007). The GLUT4 Glucose Transporter. *Cell Metabolism*, *5*(4), 237-252. http://dx.doi.org/10.1016/j.cmet.2007.03.006

Ibrahim, M., Fjære, E., Lock, E., Naville, D., Amlund, H., & Meugnier, E. et al. (2011). Chronic Consumption of Farmed Salmon Containing Persistent Organic Pollutants Causes Insulin Resistance and Obesity in Mice. *Plos ONE*, *6*(9), e25170. http://doi.org/10.1371/journal.pone.0025170

Imbeault, P. (2007). Environmental influences on adiponectin levels in humans. *Applied Physiology, Nutrition, and Metabolism = Physiologie Appliquee, Nutrition et Metabolisme*, *32*(3), 505–511. http://doi.org/10.1139/H07-017

Imbeault, P., Findlay, C., Robidoux, M., Haman, F., Blais, J., & Tremblay, A. et al. (2012). Dysregulation of Cytokine Response in Canadian First Nations Communities: Is There an Association with Persistent Organic Pollutant Levels?. *Plos ONE*, *7*(7), e39931. http://dx.doi.org/10.1371/journal.pone.0039931

Iwabu, M., Yamauchi, T., Okada-Iwabu, M., Sato, K., Nakagawa, T., Funata, M., … Kadowaki, T. (2010). Adiponectin and AdipoR1 regulate PGC-1alpha and mitochondria by Ca(2+) and AMPK/SIRT1. *Nature*, *464*(7293), 1313–1319. http://doi.org/10.1038/nature08991

Jaacks, L. M., & Staimez, L. R. (2015). Association of persistent organic pollutants and non-persistent pesticides with diabetes and diabetes-related health outcomes in Asia: A systematic review. *Environment International*, *76*, 57–70. http://doi.org/10.1016/j.envint.2014.12.001

Jin, X., Song, L., Liu, X., Chen, M., Li, Z., Cheng, L., & Ren, H. (2014). Protective Efficacy of Vitamins C and E on p,p′-DDT-Induced Cytotoxicity via the ROS-Mediated Mitochondrial Pathway and NF-κB/FasL Pathway. *PLoS ONE*, *9*(12), e113257. http://doi.org/10.1371/journal.pone.0113257

Jortay, J., Senou, M., Abou-Samra, M., Noel, L., Robert, A., Many, M.-C., & Brichard, S. M. (2012). Adiponectin and skeletal muscle: pathophysiological implications in metabolic stress. *The American Journal of Pathology*, *181*(1), 245–256. http://doi.org/10.1016/j.ajpath.2012.03.035

Kadowaki, T. (2006). Adiponectin and adiponectin receptors in insulin resistance, diabetes, and the metabolic syndrome. *Journal Of Clinical Investigation*, *116*(7), 1784-1792. http://dx.doi.org/10.1172/jci29126

Kelley, D. E., He, J., Menshikova, E. V., & Ritov, V. B. (2002). Dysfunction of mitochondria in human skeletal muscle in type 2 diabetes. *Diabetes*, *51*(10), 2944–2950. http://dx.doi.org/10.2337/diabetes.51.10.2944

Kersten, S. (2014). Integrated physiology and systems biology of PPARalpha. *Molecular Metabolism*, *3*(4), 354–371. http://doi.org/10.1016/j.molmet.2014.02.002

Kim, K.-S., Lee, Y.-M., Kim, S. G., Lee, I.-K., Lee, H.-J., Kim, J.-H., … Lee, D.-H. (2014). Associations of organochlorine pesticides and polychlorinated biphenyls in visceral vs. subcutaneous adipose tissue with type 2 diabetes and insulin resistance. *Chemosphere*, *94*, 151–157. http://doi.org/10.1016/j.chemosphere.2013. 09.066

Kim, M. J., Pelloux, V., Guyot, E., Tordjman, J., Bui, L.-C., Chevallier, A., … Barouki, R. (2012). Inflammatory pathway genes belong to major targets of persistent organic pollutants in adipose cells. *Environmental Health Perspectives*, *120*(4), 508–514. http://doi.org/10.1289/ehp.1104282

Kiviranta, H., Tuomisto, J. T., Tuomisto, J., Tukiainen, E., & Vartiainen, T. (2005). Polychlorinated dibenzo-p-dioxins, dibenzofurans, and biphenyls in the general population in Finland. *Chemosphere*, *60*(7), 854–869. http://doi.org/10.1016/j.chemosphere.2005.01.064

Krssak, M., Falk Petersen, K., Dresner, A., DiPietro, L., Vogel, S. M., Rothman, D. L., … Shulman, G. I. (1999). Intramyocellular lipid concentrations are correlated with insulin sensitivity in humans: a 1H NMR spectroscopy study. *Diabetologia*, *42*(1), 113–116. http://doi.org/10.1007/s001250051123

Kuo, C.-C., Moon, K., Thayer, K. A., & Navas-Acien, A. (2013). Environmental chemicals and type 2 diabetes: an updated systematic review of the epidemiologic evidence. *Current Diabetes Reports*, *13*(6), 831–849. http://doi.org/10.1007/s11892-013-0432-6

Kusminski, C. M., McTernan, P. G., & Kumar, S. (2005). Role of resistin in obesity, insulin resistance and Type II diabetes. *Clinical Science (London, England : 1979)*, *109*(3), 243–256. http://doi.org/10.1042/CS20050078

La Merrill, M., Karey, E., Moshier, E., Lindtner, C., La Frano, M. R., Newman, J. W., & Buettner, C. (2014). Perinatal exposure of mice to the pesticide DDT impairs energy expenditure and metabolism in adult female offspring. *PloS One*, *9*(7), e103337. http://doi.org/10.1371/journal.pone.0103337

Lahouel, A., Kebieche, M., Lakroun, Z., Rouabhi, R., Fetoui, H., & Chtourou, Y. et al. (2016). Neurobehavioral deficits and brain oxidative stress induced by chronic low dose exposure of persistent organic pollutants mixture in adult female rat. *Environmental Science And Pollution Research*, *23*(19), 19030-19040. http://dx.doi.org/10.1007/s11356-016-6913-9

Lang, I. A., Galloway, T. S., Scarlett, A., Henley, W. E., Depledge, M., Wallace, R. B., & Melzer, D. (2008). Association of urinary bisphenol A concentration with medical disorders and laboratory abnormalities in adults. *JAMA*, *300*(11), 1303–1310. http://doi.org/10.1001/jama.300.11.1303

Lee, D.-H., Steffes, M. W., Sjodin, A., Jones, R. S., Needham, L. L., & Jacobs, D. R. J. (2011). Low dose organochlorine pesticides and polychlorinated biphenyls predict obesity, dyslipidemia, and insulin resistance among people free of diabetes. *PloS One*, *6*(1), e15977. http://doi.org/10.1371/journal.pone.0015977

Lee, H. & Yang, J. (2012). PCB126 induces apoptosis of chondrocytes via ROS-dependent pathways. *Osteoarthritis And Cartilage*, *20*(10), 1179-1185. http://dx.doi.org/10.1016/j.joca.2012.06.004

Liu, Y., Turdi, S., Park, T., Morris, N., Deshaies, Y., Xu, A., & Sweeney, G. (2012). Adiponectin Corrects High-Fat Diet-Induced Disturbances in Muscle Metabolomic Profile and Whole-Body Glucose Homeostasis. *Diabetes*, *62*(3), 743-752. http://dx.doi.org/10.2337/db12-0687

Matsuda, M., & Shimomura, I. (n.d.-a). Increased oxidative stress in obesity: Implications for metabolic syndrome, diabetes, hypertension, dyslipidemia, atherosclerosis, and cancer. *Obesity Research & Clinical Practice*, *7*(5), e330–e341. http://doi.org/10.1016/j.orcp.2013.05.004

Mauger, J.-F., Nadeau, L., Caron, A., Chapados, N. A., & Aguer, C. (2016). Polychlorinated biphenyl 126 exposure in L6 myotubes alters glucose metabolism: a pilot study. *Environmental Science and Pollution Research International*, *23*(8), 8133–8140. http://doi.org/10.1007/s11356-016-6348-3

Mazulis, F., Weilg, C., Alva-Urcia, C., Pons, M. J., & Del Valle Mendoza, J. (2015). Is glucose-6-phosphate dehydrogenase deficiency more prevalent in Carrion's disease endemic areas in Latin America? *Asian Pacific Journal of Tropical Medicine*, *8*(12), 1079–1080. http://doi.org/10.1016/j.apjtm.2015.11.014

McClelland, G. B., Kraft, C. S., Michaud, D., Russell, J. C., Mueller, C. R., & Moyes, C. D. (2004). Leptin and the control of respiratory gene expression in muscle. *Biochimica et Biophysica Acta*, *1688*(1), 86–93.

McTernan, P. G., Kusminski, C. M., & Kumar, S. (2006). Resistin. *Current Opinion in Lipidology*, *17*(2). Retrieved from http://journals.lww.com/co-lipidology/Fulltext/2006/04000/Resistin.11.aspx

Midtbo, L. K., Ibrahim, M. M., Myrmel, L. S., Aune, U. L., Alvheim, A. R., Liland, N. S., … Madsen, L. (2013). Intake of farmed Atlantic salmon fed soybean oil increases insulin resistance and hepatic lipid accumulation in mice. *PloS One*, *8*(1), e53094. http://doi.org/10.1371/journal.pone.0053094

Minokoshi, Y., Kim, Y.-B., Peroni, O. D., Fryer, L. G. D., Muller, C., Carling, D., & Kahn, B. B. (2002). Leptin stimulates fatty-acid oxidation by activating AMP-activated protein kinase. *Nature*, *415*(6869), 339–343. http://doi.org/10.1038/415339a

Moon, M., Jeong, I., Jung Oh, T., Ahn, H., Kim, H., & Park, Y. et al. (2015). Long-term oral exposure to bisphenol A induces glucose intolerance and insulin resistance. *Journal Of Endocrinology*, *226*(1), 35-42. http://dx.doi.org/10.1530/joe-14-0714

Morgan, D. P., Lin, L. I., & Saikaly, H. H. (1980). Morbidity and mortality in workers occupationally exposed to pesticides. *Archives of Environmental Contamination and Toxicology*, *9*(3), 349–382. Http://doi.org/10.1007/BF01057414

Morino, K., Petersen, K. F., Dufour, S., Befroy, D., Frattini, J., Shatzkes, N., … Shulman, G. I. (2005). Reduced mitochondrial density and increased IRS-1 serine phosphorylation in muscle of insulin-resistant offspring of type 2 diabetic parents. *The Journal of Clinical Investigation*, *115*(12), 3587–3593. http://doi.org/10.1172/JCI25151

Mota, P. C., Cordeiro, M., Pereira, S. P., Oliveira, P. J., Moreno, A. J., & Ramalho-Santos, J. (2011). Differential effects of p,p'-DDE on testis and liver mitochondria: implications for reproductive toxicology. *Reproductive Toxicology (Elmsford, N.Y.)*, *31*(1), 80–85. http://doi.org/10.1016/j.reprotox.2010.09.010

Musaro, A., Fulle, S., & Fano, G. (2010). Oxidative stress and muscle homeostasis. *Current Opinion in Clinical Nutrition and Metabolic Care*, *13*(3), 236–242. http://doi.org/10.1097/MCO.0b013e3283368188

Narasimhan, T.R., Kim, H.L., & Safe, S.H. (1991). Effects of hydroxylated polychlorinated biphenyls on mouse liver mitochondrial oxidative phosphorylation. *J Biochem Toxicol 6*, 229-236. http://doi.org/10.1002/jbt.2570060309

Nishihara, Y., & Utsumi, K. (1987). 4-Chloro-4'-biphenylol as an uncoupler and an inhibitor of mitochondrial oxidative phosphorylation. *Biochem Pharmacol 36*, 3453-3457. http://doi.org/10.1016/0006-2952(87)90325-X

Neel, B. A., & Sargis, R. M. (2011). The paradox of progress: environmental disruption of metabolism and the diabetes epidemic. *Diabetes*, *60*(7), 1838–1848. http://doi.org/10.2337/db11-0153

Ngwa, E. N., Kengne, A.-P., Tiedeu-Atogho, B., Mofo-Mato, E.-P., & Sobngwi, E. (2015). Persistent organic pollutants as risk factors for type 2 diabetes. *Diabetology & Metabolic Syndrome*, *7*, 41. http://doi.org/10.1186/s13098-015-0031-6

Nieto-Vazquez, I., Fernandez-Veledo, S., de Alvaro, C., & Lorenzo, M. (2008). Dual role of interleukin-6 in regulating insulin sensitivity in murine skeletal muscle. *Diabetes*, *57*(12), 3211–3221. http://doi.org/10.2337/db07-1062

Nishihara, Y., & Utsumi, K. (1985). Effects of 1,1,1-trichloro-2,2-bis-(p-chlorophenyl)ethane (DDT) on ATPase-linked functions in isolated rat-liver mitochondria. *Food and Chemical Toxicology : An International Journal Published for the British Industrial Biological Research Association*, *23*(6), 599–602. http://doi.org/10.1016/0278-6915(85)90185-1

Pan, D. A., Lillioja, S., Kriketos, A. D., Milner, M. R., Baur, L. A., Bogardus, C., … Storlien, L. H. (1997). Skeletal muscle triglyceride levels are inversely related to insulin action. *Diabetes*, *46*(6), 983–988. http://dx.doi.org/10.2337/diab.46.6.983

Park, W.-H., Kang, Y.-C., Piao, Y., Pak, D. H., & Pak, Y. K. (2013). Causal effects of synthetic chemicals on mitochondrial deficits and diabetes pandemic. *Archives of Pharmacal Research*, *36*(2), 178–188. http://doi.org/10.1007/s12272-013-0022-9

Patsouris, D., Cao, J.-J., Vial, G., Bravard, A., Lefai, E., Durand, A., … Rieusset, J. (2014). Insulin resistance is associated with MCP1-mediated macrophage accumulation in skeletal muscle in mice and humans. *PloS One*, *9*(10), e110653. http://doi.org/10.1371/journal.pone.0110653

Pavlikova, N., Smetana, P., Halada, P., & Kovar, J. (2015). Effect of prolonged exposure to sublethal concentrations of DDT and DDE on protein expression in human pancreatic beta cells. *Environmental Research, 142*, 257-263. http://dx.doi.org/10.1016/j.envres.2015.06.046

Pellegrino, M. A., Desaphy, J.-F., Brocca, L., Pierno, S., Camerino, D. C., & Bottinelli, R. (2011). Redox homeostasis, oxidative stress and disuse muscle atrophy. *The Journal of Physiology*, *589*(Pt 9), 2147–2160. http://doi.org/10.1113/jphysiol.2010.203232

Pellegrino, M. A., Desaphy, J.-F., Brocca, L., Pierno, S., Camerino, D. C., & Bottinelli, R. (2011). Redox homeostasis, oxidative stress and disuse muscle atrophy. *The Journal of Physiology*, *589*(Pt 9), 2147–2160. http://doi.org/10.1113/jphysiol.2010.203232

Peraldi, P., & Spiegelman, B. (1998). TNF-alpha and insulin resistance: summary and future prospects. *Molecular and Cellular Biochemistry*, *182*(1-2), 169–175. http://doi.org/10.1023/A:1006865715292

Perseghin, G., Scifo, P., De Cobelli, F., Pagliato, E., Battezzati, A., Arcelloni, C., Luzi, L. (1999). Intramyocellular triglyceride content is a determinant of in vivo insulin resistance in humans: a 1H-13C nuclear magnetic resonance spectroscopy assessment in offspring of type 2 diabetic parents. *Diabetes*, *48*(8), 1600–1606. http://dx.doi.org/10.2337/diabetes.48.8.1600

Petersen, K. F., Dufour, S., Befroy, D., Garcia, R., & Shulman, G. I. (2004). Impaired mitochondrial activity in the insulin-resistant offspring of patients with type 2 diabetes. *The New England Journal of Medicine*, *350*(7), 664–671. http://doi.org/10.1056/NEJMoa031314

Pickersgill, L., Litherland, G. J., Greenberg, A. S., Walker, M., & Yeaman, S. J. (2007). Key role for ceramides in mediating insulin resistance in human muscle cells. *The Journal of Biological Chemistry*, *282*(17), 12583–12589. http://doi.org/10.1074/jbc.M611157200

Plomgaard, P., Bouzakri, K., Krogh-Madsen, R., Mittendorfer, B., Zierath, J. R., & Pedersen, B. K. (2005). Tumor necrosis factor-alpha induces skeletal muscle insulin resistance in healthy human subjects via inhibition of Akt substrate 160 phosphorylation. *Diabetes*, *54*(10), 2939–2945. http://dx.doi.org/10.2337/diabetes.54.10.2939

Preedy, Victor R and Ross J Hunter. *Adipokines*. Enfield, N.H.: Science Publishers, 2011. Print. DOI: N/A

Quinn, L. S., Strait-Bodey, L., Anderson, B. G., Argiles, J. M., & Havel, P. J. (2005). Interleukin-15 stimulates adiponectin secretion by 3T3-L1 adipocytes: evidence for a skeletal muscle-to-fat signaling pathway. *Cell Biology International*, *29*(6), 449–457. http://doi.org/10.1016/j.cellbi.2005.02.005

Rains, J. L., & Jain, S. K. (2011). Oxidative stress, insulin signaling, and diabetes. *Free Radical Biology & Medicine*, *50*(5), 567–575. http://doi.org/10.1016/j.freeradbiomed.2010.12.006

Ramadass, P., Meerarani, P., Toborek, M., Robertson, L. W., & Hennig, B. (2003). Dietary flavonoids modulate PCB-induced oxidative stress, CYP1A1 induction, and AhR-DNA binding activity in vascular endothelial cells. *Toxicological Sciences : An Official Journal of the Society of Toxicology*, *76*(1), 212–219. http://doi.org/10.1093/toxsci/kfg227

Rignell-Hydbom, A., Rylander, L., & Hagmar, L. (2007). Exposure to persistent organochlorine pollutants and type 2 diabetes mellitus. *Human & Experimental Toxicology*, *26*(5), 447–452. http://doi.org/10.1177/0960327107076886

Risebrough, R. W., Walker, W. 2nd, Schmidt, T. T., de Lappe, B. W., & Connors, C. W. (1976). Transfer of chlorinated biphenyls to Antarctica. *Nature*, *264*(5588), 738–739. http://dx.doi.org/10.1038/264738a0

Ritov, V. B., Menshikova, E. V., He, J., Ferrell, R. E., Goodpaster, B. H., & Kelley, D. E. (2005). Deficiency of subsarcolemmal mitochondria in obesity and type 2 diabetes. *Diabetes*, *54*(1), 8–14. http://dx.doi.org/10.2337/diabetes.54.1.8

Rossignol, R., Letellier, T., Malgat, M., Rocher, C., & Mazat, J. P. (2000). Tissue variation in the control of oxidative phosphorylation: implication for mitochondrial diseases. *The Biochemical Journal*, *347 Pt 1*, 45–53. http://dx.doi.org/10.1042/bj3470045

Ruzzin, J., Petersen, R., Meugnier, E., Madsen, L., Lock, E.-J., Lillefosse, H., … Froyland, L. (2010). Persistent organic pollutant exposure leads to insulin resistance syndrome. *Environmental Health Perspectives*, *118*(4), 465–471. http://doi.org/10.1289/ehp.0901321

Sabanayagam, C., Teppala, S., & Shankar, A. (2013). Relationship between urinary bisphenol A levels and prediabetes among subjects free of diabetes. *Acta Diabetologica*, *50*(4), 625–631. http://doi.org/10.1007/s00592-013-0472-z

Sader, S., Nian, M., & Liu, P. (2003). Leptin: a novel link between obesity, diabetes, cardiovascular risk, and ventricular hypertrophy. *Circulation*, *108*(6), 644–646. http://doi.org/10.1161/01.CIR.0000081427.01306.7D

Saini, V. (2010). Molecular mechanisms of insulin resistance in type 2 diabetes mellitus. *World Journal of Diabetes*, *1*(3), 68–75. http://doi.org/10.4239/wjd.v1.i3.68

Sargis, R. M. (2014). The hijacking of cellular signaling and the diabetes epidemic: mechanisms of environmental disruption of insulin action and glucose homeostasis. *Diabetes & Metabolism Journal*, *38*(1), 13–24. http://doi.org/10.4093/dmj.2014.38.1.13

Satoh, H., Nguyen, M. T. A., Miles, P. D. G., Imamura, T., Usui, I., & Olefsky, J. M. (2004). Adenovirus-mediated chronic "hyper-resistinemia" leads to in vivo insulin resistance in normal rats. *The Journal of Clinical Investigation*, *114*(2), 224–231. http://doi.org/10.1172/JCI20785

Scheepers, A., Joost, H.-G., & Schurmann, A. (2004). The glucose transporter families SGLT and GLUT: molecular basis of normal and aberrant function. *JPEN. Journal of Parenteral and Enteral Nutrition*, *28*(5), 364–371. http://doi.org/10.1177/0148607104028005364

Schmitz-Peiffer, C., Craig, D. L., & Biden, T. J. (1999). Ceramide generation is sufficient to account for the inhibition of the insulin-stimulated PKB pathway in C2C12 skeletal muscle cells pretreated with palmitate. *The Journal of Biological Chemistry*, *274*(34), 24202–24210. http://doi.org/10.1074/jbc.274.34.24202

Sell, H., Dietze-Schroeder, D., & Eckel, J. (2006). The adipocyte-myocyte axis in insulin resistance. *Trends in Endocrinology and Metabolism: TEM*, *17*(10), 416–422. http://doi.org/10.1016/j.tem.2006.10.010

Shankar, A. & Teppala, S. (2011). Relationship between Urinary Bisphenol A Levels and Diabetes Mellitus. *The Journal Of Clinical Endocrinology & Metabolism*, *96*(12), 3822-3826. http://dx.doi.org/10.1210/jc. 2011-1682

Shi, Y.-Q., Wang, Y.-P., Song, Y., Li, H.-W., Liu, C.-J., Wu, Z.-G., & Yang, K.-D. (2010). p,p'-DDE induces testicular apoptosis in prepubertal rats via the Fas/FasL pathway. *Toxicology Letters*, *193*(1), 79–85. http://doi.org/10.1016/j.toxlet.2009.12.008

Shoelson, S. E., Lee, J., & Goldfine, A. B. (2006). Inflammation and insulin resistance. *Journal of Clinical Investigation*, *116*(7), 1793–1801. http://doi.org/10.1172/JCI29069

Slim, R., Toborek, M., Robertson, L. W., Lehmler, H. J., & Hennig, B. (2000). Cellular glutathione status modulates polychlorinated biphenyl-induced stress response and apoptosis in vascular endothelial cells. *Toxicology and Applied Pharmacology*, *166*(1), 36–42. http://doi.org/10.1006/taap.2000.8944

Song, M. O., & Freedman, J. H. (2005). Activation of mitogen activated protein kinases by PCB126 (3,3',4,4',5-pentachlorobiphenyl) in HepG2 cells. *Toxicological Sciences : An Official Journal of the Society of Toxicology*, *84*(2), 308–318. http://doi.org/10.1093/toxsci/kfi084

Steinberg, G. R. (2007). Inflammation in obesity is the common link between defects in fatty acid metabolism and insulin resistance. *Cell Cycle (Georgetown, Tex.)*, *6*(8), 888–894. http://doi.org/10.4161/cc.6.8.4135

Steinberg, G. R., Dyck, D. J., Calles-Escandon, J., Tandon, N. N., Luiken, J. J. F. P., Glatz, J. F. C., & Bonen, A. (2002). Chronic leptin administration decreases fatty acid uptake and fatty acid transporters in rat skeletal muscle. *The Journal of Biological Chemistry*, *277*(11), 8854–8860. http://doi.org/10.1074/jbc.M107683200

Steinberg, G. R., Parolin, M. L., Heigenhauser, G. J. F., & Dyck, D. J. (2002). Leptin increases FA oxidation in lean but not obese human skeletal muscle: evidence of peripheral leptin resistance. *American Journal of Physiology. Endocrinology and Metabolism*, *283*(1), E187–192. http://doi.org/10.1152/ajpendo.00542.2001

Steppan, C. M., Bailey, S. T., Bhat, S., Brown, E. J., Banerjee, R. R., Wright, C. M., … Lazar, M. A. (2001). The hormone resistin links obesity to diabetes. *Nature*, *409*(6818), 307–312. http://doi.org/10.1038/35053000

Tripathy, D., Lindholm, E., Isomaa, B., Saloranta, C., Tuomi, T., & Groop, L. (2003). Familiality of metabolic abnormalities is dependent on age at onset and phenotype of the type 2 diabetic proband. *American Journal of Physiology. Endocrinology and Metabolism*, *285*(6), E1297–1303. http://doi.org/10.1152/ajpendo.00113. 2003

Turyk, M., Anderson, H., Knobeloch, L., Imm, P., & Persky, V. (2009). Organochlorine exposure and incidence of diabetes in a cohort of Great Lakes sport fish consumers. *Environmental Health Perspectives*, *117*(7), 1076–1082. http://doi.org/10.1289/ehp.0800281

Uemura, H., Arisawa, K., Hiyoshi, M., Kitayama, A., Takami, H., Sawachika, F., … Suzuki, T. (2009). Prevalence of metabolic syndrome associated with body burden levels of dioxin and related compounds among Japan's general population. *Environmental Health Perspectives*, *117*(4), 568–573. http://doi.org/10.1289/ehp.0800012

Vaag, A., Henriksen, J. E., & Beck-Nielsen, H. (1992). Decreased insulin activation of glycogen synthase in skeletal muscles in young nonobese Caucasian first-degree relatives of patients with non-insulin-dependent diabetes mellitus. *The Journal of Clinical Investigation*, *89*(3), 782–788. http://doi.org/10.1172/JCI115656

Wang, J., Lv, X., & Du, Y. (2010). Inflammatory response and insulin signaling alteration induced by PCB77. *Journal of Environmental Sciences (China)*, *22*(7), 1086–1090. http://doi.org/10.1016/S1001-0742(09) 60221-7

Wang, S., Zhao, Y., Xia, N., Zhang, W., Tang, Z., Wang, C., … Cui, S. (2015). KPNbeta1 promotes palmitate-induced insulin resistance via NF-kappaB signaling in hepatocytes. *Journal of Physiology and Biochemistry*, *71*(4), 763–772. http://doi.org/10.1007/s13105-015-0440-x

Wang, X., & Wang, W.-X. (2005). Uptake, absorption efficiency and elimination of DDT in marine phytoplankton, copepods and fish. *Environmental Pollution (Barking, Essex: 1987)*, *136*(3), 453–464. http://doi.org/10.1016/j.envpol.2005.01.004

Weigert, C., Brodbeck, K., Staiger, H., Kausch, C., Machicao, F., Haring, H. U., & Schleicher, E. D. (2004). Palmitate, but not unsaturated fatty acids, induces the expression of interleukin-6 in human myotubes through proteasome-dependent activation of nuclear factor-kappaB. *The Journal of Biological Chemistry*, *279*(23), 23942–23952. http://doi.org/10.1074/jbc.M312692200

Weisberg, S. P., McCann, D., Desai, M., Rosenbaum, M., Leibel, R. L., & Ferrante, A. W. J. (2003). Obesity is associated with macrophage accumulation in adipose tissue. *The Journal of Clinical Investigation*, *112*(12), 1796–1808. http://doi.org/10.1172/JCI19246

WHO (World Health Organization) (2011). Diabetes Programme: Facts and Figures. Available: http://www.who.int/diabetes/facts/en/ [accessed 20 April 2016].

Williams, A. A., Selvaraj, J., Srinivasan, C., Sathish, S., Rajesh, P., Balaji, V., … Balasubramanian, K. (2013). Protective role of lycopene against Aroclor 1254-induced changes on GLUT4 in the skeletal muscles of adult male rat. *Drug and Chemical Toxicology*, *36*(3), 320–328. http://doi.org/10.3109/01480545.2012.720991

Wilson, J., Berntsen, H. F., Zimmer, K. E., Frizzell, C., Verhaegen, S., Ropstad, E., & Connolly, L. (2016). Effects of defined mixtures of persistent organic pollutants (POPs) on multiple cellular responses in the human hepatocarcinoma cell line, HepG2, using high content analysis screening. *Toxicology and Applied Pharmacology*, *294*, 21–31. http://doi.org/10.1016/j.taap.2016.01.001

Wu, H., Bertrand, K. A., Choi, A. L., Hu, F. B., Laden, F., Grandjean, P., & Sun, Q. (2013). Persistent organic pollutants and type 2 diabetes: a prospective analysis in the nurses' health study and meta-analysis. *Environmental Health Perspectives*, *121*(2), 153–161. http://doi.org/10.1289/ehp.1205248

Xiong, X., Wang, A., Liu, G., Liu, H., Wang, C., Xia, T., … Yang, K. (2006). Effects of p,p'-dichlorodiphenyldichloroethylene on the expressions of transferrin and androgen-binding protein in rat Sertoli cells. *Environmental Research*, *101*(3), 334–339. http://doi.org/10.1016/j.envres.2005.11.003

Yamauchi, T., Kamon, J., Minokoshi, Y., Ito, Y., Waki, H., Uchida, S., … Kadowaki, T. (2002b). Adiponectin stimulates glucose utilization and fatty-acid oxidation by activating. *Nature Medicine*, *8*(11), 1288–1295. http://doi.org/10.1038/nm788

Yang, J. (2014). Enhanced skeletal muscle for effective glucose homeostasis. *Progress in Molecular Biology and Translational Science*, *121*, 133–163. http://doi.org/10.1016/B978-0-12-800101-1.00005-3

Yoon, M. J., Lee, G. Y., Chung, J.-J., Ahn, Y. H., Hong, S. H., & Kim, J. B. (2006). Adiponectin increases fatty acid oxidation in skeletal muscle cells by sequential activation of AMP-activated protein kinase, p38 mitogen-activated protein kinase, and peroxisome proliferator-activated receptor alpha. *Diabetes*, *55*(9), 2562–2570. http://doi.org/10.2337/db05-1322

Yoshida, T., Okuno, A., Izumi, M., Takahashi, K., Hagisawa, Y., Ohsumi, J., & Fujiwara, T. (2008). CS-917, a fructose 1,6-bisphosphatase inhibitor, improves postprandial hyperglycemia after meal loading in non-obese type 2 diabetic Goto-Kakizaki rats. *European Journal of Pharmacology*, *601*(1-3), 192–197. http://doi.org/10.1016/j.ejphar.2008.10.050

Zabielski, P., Lanza, I. R., Gopala, S., Heppelmann, C. J. H., Bergen, H. R. 3rd, Dasari, S., & Nair, K. S. (2016). Altered Skeletal Muscle Mitochondrial Proteome As the Basis of Disruption of Mitochondrial Function in Diabetic Mice. *Diabetes*, *65*(3), 561–573. http://doi.org/10.2337/db15-0823

Zhang, L., Keung, W., Samokhvalov, V., Wang, W., & Lopaschuk, G. D. (2010). Role of fatty acid uptake and fatty acid beta-oxidation in mediating insulin resistance in heart and skeletal muscle. *Biochimica et Biophysica Acta*, *1801*(1), 1–22. http://doi.org/10.1016/j.bbalip.2009.09.014

9

Assessment of Water Resources and Analysis of Safe Yield and Reliability of Surface Water Reservoirs of Asmara Water Supply System

Kahsay N. Zeraebruk[1], Alfred O. Mayabi[1] & John M. Gathenya[1]

[1] College of Engineering and Technology, Jomo Kenyatta University of Agriculture and Technology, Kenya

Correspondence: Kahsay Zeraebruk, Department of Civil Construction and Environmental Engineering, JKUAT, P.O. Box 62000–00200 Nairobi. E-mail: kahsayneg2000@gmail.com

Abstract

In a water supply system safe yield is the average annual volume that can be supplied to the system subject to an adopted set of operational rules and a typical demand pattern without violating a given level of service standard. It is dependent upon storage and hydrologic (rainfall/runoff/evaporation) characteristics of the sources, the source facilities, upstream and downstream permitted withdrawals and minimum in-stream flow requirements.

For effective operation and management of a water supply system, it is important to have knowledge of water balance of the reservoirs and estimate their safe yield at a certain level of reliability. In this study, to assess water resources potential of existing surface water sources and new catchments and estimate the water balance of the water supply system, the hydrologic simulation model, SWAT (Soil and Water Assessment Tool) was utilized. The model was calibrated and validated successfully. The safe yield and corresponding reliability of reservoirs were estimated using a deterministic water balance model. The results of the water balance analysis and projected water demand were used to assess existing water supply situation and challenges in future. The assessment indicated that the gap between demand and supply at high population growth rate scenario is wide and very critical.

To close the gap between the available water supply and the increasing water demand in the study area, utmost attention is needed by the decision making authorities and the management of the water utility to improve performance efficiency of the water supply system by instituting effective water governance and reducing leakage losses.

Keywords: Safe yield, SWAT, reliability, water demand, population growth and projection, decision support tool

1. Introduction

In a water distribution system, the purpose of a water supply reservoir is to balance differences between supply and demand. The temporal variation of storage is a function of several variables such as inflow, outflow and losses. The yield or outflow from a reservoir or a group of reservoirs depends on the active storage capacity of each reservoir and the water that flows into each reservoir, i.e., the inflow (Ahmed, Srivastava, & Rani, 2013). Hence, in order to effectively utilize the water reservoirs for different purposes such as water supply, hydropower generation, irrigation, and flood control, knowledge of water quantity within the reservoir is required (Yunus & Fidelia, 2012). In other words, the balancing or buffering ability of a reservoir is limited by its maximum capacity and its current storage volume at the beginning of the cycle. Under normal circumstances, the configuration and size of a reservoir system impose the primary constraints on how much surface runoff can be controlled in a river basin. In addition, regional hydrology plays a critical role in determining how well the water stored in reservoirs may be regulated.

Reservoir safe yield is generally defined as the reliable withdrawal rate that can be provided by a combination of streamflows and reservoir storage through a defined critical drought period. In a water supply system the safe yield is the average annual volume that can be supplied to the system subject to an adopted set of operational rules and a typical demand pattern without violating a given level of service standard. Safe yield is dependent upon storage and hydrologic (rainfall/runoff/evaporation) characteristics of the sources, the source facilities, upstream and downstream permitted withdrawals and minimum in-stream flow requirements. However, as the relationship between storage size and reservoir yield needs to be specified in a reservoir storage–yield analysis it is a straightforward exercise to interchange these two variables (McMahon, Pegram, Vogel, & Peel, 2007).

Storage reservoirs tend to be large and complex systems requiring equally complex mathematical models to simulate their behaviour (Vogel & Bolognese, 1995). Quantitatively, reliability of a water reservoir is defined as the percentage of time that the water supply system is able to meet the full demand. Therefore, deficiency is defined as not meeting the full demand. An adaptive management strategy is needed when one has to make a decision on the face of uncertainty and have a chance to learn from management decisions through careful monitoring of key variables (Asefa, Adams, & Wanakule, 2015). Such condition would result in water use restrictions until the water supply system is again able to meet all quantitative criteria. To fully understand the ecologic, geomorphologic, and hydrologic impacts of dams, it is necessary to integrate our knowledge of all factors that influence their operational behaviour (Vogel et al., 2007).

One of the main aspects of the safe yield and reliability analysis is water demand. Estimating and forecasting water demand become necessary as the urban population dependent on public water supplies increases rapidly and new demands for water are not easily met. Considerable efforts have been put into the development of urban water supply projections in the last four decades resulted in a wealth of understanding and sophisticated forecasting techniques in this field (Okeola, Balogun, & Bell, 2015). Different kinds of data-sets have been used and the quantity and type of data available determine which forecasting method should be considered for application. There is no absolute level of accuracy that is appropriate in all demand forecasting situations (Okeola et al., 2015).

Analysis of demand for water in Least Developed Countries (LDCs) is complicated by abundant evidence that, contrary to what is observed in most developed countries, households in LDCs have access to and may use more than one of several types of water sources, such as in-house tap connections, public or private wells, public or (someone else's) private taps, water vendors or resellers, tank trucks, water provided by neighbours, rainwater collection, or water collected from rivers, streams, or lakes (Nauges & Whittington, 2009).

Scarcity and unsustainable use of water resources have become the major critical developmental issues throughout the world. Population growth increases demand and competition for water resources and compounds water scarcity. Sub Saharan Africa is projected to be severely impacted by water scarcity because of climate change (Mutunga, Zulu, & De Souza, 2012). The problem of water is further exacerbated by relatively high annual population growth rates. Mutunga et al. (2012) stated that, fifteen countries in Sub Sahara Africa, which are home to about 400 million people, are suffering from water scarcity or water stress. Further the report by Mutunga et al. (2012) indicated that most of the water stressed and scarce countries in Africa had recorded high population growth rates (including Eritrea) showing growth rates as high as 3.2 percent per year. The high population growth rates combined with anticipated changes in climate would result in water shortages to likely become even more severe, especially in areas where economic and political problems impede access to fresh water.

Eritrea had the third highest urban population growth rate in Africa for the years 2000–2005, where the annual urban population growth was 6% (Tewolde & Cabral, 2011). Reports indicate that Asmara was one of the fastest growing cities in Africa. The growth of the city has also been witnessed through surface area growth by incorporating most of the villages around the city. As a result, the Greater Asmara Area (GAA) has experienced high population growth in the past two decades, particularly, after the independence of the country in 1991 (Tewolde & Cabral, 2011).

Ensuring a water supply utility's ability to be there when needed in the future is far from exact science; consequently one of the most challenging tasks of water supply utilities is planning the timing and quantity of new water supply sources as demand for water consumption grows (Asefa et al., 2015).

2. Description of the Study Area

This study is based on Asmara water supply system. Asmara is the capital of the state of Eritrea and the administrative centre of the Central Region known as Administration of Maekel Zone (AMZ). Asmara is located at an average elevation of 2350 m above mean sea level in the central highlands of Eritrea, Figure 1.

As a capital city and major business centre, Asmara has been growing relatively faster in recent years in surface area and population. Records indicate that the current population of Asmara, including the satellite villages, to be around 450,000 (Database of AMZ, 2012). Asmara Water Supply and Sewerage Department (AWSD) is the only public utility responsible for water and sewerage services in and around Asmara.

Systematic development of water supply and management in Eritrea dates back to the colonial times. The Italian colonial administration developed domestic water supply as part of overall programme to improve the level of personal hygiene and environmental sanitation in the major towns of Eritrea, especially in the capital city, Asmara. This is evident as some of the water supply infrastructure facilities currently in use for supplying water to the capital were built during the colonial period.

The major source of water supply for Asmara is surface water produced from rainfall occurring mainly during the rainy months of July-September. The runoff created over the drainage areas joins streams and rivers and finally collected into reservoir dams located around the city. The current water supply and management situation in Asmara and the surrounding villages are constrained due to several problems attributed to drought as well as low technical and financial performance of the water utility.

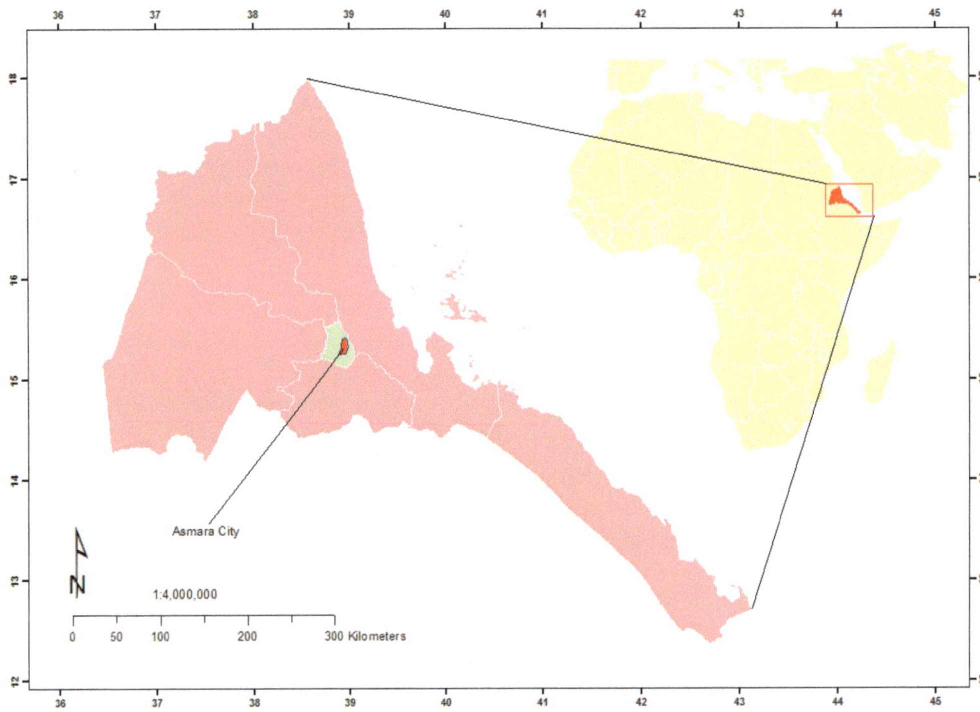

Figure 1. Location of the study area (Asmara city)

Dams have been constructed across a number of streams and rivers to impound the seasonal runoff. The surface water reservoirs currently supplying water to Asmara and surrounding areas are: Mai-Nefhi, Toker, Adi-Sheka, Mai-Sirwa, and Struatta-Vaudetto (Figure 2).

There are three water treatment plants currently under operation. Mai-Nefhi treatment plant is situated just downstream of the Mai-Nefhi reservoir. The raw water from Adisheka, Stretta-Vaudetto and Maiserwa is treated at the oldest treatment plant known as Stretta-Vaudetto WTP. The water from Toker Dam is pumped to a distance of 10 km to the new treatment plant of Adi-Nifas located next to Stretta-Vaudetto WTP.

Ground water has not been exploited as a direct source of the municipal water. This could be attributed to the quality problem which is mostly brackish in nature. Upon detailed assessment and quantity and quality investigation ground water may be utilized for different purposes which require lesser quality of water.

Small quantities of water are currently obtained from ground water aquifers in the city. Many private wells exist in Asmara serving as either alternative or supplemental sources to the municipal supply. The well fields are located in the eastern and southern areas of Asmara, and abstract water from the zone between relatively recent surface basalts and the underlying weathered laterite (Ethiopia Water Supply and Sewerage Authority, 1985). The quality of this water is variable, having brackish taste at times. The water from the ground water is mostly used for irrigation of small areas of land for vegetable growth.

Three new catchments, within the Central Regional Administration, were identified as potential water supply sources for the future. These are: Ungula catchment, Dem-Sebay Catchment and Aba-Buruk catchment (Figure 3).

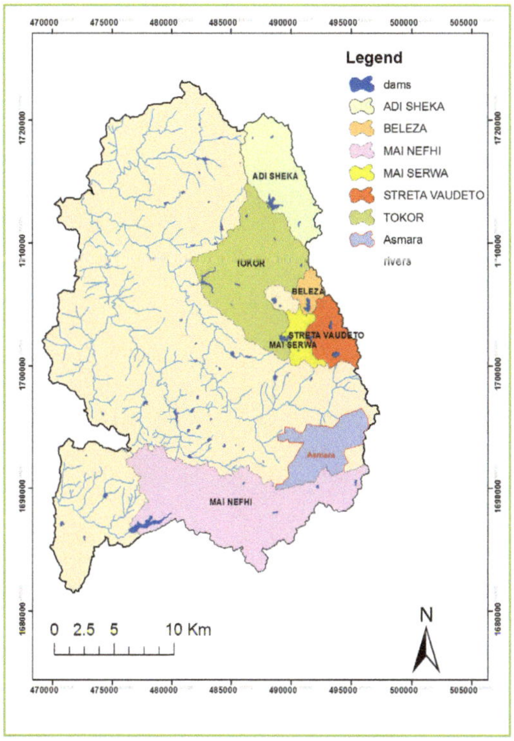

Figure 2. Catchment areas developed for Asmara water supply

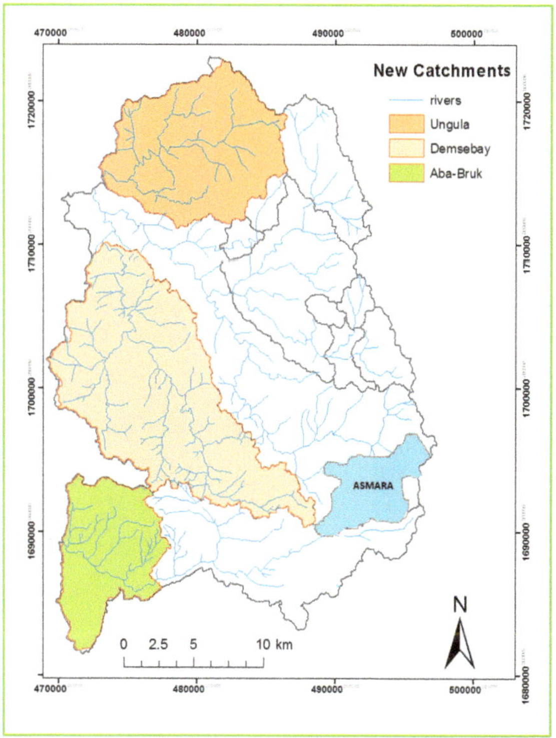

Figure 3. Catchment areas for future water sources development

The surface water sources are dependent on streamflow created from rainfall during the rainy months of mostly July to September. The recurrent drought, evaporation loss during the long dry months and seepage loss from the reservoirs must be taken into account when trying to estimate the water budget of the catchment areas and reliability of safe-yield of the water supply system.

3. Methodology

3.1 Water Resources Analysis

For planning and decision making purposes and to get reliable information of the water resources potential of catchment areas, hydrological analysis was carried out on existing surface sources with hydrological simulation model, Soil Water Assessment Tool (SWAT). It is a basin scale, continuous time model that operates on a daily time step and is designed to predict the impact of land management practices on water, sediment and agricultural chemical yields in large complex both gauged and un-gauged watersheds with varying land use and management conditions over a long periods of time (Neitsch et al., 2011). Digital elevation model (DEM), weather data, soil data and land use/land cover data are the most important data input for the setup of the SWAT model and for the simulation of the hydrological components (Fiseha, Setegn, Melesse, Volpi, & Fiori, 2010). The calibrated hydrological model, SWAT, was used to estimate the streamflow hydrograph of the catchment areas.

For purposes of planning and future development schemes, hydrological analysis using SWAT was also carried out on the catchments identified as future sources of water for Asmara. Field investigation by the researchers and experts from Eritrea Water Resources Department indicated that the catchments are well qualified technically for development by constructing dams. Annual stream inflow volumes for each of the catchment areas were estimated using the hydrological model, SWAT.

3.2 Safe Yield Estimation

For each of the surface water supply reservoir, safe yield was estimated through the use of a deterministic computer model developed with Microsoft Excel software. The spread-sheet model is a mass balance model developed on the basis of simple accounting (addition and subtraction) of the change in reservoir volume or storage as a function of surface water inflow, withdrawal (outflow), and losses due to evaporation. The model performs calculations and logical functions in a sequential manner temporally on monthly time-step. The calculations and logical functions normally begin using values from the first month of the period of record and are repeated for every month until the end of the period of record, with each time beginning with the previous time's ending storage values.

The attainable safe yield during the period of analysis was found by iteration of the monthly mass balance equation for each of the reservoirs. Monthly accounting of water volumes for each reservoir adheres to the following general algorithm:

Ending Storage = Beginning of storage + Natural inflow + Pumped/diverted inflow –water supply release/deliveries – Evaporation – Minimum instream flow

For the surface water reservoirs of Asmara water supply system, the model calculation uses the storage equation concept as developed from the basic mass balance storage equation as:

$$I_t - O_t = \Delta S_t \quad (1)$$

Where:

I_t = total inflow volume to the reservoir during a specified time period

O_t = the total outflow which is the sum of the volume withdrawal (demand) and the volume losses due to evaporation from the reservoir during the time step t.

ΔS_t = change in storage volume of the reservoir during the specified time period

The mass balance equation can be converted to a form that can analyse the cumulative effects of inflows and outflows over consecutive time steps as follows:

$$\Delta S_t = S_{t+1} - S_t \quad (2)$$

Where:

ΔS_t = change in storage of the reservoir during time step t

S_{t+1} = storage of the reservoir volume at the end of time step t+1

S_t = storage of the reservoir volume at the end of time step t

By substituting this relationship, equation 3.2 is transformed to:

$$I_t - O_t = S_{t+1} - S_t \quad (3)$$

$$O_t = D_t + E_t \quad (4)$$

D_t = the total volume of water withdrawn to satisfy the demand during the time step t

E_t = the total volume of evaporation loss form the reservoir surface area during the time step t.

Equation 3.6 and 3.7 can be combined to a more useful form of the mass balance model as:

$$S_{t+1} = S_t + I_t - D_t - E_t \quad (\text{for } S_{min} \leq S_{t+1} \leq S_{max}) \quad (5)$$

S_{min} = the minimum capacity of the reservoir below which withdrawal is not allowed

S_{max} = the maximum capacity of the reservoir

Estimation of the rate of evaporation from the reservoirs was done using the FAO CROPWAT 8.0 Software. CROPWAT 8.0 for Windows is a computer programming model for the calculation of important irrigation parameters such as evapotranspiration.

The evaporation, E_t, was estimated based on the original Penman-Monteith equation and the equations of the aerodynamic and surface resistance. The Penman-Monteith method to estimate E_t is expressed as:

$$E_t = \frac{0.408\Delta(R_n - G) + \gamma\frac{900}{T+273}U_2(e_s - e_a)}{\Delta + \gamma(1 + 0.34U_2)} \quad (6)$$

where:

E_t = reference evapotranspiration [mm day^{-1}], R_n = net radiation at the crop surface [MJ m^{-2} day-1],

G = soil heat flux density [MJ m^{-2} day-1], T = mean daily air temperature at 2 m height [°C],

u_2 = wind speed at 2 m height [m s^{-1}], e_s = saturation vapour pressure [kPa],

e_a = actual vapour pressure [kPa], e_s - e_a = saturation vapour pressure deficit [kPa],

Δ = slope of vapour pressure curve [kPa °C^{-1}], γ = psychrometric constant [kPa °C^{-1}].

The monthly volume of reservoir evaporation $\mathbf{Ev_t}$ is a function of the rate of evaporation and the free surface area of the reservoir, **A**, exposed to the atmosphere; and it was estimated as:

$$Ev_t = E_t \, x \, A \tag{7}$$

The storage of the reservoir control point is repetitively calculated on a monthly basis by adding total monthly inflows and subtracting total monthly outflows. The total monthly inflows to reservoir are obtained from historical records if the river is gauged otherwise estimated with the hydrological simulation model SWAT. The total monthly outflows from a reservoir are typically calculated by summing monthly withdrawals or releases from the reservoir to serve customer demands.

The model can be beneficial for regulators and managers of Asmara Water Supply utility as a decision-making tool for effective and efficient management of the water supply system.

3.3 Reservoir Reliability Model

The simulation was constructed around a flow distribution logic that defines how the system will operate and provides criteria that defines how monthly water demands will be met based on the monthly streamflow, evaporation losses, and previously stored water. The logic is applied to each time step in the simulation as shown in Figure 4.

Periods when the reservoir was less than or equal to the minimum storage capacity ($S_{t+1} > S_{mi}$) were noted and the time based reliability is obtained using:

$$R_e = \left[\frac{\sum DEi}{N}\right] 100 \tag{8}$$

Where R_e (%) is the reliability of the reservoir, ΣDE_i is the sum of months when the reservoir storage is greater than S_{min} (demand-met), and N is number of months in the simulation period.

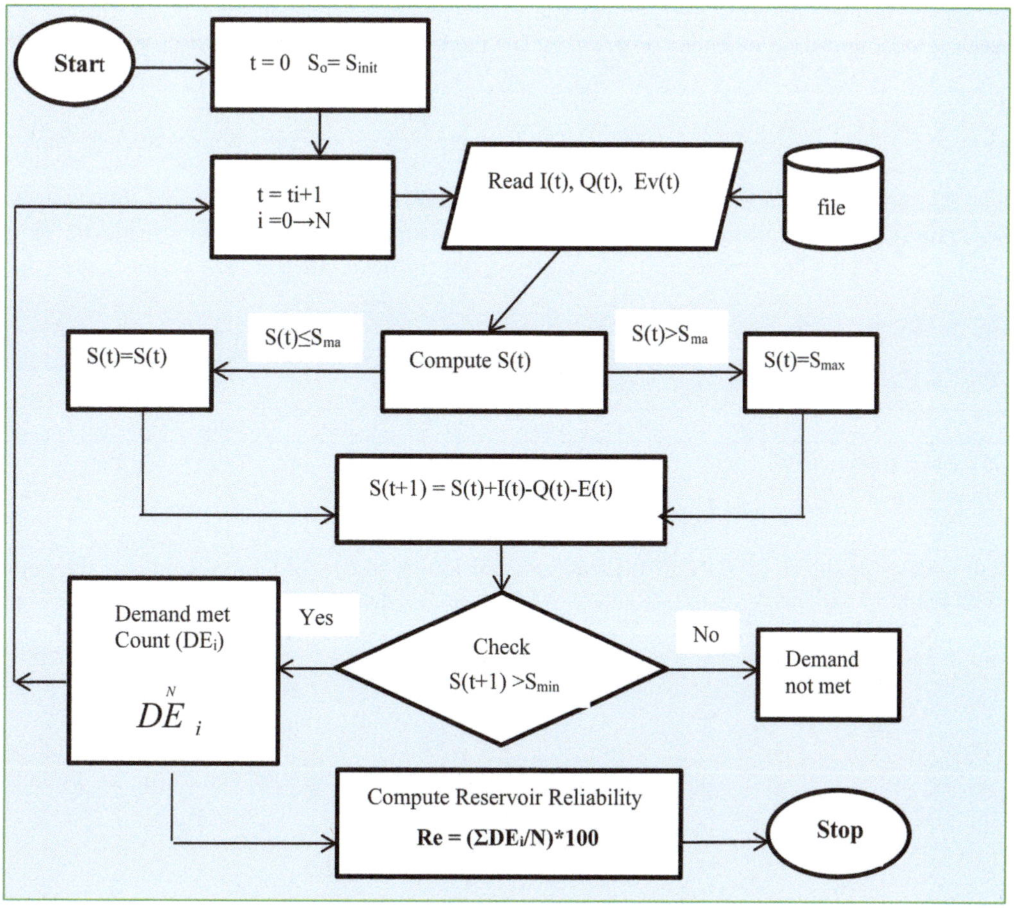

Figure 4. Reservoir Reliability analysis flow diagram

3.4 Water Demand Estimation

Accurate water demand forecasting is an essential ingredient in water resources planning. Future growth in the water demand depends on many factors. The traditional approach of hydrologists and water resources engineers has been to focus on the supply side and the assessment of available water resources

The model developed here to estimate municipal water demand uses a simple purpose-specific process considering the population growth (the number of potential subscribers) and growth per capita water demand as major factors affecting the water requirements in the study area without considering the socio-econometric factors.

3.4.1 Population Estimation and Forecasting

Government policymakers and planners around the world use present population sizes and population projections to gauge demand for food, water, energy, and other public services for now and for the future. The size, composition, and distribution of populations and forecasting future demographic characteristics are part of any analysis of sustainable development challenges and opportunities (Mutunga et al., 2012). Knowledge of the current size of the population, its characteristics and the expected growth in the future is one of the important tools required for decision making process.

The report by UN-Habitat 2014 shows that the decade average rate of change in urban population for Eritrea is relatively high, estimated: at 4.22% (1970-80), 5.30% (2000-10), 4.75% (2010-20), and projected at 4.03% (2020-30), 3.56% (2030-40) and 3.07% (2040-2050) (UN-Habitat, 2014).

In order to plan for the future water demands of the population of Asmara (the City and satellite villages), it is important to define the demographics and how the population trends are anticipated to change over time.

The population data collected from the database of Administration of Central Region for 2005 and 2010 indicates that the population of Asmara and the satellite villages was around 420,000 and 460,000 respectively (K. Zeraebruk, Mayabi, Gathenya, & Zemenfes, 2014). These data were used as baseline for future projections. In this study the Geometric Increase Method was used as a model for population projection. In this method, the annual percentage increase or percentage growth rate is assumed to be constant for a given forecast period, and the increase is compounded over existing population. The model is based on the following equation.

$$P_t = P_o (1 + r)^t \tag{9}$$

Where:

P_t = Population projected after t years, P_o = Base year population, r = Growth rate (%/), t = time in years;

The growth rate (r) has been acquired from UN Population estimation and projection made for Eritrea, from 1950 to 2100 for two growth rate scenarios: Moderate-variance growth rate and high-variance growth rate (UN, 2013).

Table 1. Population growth rate for Asmara (UN forecast 2013)

Year	2015	2020	2025	2030	2040	2050	2060
Medium Variance growth rate (%)	3.3%	2.8%	2.6%	2.4%	2.1%	1.7%	1.4%
High Variance growth rate (%)	3.4%	3.0%	2.6%	2.7%	2.4%	2.2%	1.8%

The growth rates as predicted by (UN, 2013), and adopted in this study, for the projection of the population in Asmara with medium-variance growth rate and high-variance growth rate are summarised in Table 1 and Figure 5.

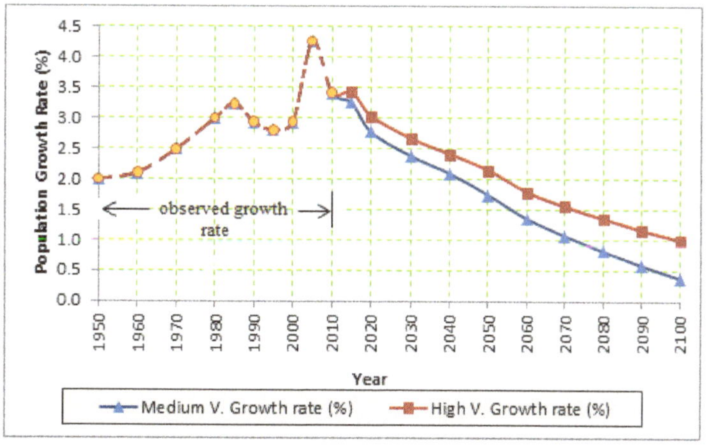

Figure 5. Projected growth rate of population for Asmara (Source: UN, 2013)

Using the population-growth model and the annual growth rate extracted from the UN data, the population of Asmara (including the village around the city) has been estimated and projected up to the year 2060 under moderate-variance and high-variance growth rates. The results of the model are shown in Figure 6.

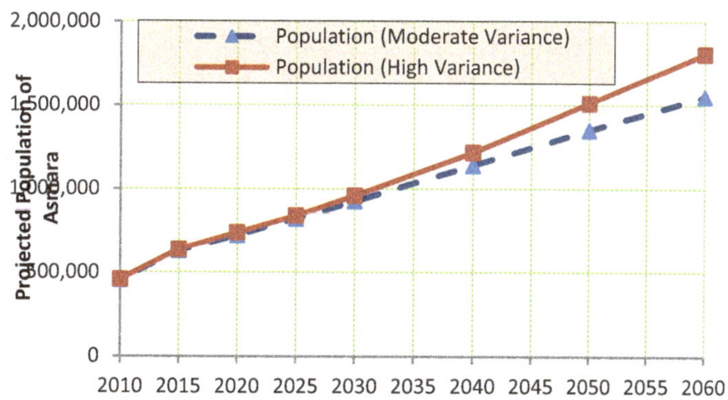

Figure 6. Population projection for Moderate and high variance scenario

3.4.2 Water Demand Analysis

The model developed here to estimate municipal water demand uses a simple purpose-specific process considering the population growth (the number of potential subscribers) and growth per capita water demand as major factors affecting the water requirements in the study area without considering the socio-econometric factors.

The study report under 'Technical and Institutional Rehabilitation Study of Asmara Water Supply System', SAUR Intenational and BRL ingenierie (1997) devised a conceptual water demand model, which calculates consumption at each unit zone of the project area as follows:

$$d_i = dom_i + non\text{-}dom_i \qquad (10)$$

$$D = \sum d_i\,(1 + \%\,Leakage) \qquad (11)$$

Where

d_i = demand of a unit zone, dom_i = domestic consumption of zone I

% Leakage = percentage of physical losses, D = overall demand to be supplied (production)

$non\text{-}dom_i$ = non-domestic consumption of zone i i.e. small industries and large consumers

From these definitions, we underline that the water drawn at a connection point is the consumption and that the corresponding demand is the consumption plus the losses.

4. Results and Discussion

4.1 Assessment of the Water Supply Sources

Urban water supply must be operated with high degree of reliability in most cases between 80 to 90 per cent. Since demand for water is always increasing with time, keeping such a high reliability of water supply system requires effective water resources management and strategic planning.

The river catchment areas currently used for Asmara water supply are Adi-Sheka, Stretta-Vaudetto, Mai-Sirwa, Mai-Nefhi and Toker all of them located within the boundary of Central Regional Administration.

The outcome of simulation analysis with the yield-reliability model of existing surface water reservoirs of Mai-Nefhi, Toker Adi-Sheka, Mai-Sirwa, and Stretta-Vaudetto and are shown in Figure 7.

The water supply from the reservoirs has been estimated considering safe yield of 80% reliability. This means that there is a risk of 1 out of 5 for the reservoirs to fail meeting the water demand. The safe yield of each of the existing water supply reservoirs at 80% reliability is summarised in Table 2. With this level of reliability, the results showed that the total water supply capacity of the existing reservoirs was around 43,000 m^3/day.

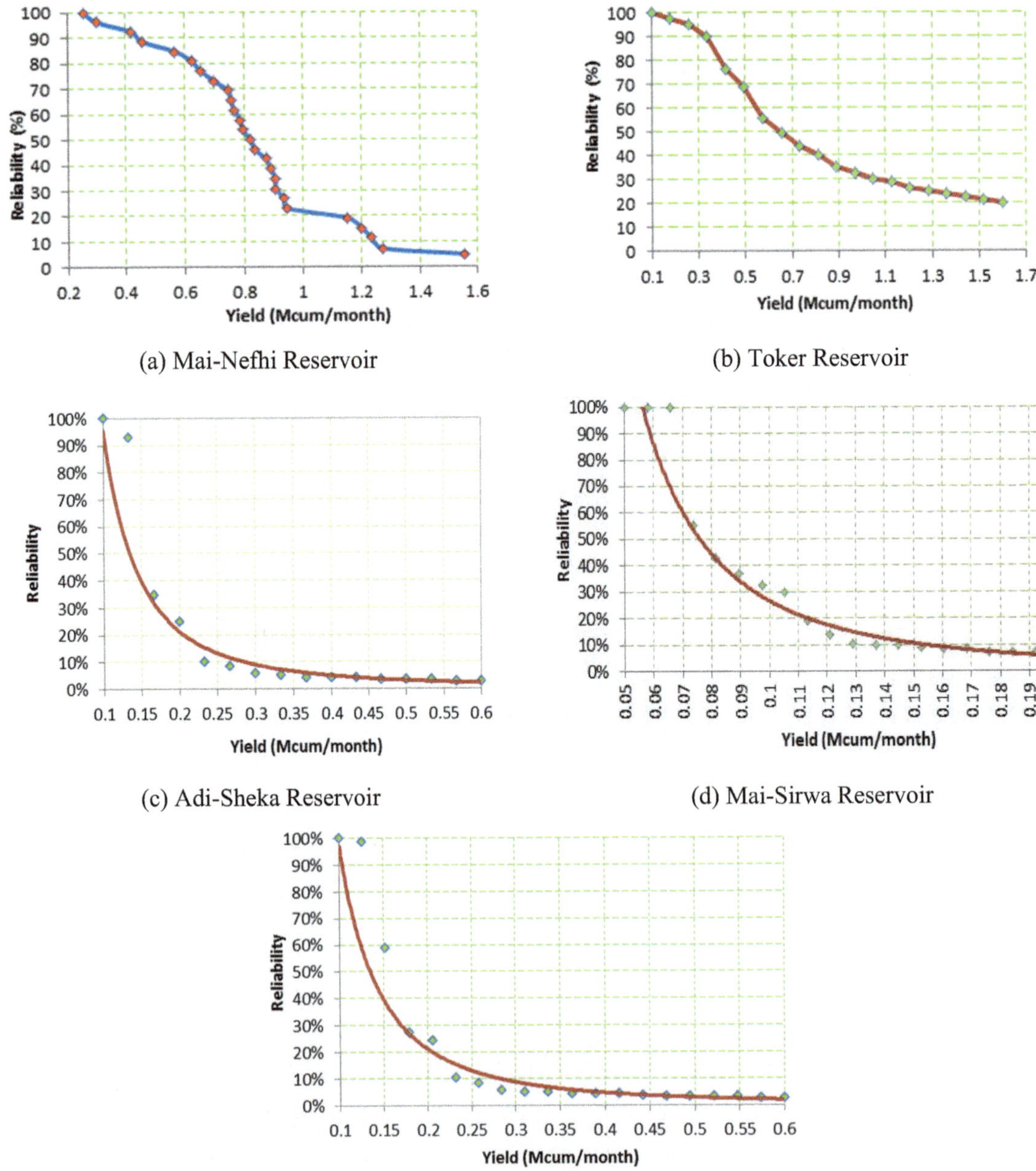

(a) Mai-Nefhi Reservoir

(b) Toker Reservoir

(c) Adi-Sheka Reservoir

(d) Mai-Sirwa Reservoir

(e) Stretta-Vaudetto reservoir

Figure 7. Yield- reliability curves for the surface water reservoirs of Asmara

Table 2. Summary of water supply yield of the existing reservoirs

No	Reservoir Name	Safe Yield (80% Reliability)	
		m^3/month	m^3/day
1	Toker	400,000	13,300
2	Adi-Sheka	117,000	3,900
4	Stretta-Vaudetto	112,000	3,733
5	Mai-Nefhi	600,000	20,000
6	Mai-Sirwa	62,000	2,067
SUM		1,291,000	43,000

4.2 Assessment of Future Water Supply Sources

Three new catchment areas, within the Central Regional Administration, were identified as potential water supply sources for the future. These are: Ungula catchment, Dem-Sebay Catchment and Aba-Buruk catchment (see Figure 3).

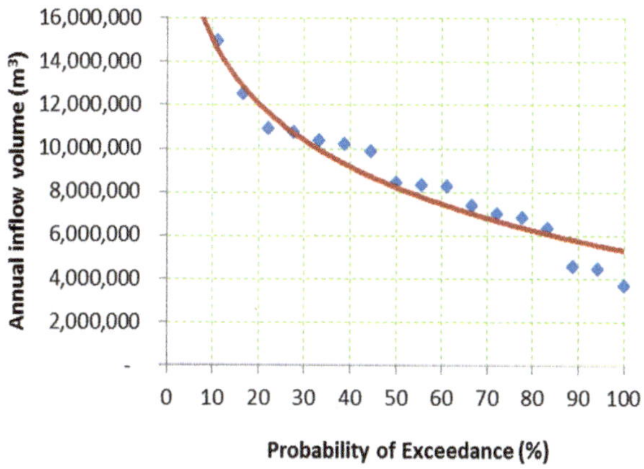

(a) Ungula catchment area

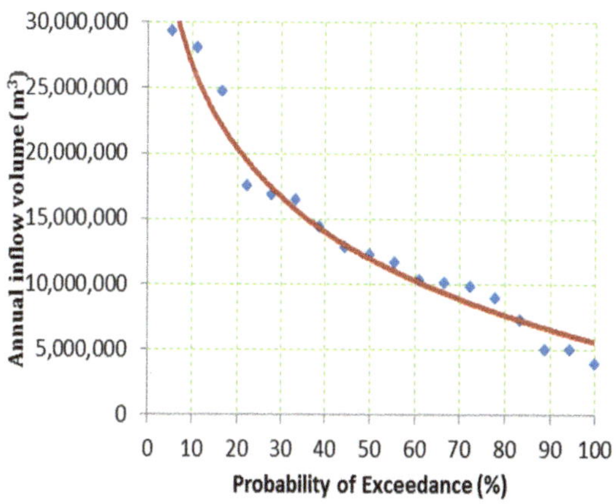

(b) Dem-Sebay catchment area

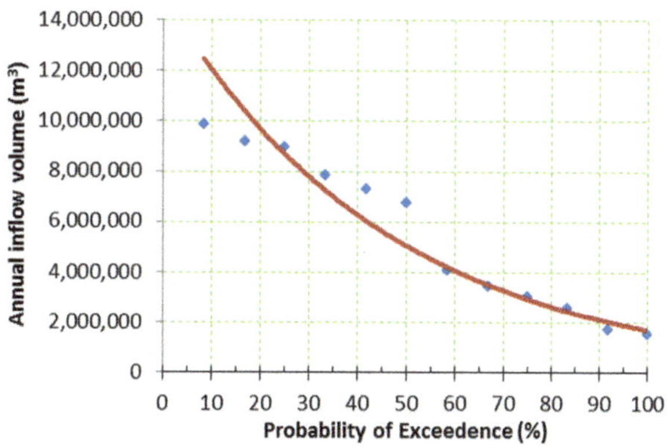

(c) Aba-Buruk catchment area

Figure 8. Flow Duration Curve (FDC) for future water supply sources for Asmara

The calibrated hydrological model, SWAT, developed in this study was used to estimate the streamflow hydrograph for these catchment areas. Each area was delineated with DEM data and the other input data required to running the model such as, weather, soil, land-use/cover, were used to predict the monthly streamflow expected in each of the catchment areas, as discussed in the following sections.

For each of the three catchments the flow duration analysis was conducted and the results of the probability of exceedance analysis for Ungula, Dem-Sebay and Aba-Bruk are summarised in Figure 8.

Streamflow analysis with the hydrological model (SWAT) indicated that the Ungula, Dem-Sebay and Aba-Bruk catchment areas could deliver around 6.0, 8.0 and 3.0 Mcum of water every year at 80% of probability of exceedance (Figure 8).

4.3 Water Demand Projection

The amount of water needed by consumers (water demand) depends on the population to be served and the rate of per capita water consumption. Both this parameters are time dependent and in most cases they tend to grow with time which is true for cities in developing countries such as Asmara.

The conceptual water demand model was used to simulate the future water demands projected up to 2060 for the moderate –variance and high-variance growth rates. The growth in per capita demand assumed and the corresponding daily per capita demand (l/c/d) for the moderate and high population growth scenarios are shown on Figure 9.

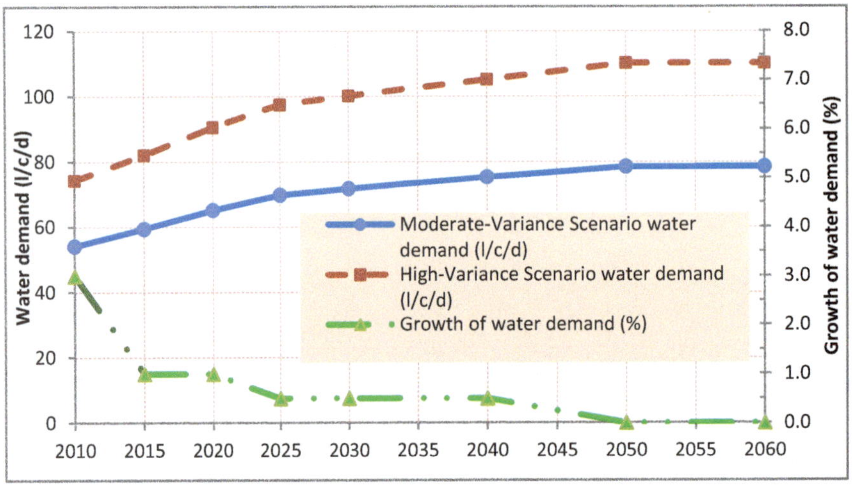

Figure 9. Projected daily per capita demand and water demand growth rate

The projected daily water demand for the moderate growth rate scenario and high growth rate scenario are shown in Figure 10.

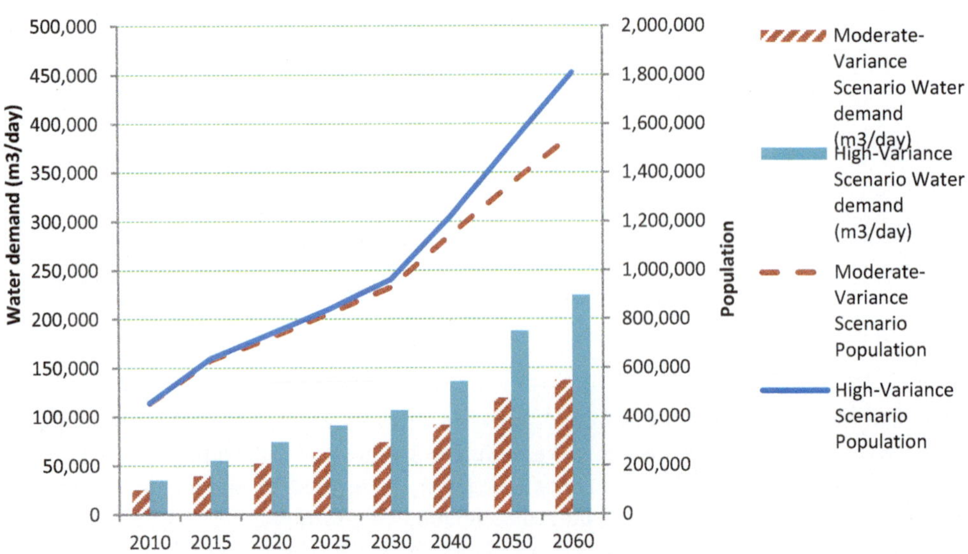

Figure 10. Asmara City projected population and daily water demand

4.4 Comparison of Safe Water Supply Yield and Demand

The water supply potential of the surface water sources has been analysed in detail. The yield of each of the existing surface water reservoirs were analysed considering 80% reliability. Besides, analyses of the annual average water resources potential of the catchment areas that may be developed as additional sources of water supply to cater for the increasing water demand of the city were also conducted.

Comparison of water demand and supply were carried out, to understand the current situation and future water supply challenges. Such information could help the water authorities to make informed decisions and device a roadmap for sustainable management and development of the water resources. The comparison process was based on the two population growth scenarios i.e. moderate-growth rate and high–growth-rate and also on certain development scenario of the new catchments identified as potential sources of water for the city.

The results of the analysis showed that for the moderate growth rate scenario the water crisis could be managed through systematic development of the new catchments areas up to 2040 as shown in Table 3 and Figure 11. Minimizing leakage losses and introduction of effective and efficient water governance from source up to the point of end-users would play crucial role to close the gap between the available water resources for supply and water demand.

Table 3. Water demand and supply situation for moderate-growth rate

Year	Annual Demand (Mcum)	Annual Supply (Mcum)	Assumed Water Supply Sources Development Scenario
2010	9.2	15.5	Existing Reservoirs
2015	14.4	15.5	Existing Reservoirs
2020	19.0	23.5	Existing Reservoirs + Demsebay
2025	23.1	26.5	Existing Reservoirs + Demsebay + Ababruk
2030	26.8	32.5	Existing Reservoirs + Demsebay + Ababruk + Ungula
2040	33.3	32.5	Existing Reservoirs + Demsebay + Ababruk + Ungula
2050	43.5	32.5	Existing Reservoirs + Demsebay + Ababruk + Ungula
2060	50.0	32.5	Existing Reservoirs + Demsebay + Ababruk + Ungula

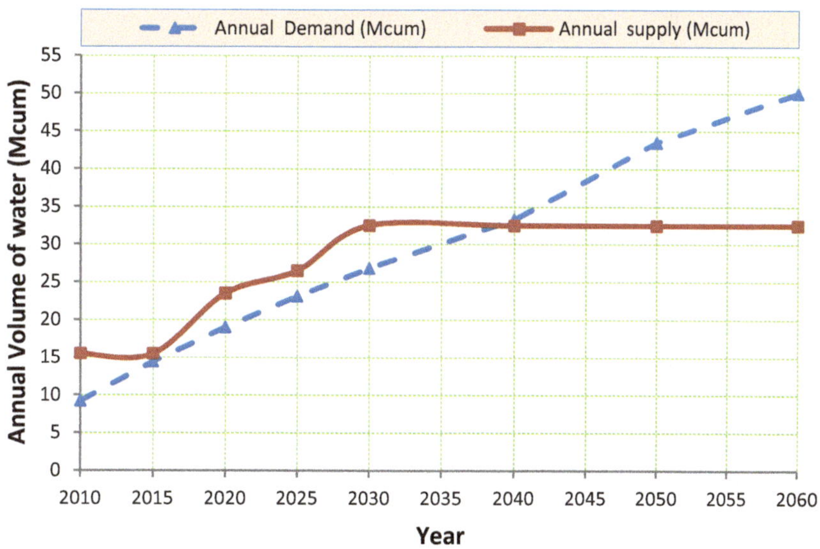

Figure 11. Water demand and supply comparison for moderate-growth rate

Assuming the high growth rate scenario of population as predicted by UN 2013, the gap between demand and supply will be wider and more critical. The analysis as indicated in Table 4 and Figure 12 predicted that the water crisis would be so critical. The water authorities should plan and consider systematic development of the new catchments areas identified for water supply immediately.

Table 4. Water demand and supply comparison for high-growth rate scenario

Year	Annual Demand (Mcum)	Annual supply (Mcum)	Development Scenario of Water Supply Sources
2010	12.6	15.5	Existing Reservoirs
2015	20.0	15.5	Existing Reservoirs
2020	27.0	23.5	Existing Reservoirs + Demsebay
2025	33.0	26.5	Existing Reservoirs + Demsebay + Ababruk
2030	38.8	32.5	Existing Reservoirs + Demsebay + Ababruk + Ungula
2040	49.6	32.5	Existing Reservoirs + Demsebay + Ababruk + Ungula
2050	68.4	32.5	Existing Reservoirs + Demsebay + Ababruk + Ungula
2060	81.7	32.5	Existing Reservoirs + Demsebay + Ababruk + Ungula

Exploration of new water sources outside the Central Regional administration but not very far from Asmara may also be considered as a long term solution to the water crisis. Another important issue that must be considered in this context would be to modernize the overall management system of the water utility and improve the efficiency of the water supply distribution system by controlling and minimizing leakage losses and other forms of Non-revenue water.

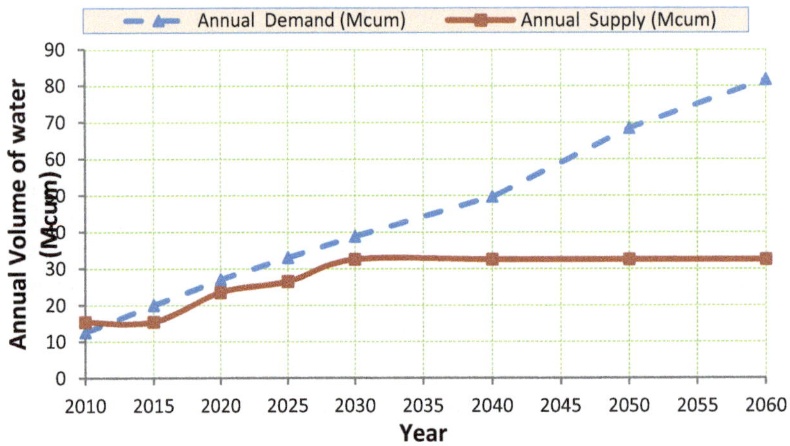

Figure 12. Water demand and supply comparison for high-growth rate

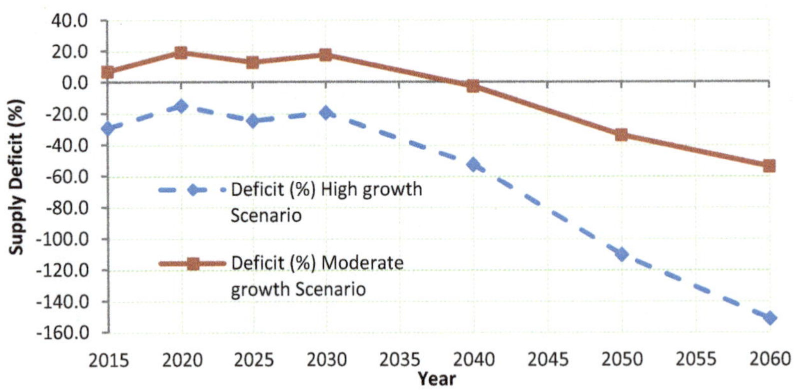

Figure 13. Prediction of water supply deficit in Asmara

The comparative analysis results between water available for supply and demand in the time period from 2015 to 2060 predicted very critical shortage of water in the Greater Asmara Area. Figure 13 shows the gap between supply and demand in terms of water supply deficit for the high and moderate population growth rate and based on the assumption of the development of new sources.

The UN-population report (2013) indicated that Eritrea is one of countries in Africa with high population growth rate. Hence, in this case the high growth rate is the most likely scenario and creates the most critical situation for planning, design and development. It is also the most critical condition in terms of water demand, development of infrastructure, and financial investment and as such requires serious consideration by policy and decision makers in Eritrea in general and in Asmara in particular.

5. Conclusion

Water has become increasingly scarce and less reliable due to factors such as population growth, economic growth, increased urbanization and inefficient water management system and the supply is shrinking due to changing climatic patterns. The challenge of servicing more people with scarcer water resources is going to become a norm rather than an exception for AWSD and the situation requires critical appraisal of current perspective on urban water management. Policy and decision makers in the region, and managers of AWSD need to consider efficient water use practices and develop technical and management strategies that can respond to all these challenges simultaneously.

Population projections can alert policymakers to major trends that may affect economic development and help them develop policies that can be adapted for various projection scenarios. However, it is not possible to tell the future condition accurately and that is why users of population projections need to understand the reliability and the limitations of projection scenarios. Awareness of how projections are prepared and the possible sources of uncertainty in the numbers can help policymakers more effectively incorporate projections in their planning process.

In this study, the water supply yield of each of the existing surface water reservoirs were analysed with a water-balance model considering 80% reliability. Besides, analyses of the annual average water supply potential of the new catchment areas were conducted at 80% of probability of exceedance. The water demand and supply comparison was made on the basis of two population growth scenarios (moderate-growth rate and high–growth-rate) and on assumed development scenarios of the new catchments.

The following conclusion can be made:

- The water-balance model developed with a Visual Basic programming language can be used successfully as a management and decision making tool to predict the safe yield and corresponding reliability of the existing water supply reservoirs
- For the moderate growth rate scenario the predicted gap between water demand and supply may be managed through systematic development of the new catchment areas up to 2040 (annual supply of 32 x 10^6 m^3 against annual demand of 33 x 10^6 m^3).
- For the high growth rate scenario the water crisis is predicted to reach a serious level in the near future (annual supply of 24 x 10^6 m^3 against annual demand of 27 x 10^6 m^3 predicted in 2020). This scenario calls for development of the new catchment areas and treatment facilities much earlier as well as introduction of appropriate demand management measures.
- To close the gap between the available water supply and increasing water demand, utmost attention is needed by the decision making authorities and the management of the water utility to improve performance efficiency of the water supply system by instituting effective water governance and reducing leakage losses.

Reference

Ahmed, G. L., Srivastava, D. K., & Rani, D. (2013). Optimization – Simulation Models For Yield Assessment of a Single Reservoir System. *Journal of Indian Water Resources Society, 33*(4), 9–16.

Asefa, T., Adams, A., & Wanakule, N. (2015). A Level-of-Service Concept for Planning Future Water Supply Projects under Probabilistic Demand and Supply Framework. *JAWRA Journal of the American Water Resources Association, 51*(5), 1272–1285. https://doi.org/10.1111/1752-1688.12309

Ethiopia Water Supply and Sewerage Authority. (1985). *Asmara Water Supply Feasibility Study (Volume 1: Main report).*

Fiseha, B. M., Setegn, S. G., Melesse, A. M., Volpi, E., & Fiori, A. (2010). Hydrological analysis of the Upper Tiber River Basin, Central Italy: a watershed modelling approach. *HYDROLOGICAL PROCESSES Hydrol. Process. (2012), 2274*(November 2008), 2267–2274. https://doi.org/10.1002/hyp

McMahon, T. a., Pegram, G. G. S., Vogel, R. M., & Peel, M. C. (2007). Revisiting reservoir storage-yield relationships using a global streamflow database. *Advances in Water Resources, 30*(8), 1858–1872. https://doi.org/10.1016/j.advwatres.2007.02.003

Mutunga, C., Zulu, E., & De Souza, R.-M. (2012). *Population Dynamics , Climate Change , and Sustainable Development in Africa.*

Nauges, C., & Whittington, D. (2009). Estimation of Water Demand in Developing Countries: An Overview. *The World Bank Research Observer, 25*(2), 263–294. https://doi.org/10.1093/wbro/lkp016

Neitsch, S., Arnold, J., Kiniry, J., & Williams, J. (2011). Soil & Water Assessment Tool Theoretical Documentation Version 2009, 1–647.

Okeola, O. G., Balogun, S. O., & Bell, S. (2015). Estimating a municipal water supply reliability. *Cogent Engineering, 2*(1), 10. https://doi.org/10.1080/23311916.2015.1012988

SAUR Intenational and BRL ingenierie. (1997). *Technical and Institutional Rehabilitation Study of Asmara Water Supply System*. Administration of Central Region (Department of Infrastructure) Asmara 1997.

Tewolde, M. G., & Cabral, P. (2011). Urban sprawl analysis and modeling in Asmara, Eritrea. *Remote Sensing, 3*(10), 2148–2165. https://doi.org/10.3390/rs3102148

UN. (2013). *United Nations, Department of Economic and Social Affairs, Population Division (2013). World Population Prospects: The 2012 Revision, Volume I: Comprehensive Tables ST/ESA/SER.A/336.* (Vol. I).

UN-Habitat. (2014). *THE STATE OF AFRICAN CITIES 2014: Re-imagining sustainable urban transitions.*

Vogel, R. M., & Bolognese, R. A. (1995). Storage-reliability-resilience-yield relations for over-year water supply systems. *Water Resources Research.* https://doi.org/10.1029/94WR02972

Vogel, R. M., Sieber, J., Archfield, S. a., Smith, M. P., Apse, C. D., & Huber-Lee, A. (2007). Relations among storage, yield, and instream flow. *Water Resources Research, 43*(5), 1–12. https://doi.org/10.1029/2006 WR005226

Yunus, S. D., & Fidelia, N. N. (2012). Reservoir storage variations from hydrological mass balance and satellite radar altimetry. *International Journal of Water Resources and Environmental Engineering, 4*(6), 201–207. https://doi.org/10.5897/IJWREE11.140

Zeraebruk, K., Mayabi, A., Gathenya, J., & Zemenfes, T. (2014). Assessment of Level and Quality of Water Supply Service Delivery for Development of Decision Support Tools: Case Study Asmara Water Supply. *International Journal of Sciences: Basic and Applied Research (IJSBAR), 14*(1), 93–107.

Zeraebruk, K. N., Mayabi, A. O., Gathenya, J. M., & Tsige, Z. (2014). Assessment of Water Supply Services and Operational Performance of Asmara Water Supply Department (AWSD) for Development of Decision Support Tools. *Environment and Natural Resources Research, 4*(4), 208–222. https://doi.org/10.5539/enrr.v4n4p208

10

Heavy Precipitation Events in Marmara Region and connections with the North Atlantic and Arctic Oscillation Patterns

Sevinc A. Sirdas[1], Ali Behzat Diriker[1] & Vinay Kumar[2]

[1] Department of Meteorological Engineering, Faculty of Aeronautics and Astronautics, Istanbul Technical University, Turkey

[2] Department of Earth, Ocean & Atmospheric Sciences, Florida State University, Tallahassee, FL, 32306, USA

Correspondence: Sevinc A. Sirdas, Department of Meteorological Engineering, Faculty of Aeronautics and Astronautics, Istanbul Technical University, Turkey. E-mail: sirdas@itu.edu.tr

Abstract

The prediction and understanding of the extreme weather events became one of the top priorities of the forecasting centers. To analyze the heavy rainfall events, the data from reanalysis and station datasets are utilized. The linear regression analysis and principle component analysis (PCA) are applied on precipitation, temperature, Arctic Oscillation Indices (AO) and North Atlantic Oscillation Indices (NAO) datasets. The results indicate that the relationship between these variables can be expressed neither by linear regression nor by PCA, which falls in short in capturing the complexity of the problem. The results indicate the importance of low-level jet in two cases out of three cases considered here.

Keywords: Heavy Rain, Extreme Weather Events, Low Level Jet

1. Introduction

Extreme weather events have significant effects on both the society and the individual which makes it important to understand and predict them correctly. These events are affected by a wide range of factors from topography, local forcing to remote teleconnection patterns such as El Niño Southern Oscillation (ENSO) and Arctic Oscillation (AO). On the other hand, extreme precipitation events such as flash floods, heavy rain may sometimes intensify low-level jets in the troposphere (Diaz & Murnane, 2008).

The climate of a region is affected by its geographical location and the geographical features of its surroundings. Large surfaces of land and water such as continents and oceans affect the air mass above it via their radiation balance (incoming, absorbed & reflected radiations). The climate of Turkey is affected by four air masses, namely: cP (Continental Polar, originates from Siberia), mP (Maritime Polar; originates from the northern part of Atlantic Ocean), mT (Maritime Tropical; originates from the equatorial part of Atlantic Ocean, Mediterranean and Aegean seas), cT (Continental Tropical; originates from North Africa) (Türkes, 1996; Sirdas, 2003). Precipitation is an important parameter which is directly affected by the climate of a region. In other way, by using the precipitation data, the climatological characteristics of a region can be determined. Due to its geographical location, Turkey receives rainfall throughout the year with varying intensity of precipitation over the country e.g. a city in the East Black sea region of Turkey receives 2200 mm of precipitation annually, while another one in Central Anatolia receives only 320 mm of precipitation (Sirdas, 2003; Sirdas & Sen, 2003).

Extreme weather events such as floods, droughts, heat waves and tropical cyclones are related to the climate and somehow linked with the recent global climate change. Furthermore, it is difficult to state that each and every one of the event is an indication of a change in the climate. As Mark Twain once said "The climate is what we expect and the weather is what we get". Extreme weather events can be separated in two groups: firstly, weather-based events relatively short-termed events which can be predicted 1-2 weeks ahead such as tropical cyclones, severe floods etc; secondly, climate-based events such as drought, season-long heat waves, multiple occurrences of severe storm events and record wildfire (Diaz & Murnane, 2008). Extreme weather events occur due to the complex interactions between large-scale atmosphere-ocean circulation patterns such as the Arctic Oscillation, El-Niño-Southern Oscillation with local weather (Khandekar, 2013). Most of Indian summer monsoonal droughts are associated with an El-Niño of the central-eastern Pacific Ocean (Khandekar & Neralla, 1984; Yarav, Ramu, & Dimri, 2013; Kumar, 2016). ENSO has also been correlated to the droughts in Amazonia region, while floods in south-eastern region of South America (Cavalcanti, 2012). Martin & Schumacher (2011) found negative correlation between the Caribbean Low-Level Jet (CLLJ) and Caribbean precipitation based on model study. Although, in many other cases as well low-level jets have been found to relate to the extreme precipitation events, somehow the correlation between the low-level jets and precipitation events is hard to establish. Lazarus (2009) states that during El Niño years, the precipitation observed in Florida increases where suppresses during a La Nina years. There are many studies, those explain rainfall prediction and variability on the scale of time and space over Euro-Mediterranean region (Sirdas et al., 2007), Iran (Amiri, Amerian, & Mesgari, 2016); Amiri & Mesgari 2016) and India (Kumar & Krishnamurti, 2012; Kumar & Krishnamurti, 2015).

Precipitation extremes are statistical anomalies of a log-normal probability density function over a region (Sirdas et al., 2013). These extremes have been found to be affected by a plethora of factors, such as sea surface temperatures (SST) anomalies (Grimm & Tedeschi, 2009), teleconnection patterns (Vasconcellos & Cavalcanti, 2010), synoptic systems like frontal systems (Vasconcellos & Cavalcanti, 2010), persistent systems (Carvalho, et al. 2002) and large mesoscale convective systems (MCSs) (Durkee, et al. 2009) in their study of South American climate. Haylock & Goodess (2004) demonstrated a link between NAO and extreme precipitation in winter (DJF) by examining two indices of dry-days and wet-day from the data of 347 stations over Europe. A sign of negative NAO has an impact on the storm track in eastern tropical Pacific and western Atlantic (Cassou, 2008). Furthermore, this study demonstrates a link between MJO (Madden-Julian Oscillation) and NAO, based on daily geopotential height maps from 1974 to 2007.

This study begins by summarizing elementary information on the difference between weather and climate and gives what are the elementary parts of the climate extremes in Turkey. Further an attempt is being made to introduce the concept of extreme weather and its association with certain mesoscale convective systems such as ENSO and NAO. The purpose of this study is to analyse the relationship between teleconnection patterns (namely the AO and the NAO) and statistically extreme precipitation events (such as heavy rains). In the last section the manuscript comprises, 3 case studies to examine a physical relationship between the low-level jets and extreme precipitation events.

2. Data & Distribution Analysis

2.1 Locations of Meteorological Stations

The following section focuses on the analysis of the monthly precipitation and daily temperature data from 23 meteorological observation stations in the Marmara region. The duration of this dataset is 40 years (1970-2010). This study comprises the average precipitation received by each station, total precipitation received by each station, and total precipitation received during 1970 to 2010. The meteorological observation stations mentioned above are: Ayvalık, Bandırma, Bursa, Balıkesir, Bozcaada, Çanakkale, Çorlu, Dursunbey, Edirne, Edremit, Florya, Geyve, Gönen, Kartal, Keleş, Kireçburnu, Kırklareli, Kocaeli, Kumköy, Lüleburgaz, Sakarya, Şile and Tekirdağ. Figure 1 shows the geographical locations of 23 stations.

2.2 Data Analysis of Meteorological Station Data

Datasets from 23 stations of Marmara region have been analysed to find the good cases those would show a dynamical relationship between the oscillations and the extreme weather events. The temperature and total rainfall data have been studied on a monthly basis to understand their variability. Some of the special cases of temperature are also studied on a daily basis to layout the relationship between the cases and the oscillations.

The amount of precipitation received by Marmara Region varies from January to December months and is lowest for summer season. Majority of stations show maximum annual precipitation is being received in autumn (SON) and winter (DJF) months. Generally, the amount of precipitation ranges between 0-100 mm in summer (JJA) months to a peak of 400-450 mm in winter (DJF) months, while the spring (MAM) months and autumn (SON) months lying in between. In general, autumn months having more precipitation than spring months. Turkey receives most of the precipitation during autumn (SON) and winter (DJF) months (Figure 2). There are some extreme cases in which Şile received the highest precipitation, more than 300 mm of precipitation in the month of August e.g. 1997. Such a heavy rainfall in the month of August is anomalous. The high peaks (Fig 2) indicate to an extreme weather event as compared to lowest peak in the summer months. However, the coastal provinces such as Edremit, Çanakkale, Ayvalık and Bandırma show gradual change in the precipitation, while the provinces away from sea such as Sakarya, Balıkesir, Keleş and Dursunbey show a dramatic change in the precipitation regime, between spring (MAM) and summer (JJA) months.

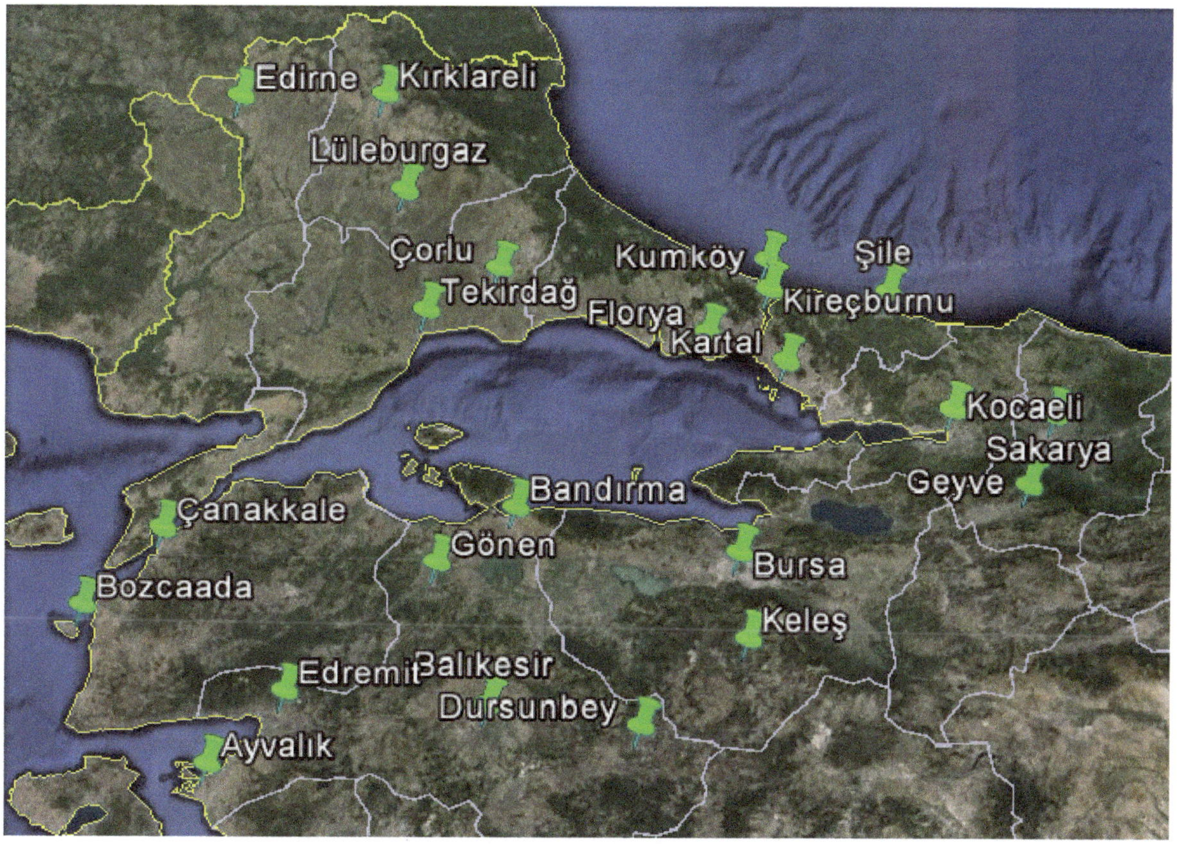

Figure 1. Map showing the locations of the Meteorological Stations

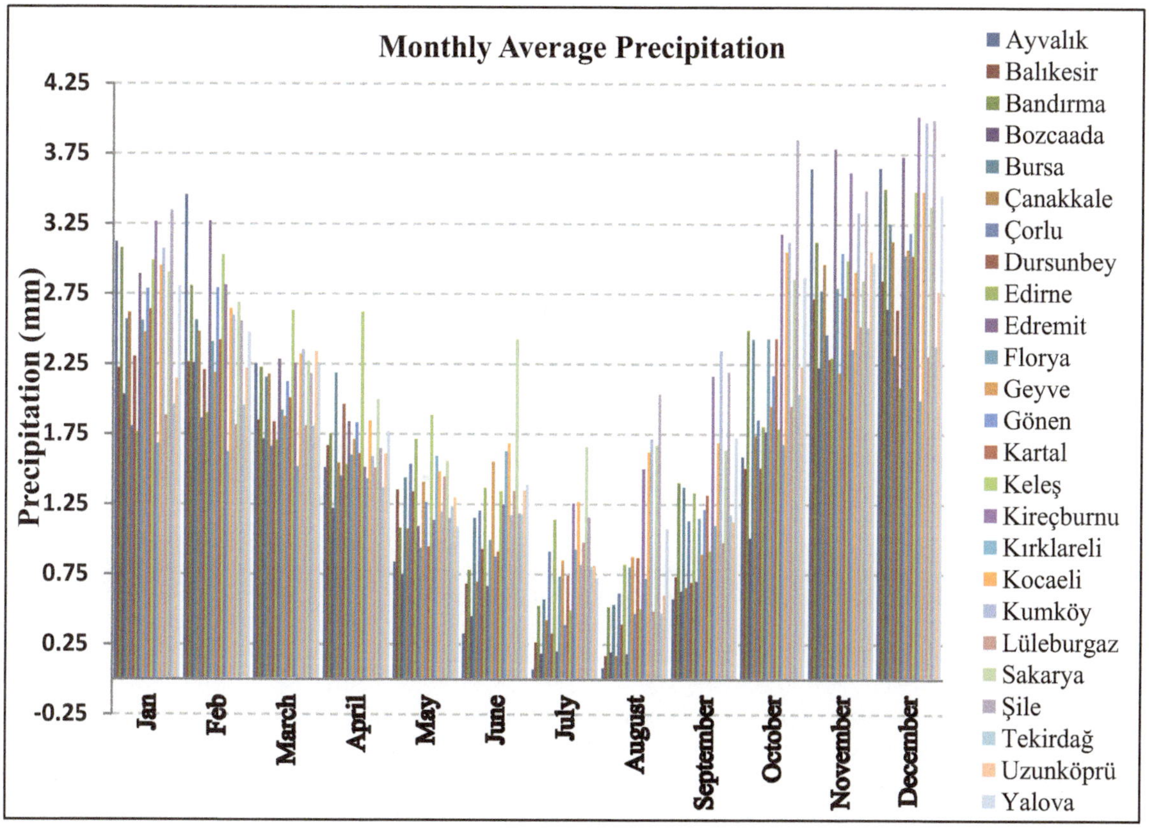

Figure 2. Climatology of monthly average precipitation based on the period of 1970-2010

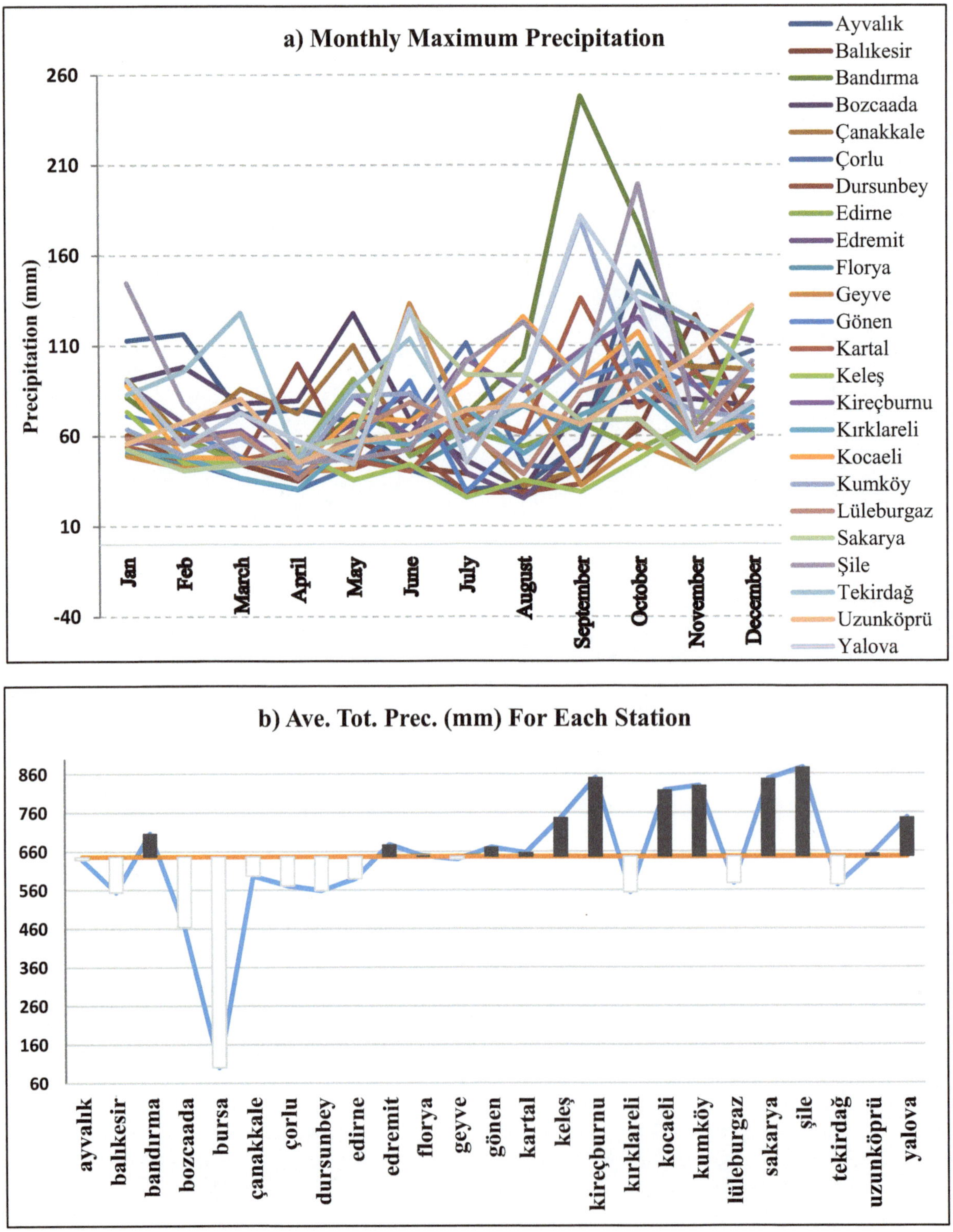

Figure 3. (a) Monthly maximum and, (b) average annual total precipitation (mm) observed by each station over 1970-2010

Figure 3 shows the amount of average annual total precipitation received by each station over 1970-2010. This is used as a safety assessment showing the reliability of the data. In this case the precipitation observed by Bursa station is around 100 mm per year which doesn't fit well with the mean value and standard deviation of the data which are 645mm and 104mm respectively. It is an error in either the precipitation data or in the observation equipment of the station. Another case for these lowest precipitation values of Bursa station could reveal the climate change impact on rainfall pattern.

Figure 4 is obtained by averaging the total amount of precipitation received by each station, every year. It is showing the increasing trend overall.

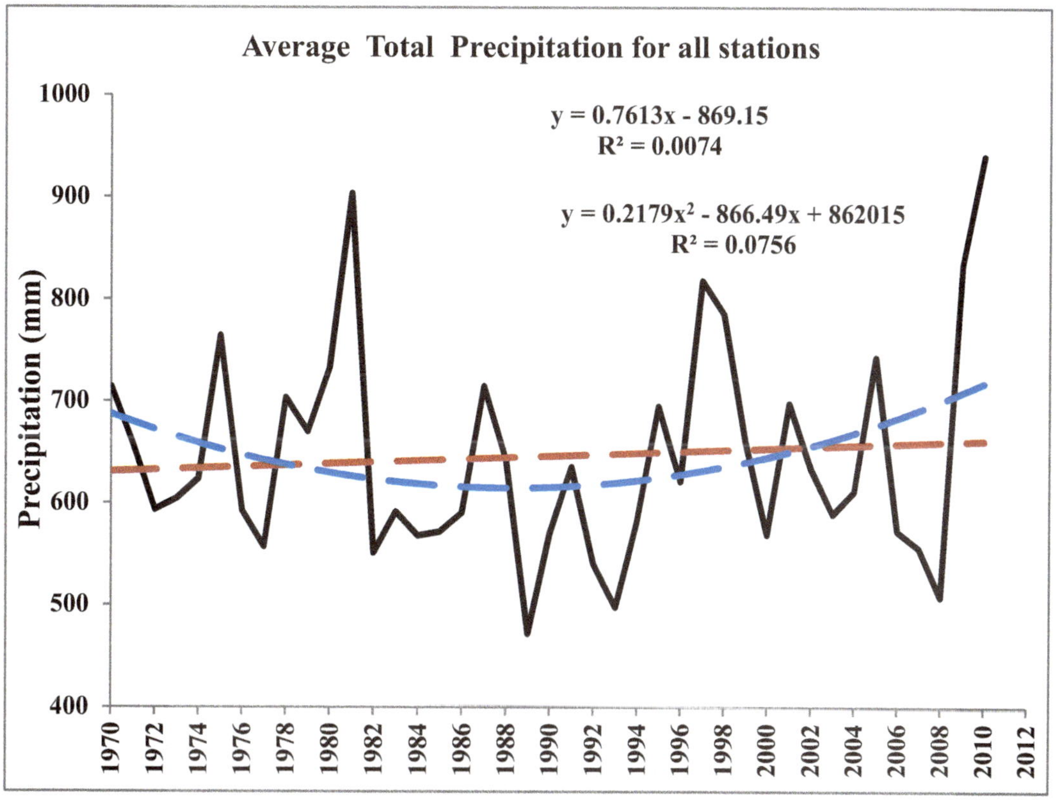

Figure 4. Average total precipitation observed by all stations and the trend line throughout 1970-2010

2.3 ERA-Interim Data

For the purpose of case study, 3 cases/dates were selected for the low-level jets and their effects on extreme weather events. The case study on these dates has been done by examining the wind velocity at 1000, 850 and 200 mb levels and surface data of 2m temperature, 2m dew point temperature, sea surface temperature (SST), mean sea level pressure (MSLP), evaporation, surface precipitation. The data has been acquired from ERA-Interim and consists of the years 1985, 1999 and 2009.

2.4 The Arctic Oscillation (AO) and the North Atlantic Oscillation (NAO)

The AO and the NAO data used in this study begins from the January of 1950 and ends with February of 2014 and consists of monthly values. The data used in this study has been obtained from NCEP reanalysis. Due to the discontinuity of the precipitation data and the limited areal coverage of the temperature data, linear regression and principle component calculations were calculated for 8 stations only. The linear regression analysis used in this study to find the relationship between precipitation and Arctic and North Atlantic oscillations uses precipitation as the independent variable and Arctic and North Atlantic oscillations as the dependent variables.

The results of the linear regression analyses provide a correlation coefficient ranging between 0.00002 and 0.0054 (only selected stations are shown in Figure 5 and 6), which indicates a negligible linear correlation between precipitation-AO and precipitation-NAO. The minuscule correlation coefficient may be the result of the precipitation data comprises too much low values between 0.001mm and 1mm. Days with 0mm of precipitation were not included in these calculations.

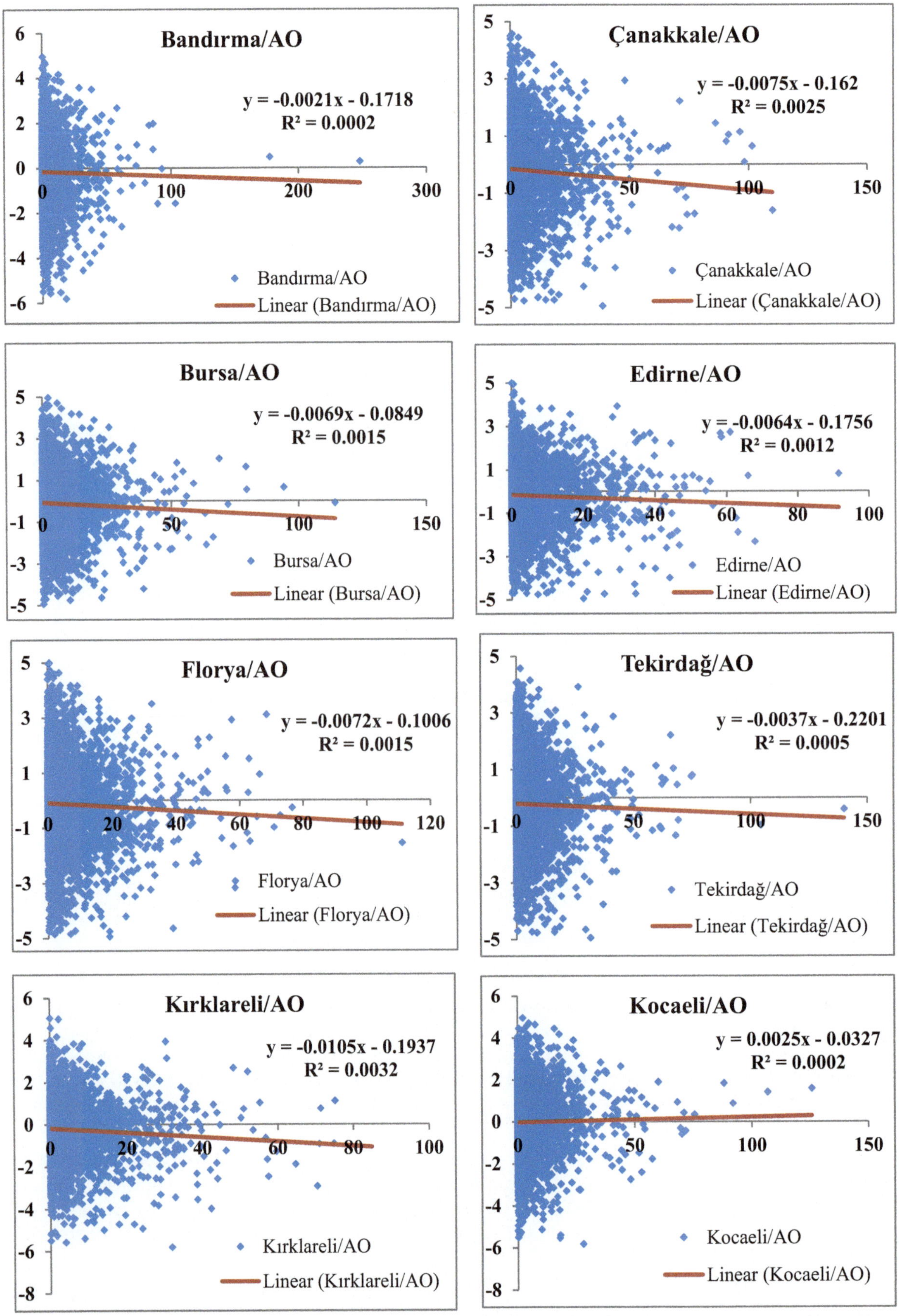

Figure 5. The Linear Regression Graphs for Precipitation data taken from 8 Meteorological Observation Stations and AO

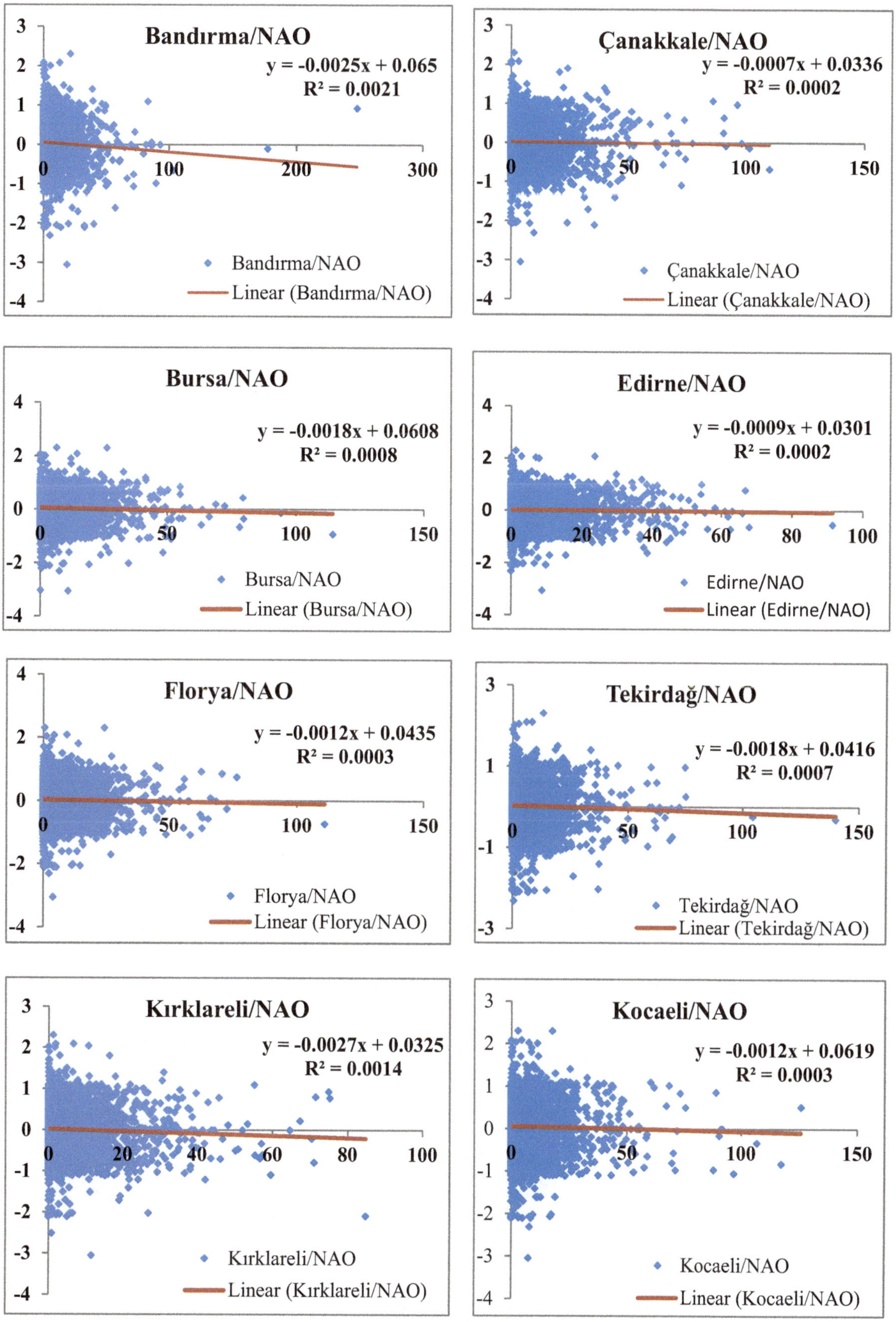

Figure 6. The Linear Regression Graphs for Precipitation data taken from 8 Meteorological Observation Stations and NAO

Table 1. Results of PCA for meteorological stations of Bandırma, Bursa, Edirne, Çanakkale, Florya, Kırklareli, Kocaeli and Tekirdağ.

```
Bandırma

      Prec.   AO   NAO Temp.
Prec.  1.00 0.00 -0.05 -0.06
AO     0.00 1.00  0.38  0.00
NAO   -0.05 0.38  1.00 -0.09
Temp. -0.06 0.00 -0.09  1.00

Standard deviations:
[1] 1.1815150 1.0307160 0.9703704 0.7746147

Rotation:
              PC1          PC2         PC3        PC4
Prec.  0.06605102 -0.707600675  0.69501210  0.1090721
AO    -0.68759208  0.075113799  0.24825277 -0.6781929
NAO   -0.70842344  0.004809511 -0.03837458  0.7047272
Temp.  0.14488140  0.702592392  0.67368846  0.1775308

Importance of components:
                         PC1   PC2   PC3   PC4
Standard deviation     1.182 1.031 0.970 0.775
Proportion of Variance 0.349 0.266 0.235 0.150
Cumulative Proportion  0.349 0.615 0.850 1.000
```

```
Bursa

      Prec.    AO   NAO Temp.
Prec.  1.00 -0.04 -0.03 -0.09
AO    -0.04  1.00  0.38 -0.02
NAO   -0.03  0.38  1.00 -0.10
Temp. -0.09 -0.02 -0.10  1.00

Standard deviations:
[1] 1.1843328 1.0443919 0.9483922 0.7792007

Rotation:
              PC1         PC2         PC3         PC4
Prec.  0.07830035 -0.71428912 -0.69511435 -0.02182078
AO    -0.68424735  0.11611188 -0.21793109  0.68617027
NAO   -0.70219710 -0.01539596 -0.04096949 -0.71063612
Temp.  0.18053767  0.68997975 -0.68384470 -0.15391725

Importance of components:
                         PC1   PC2   PC3   PC4
Standard deviation     1.184 1.044 0.948 0.779
Proportion of Variance 0.351 0.273 0.225 0.152
Cumulative Proportion  0.351 0.623 0.848 1.000
```

```
Edirne

      Prec.    AO   NAO Temp.
Prec.  1.00 -0.03 -0.01 -0.06
AO    -0.03  1.00  0.39  0.06
NAO   -0.01  0.39  1.00 -0.07
Temp. -0.06  0.06 -0.07  1.00

Standard deviations:
[1] 1.1786934 1.0332799 0.9743680 0.7704684

Rotation:
               PC1        PC2         PC3          PC4
Prec. -0.084700660 -0.6334758 -0.76909136  0.005715628
AO     0.704861487  0.0874210 -0.14451284  0.688944040
NAO    0.704201113 -0.1531668  0.04346218 -0.691918907
Temp. -0.009842515  0.7533962 -0.62106808 -0.215804697

Importance of components:
                         PC1   PC2   PC3   PC4
Standard deviation     1.179 1.033 0.974 0.770
Proportion of Variance 0.347 0.267 0.237 0.148
Cumulative Proportion  0.347 0.614 0.852 1.000
```

```
Çanakkale

      Prec.    AO   NAO Temp.
Prec.  1.00 -0.04 -0.01 -0.09
AO    -0.04  1.00  0.38  0.00
NAO   -0.01  0.38  1.00 -0.09
Temp. -0.09  0.00 -0.09  1.00

Standard deviations:
[1] 1.1813632 1.0467354 0.9493654 0.7793788

Rotation:
             PC1         PC2         PC3         PC4
Prec.  0.0683581 -0.70455292 -0.70614896 -0.01691181
AO    -0.6891624  0.13196237 -0.18187953 -0.68888381
NAO   -0.7044381 -0.04395618 -0.04127263  0.70719965
Temp.  0.1554002  0.69588719 -0.68305934  0.15818264

Importance of components:
                         PC1   PC2   PC3   PC4
Standard deviation     1.181 1.047 0.949 0.779
Proportion of Variance 0.349 0.274 0.225 0.152
Cumulative Proportion  0.349 0.623 0.848 1.000
```

```
Florya

      Prec.    AO   NAO Temp.
Prec.  1.00 -0.04 -0.02 -0.06
AO    -0.04  1.00  0.38  0.00
NAO   -0.02  0.38  1.00 -0.10
Temp. -0.06  0.00 -0.10  1.00

Standard deviations:
[1] 1.1827241 1.0314590 0.9658862 0.7773801

Rotation:
              PC1         PC2        PC3          PC4
Prec. -0.08119985  0.69903739 0.71044341  0.004844073
AO     0.68444675 -0.13426493 0.21499881 -0.683579616
NAO    0.70461321  0.04313535 0.03326668  0.707497624
Temp. -0.16869930 -0.70104136 0.66928245  0.179283452

Importance of components:
                        PC1   PC2   PC3   PC4
Standard deviation     1.18 1.031 0.966 0.777
Proportion of Variance 0.35 0.266 0.233 0.151
Cumulative Proportion  0.35 0.616 0.849 1.000
```

```
Kırklareli

      Prec.    AO   NAO Temp.
Prec.  1.00 -0.06 -0.03 -0.01
AO    -0.06  1.00  0.39  0.02
NAO   -0.03  0.39  1.00 -0.08
Temp. -0.01  0.02 -0.08  1.00

Standard deviations:
[1] 1.1846487 1.0111072 0.9892289 0.7718134

Rotation:
             PC1        PC2        PC3         PC4
Prec.  0.1542988 -0.5204520 0.83899057 -0.03763546
AO    -0.6906083  0.1295493 0.17645241 -0.68930524
NAO   -0.6994522 -0.1232376 0.08354388  0.69899896
Temp.  0.1000935  0.8349606 0.50791712  0.18666093

Importance of components:
                         PC1   PC2   PC3   PC4
Standard deviation     1.185 1.011 0.989 0.772
Proportion of Variance 0.351 0.256 0.245 0.149
Cumulative Proportion  0.351 0.606 0.851 1.000
```

```
Kocaeli

      Prec.    AO   NAO Temp.
Prec.  1.00  0.00 -0.02 -0.05
AO     0.00  1.00  0.39 -0.03
NAO   -0.02  0.39  1.00 -0.10
Temp. -0.05 -0.03 -0.10  1.00

Standard deviations:
[1] 1.1872974 1.0200783 0.9725015 0.7771782

Rotation:
              PC1         PC2        PC3         PC4
Prec. -0.01210297 -0.76471220 0.64137895 -0.06084259
AO     0.67798644  0.12918453 0.23184649  0.68548738
NAO    0.69930392  0.06463622 0.02276392 -0.71153214
Temp. -0.22619005  0.62797195 0.73100072 -0.14187049

Importance of components:
                         PC1   PC2   PC3   PC4
Standard deviation     1.187 1.020 0.973 0.777
Proportion of Variance 0.352 0.260 0.236 0.151
Cumulative Proportion  0.352 0.613 0.849 1.000
```

```
Tekirdag

      Prec.    AO   NAO Temp.
Prec.  1.00 -0.02 -0.03 -0.05
AO    -0.02  1.00  0.38  0.02
NAO   -0.03  0.38  1.00 -0.09
Temp. -0.05  0.02 -0.09  1.00

Standard deviations:
[1] 1.1788126 1.0273793 0.9753661 0.7768872

Rotation:
              PC1         PC2         PC3         PC4
Prec.  0.06971922 -0.67502766  0.73265800  0.05185701
AO    -0.69234181  0.10147712  0.20775696 -0.68352195
NAO   -0.70903408 -0.04787393 -0.02639852  0.70305183
Temp.  0.11433598  0.72921063  0.64757426  0.18927948

Importance of components:
                         PC1   PC2   PC3   PC4
Standard deviation     1.179 1.027 0.975 0.777
Proportion of Variance 0.347 0.264 0.238 0.151
Cumulative Proportion  0.347 0.611 0.849 1.000
```

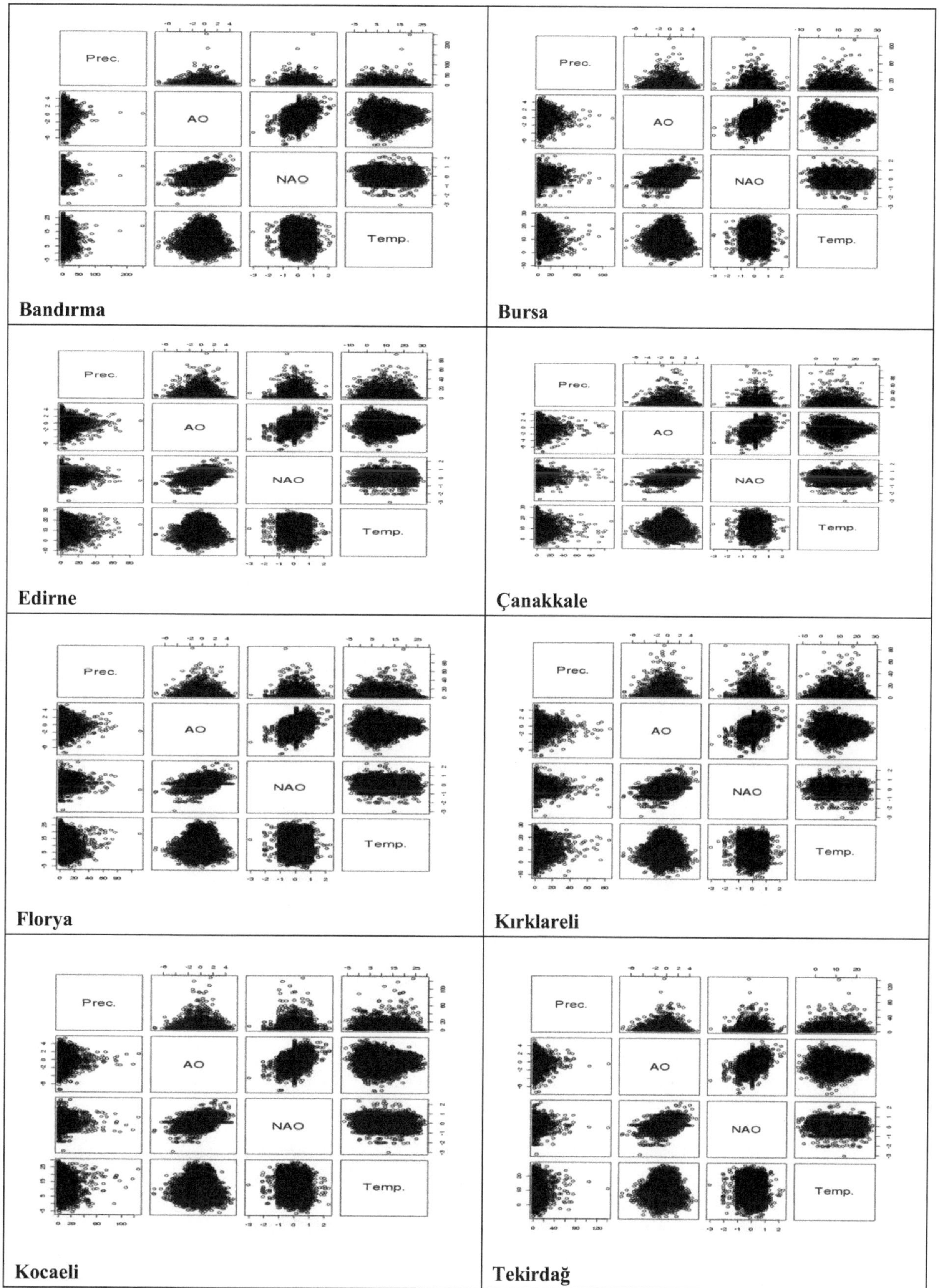

Figure 7. The Linear Regression Graphs for 8 stations. The variables are: Precipitation, AO, NAO, and Temperature

The following calculations have been made in the program R. The following graphics are the linear regression plots of precipitation, temperature, AO and NAO for each of the stations in Figure 7. PCA involves the analysis of correlation coefficients of the linear combinations of these variables and the components in terms of standard deviation, proportion of variances, cumulative proportion of variances. The results for the 8 stations considered in this study are as follows in Table 1.

According to these PCA Table 1 the highest value of the PCA-1 stands for precipitation. The high value of variance and standard deviation for precipitation, which are both important factors for PCA due to its assumptions of linearity and its assumption of large variances being more important dynamically.

3. Evaluation of Case Studies with ERA-Interim Data

3.1 The Selection of the Cases

The third section includes 3 case studies in order to examine the physical relationship between the low-level jets and extreme precipitation events. For the purpose of this study, the cases were selected primarily from the precipitation data by finding the days, those have the maximum precipitation between the years 1970-2010 for each station. The dates of these events were then compared with those ones indicate a pattern due to either the proximity of the stations or the events happening in relatively close time intervals indicate that multiple stations were affected by the same event. Somehow, in some extreme events no such relation could've been found. Such dates are mentioned in the Table 2.

Table 2. Extreme Weather Cases and Locations

Name of Stations	Number of Stations	Year	Month	Day	Amount of Precipitation (mm)
Kirecburnu	17061	1985	10	16	125.5
Sile	17610	1985	10	17	199.3
Florya	17636	1997	10	14	111.0
Tekirdag	17056	1997	10	16	140.1
Geyve	17662	1999	6	26	133.2
Sakarya	17069	1999	6	26	127.7
Bandırma	17114	2009	9	9	248.0
Kumkoy	17059	2009	9	13	179.4

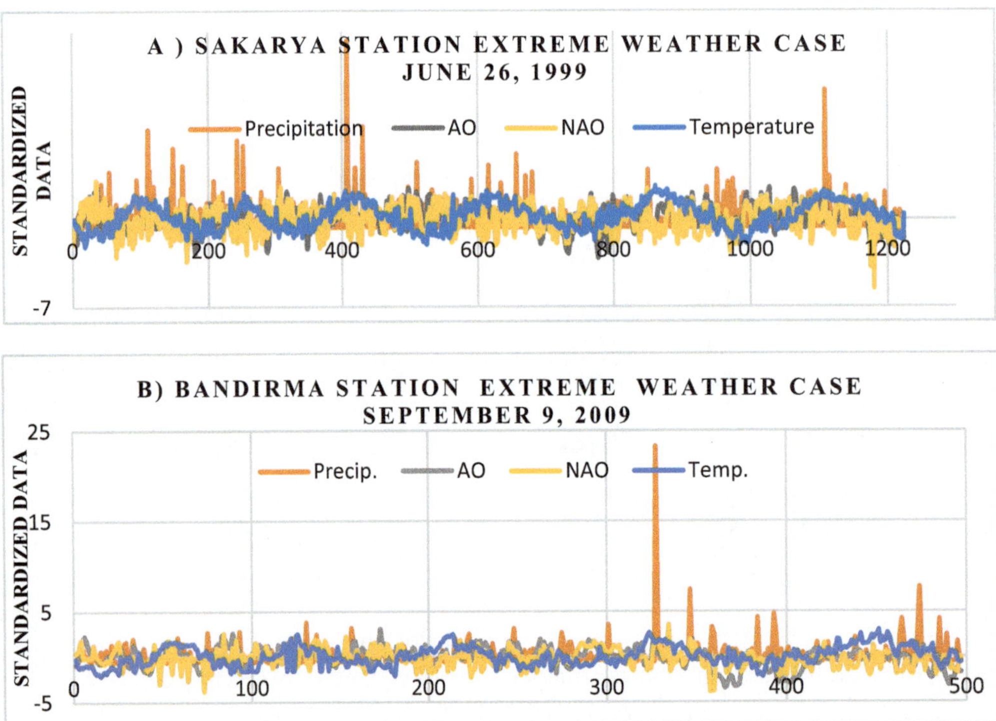

Figure 8. Standardized time series of two cases for precipitation, temperature, AO and NAO

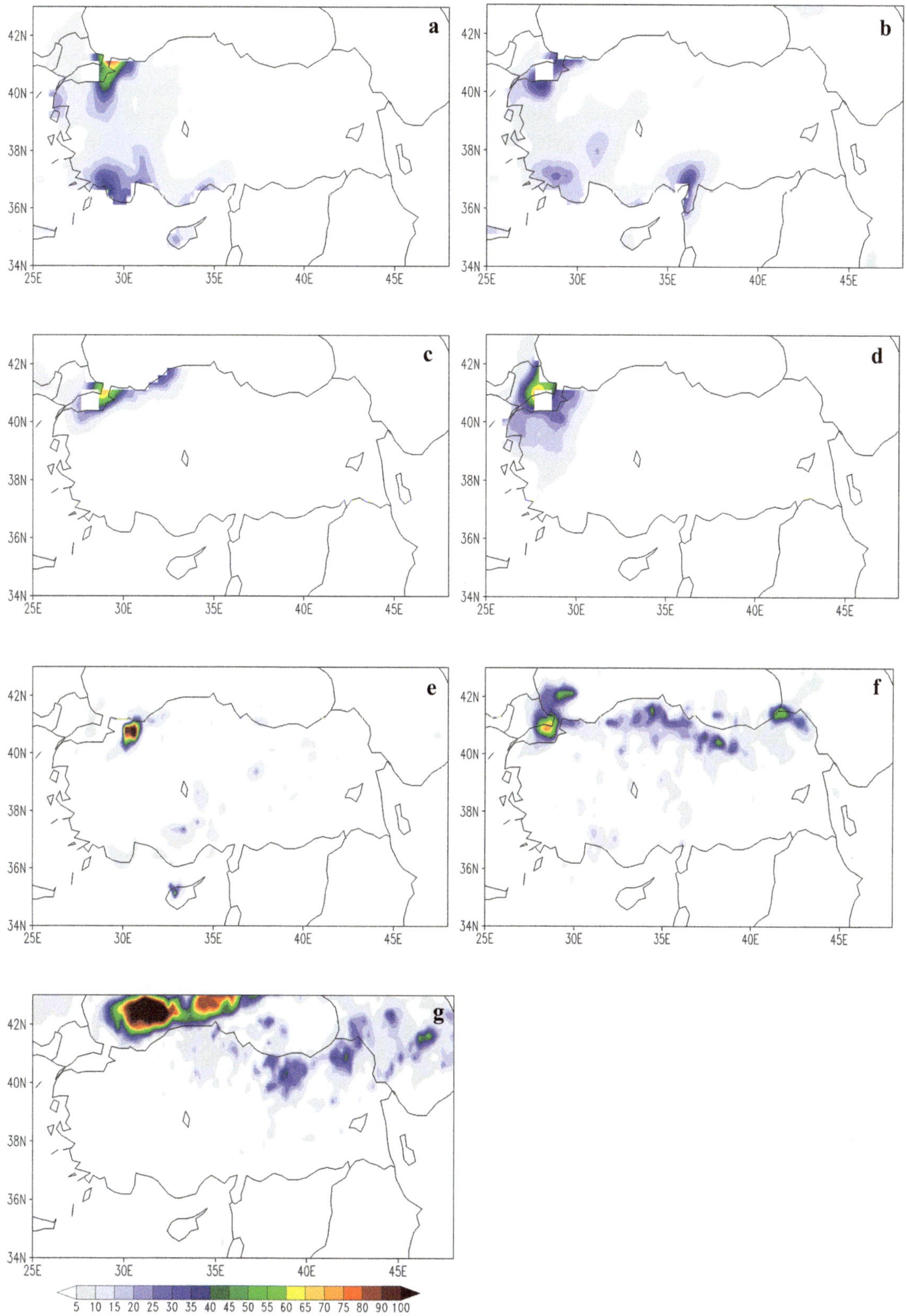

Figure 9. Daily precipitation from APHRODITE and TRMM3B42 in mm for (a) October 16, 1985 (b) October 17, 1985 (c) October 14, 1997 (d) October 16, 1997 (e) June 26, 1999 (f) September 09, 2009 and (g) September 13, 2009

As a result, October 16, 1985; June 26, 1999 and September 11, 2009 were selected for this study, because of the severity of the amount of precipitation. Extreme heavy rain cases of 2009 and 1999 are chosen to relate the interconnection of precipitation with NAO, AO and temperature. Figure 8 displays the standardized time series of precipitation, AO, NAO and temperature for two cases, those were observed in Sakarya and Bandırma regions on June 26, 1999 and September 9, 2009. The precipitation and temperature data sets were standardized to make them dimensionless to relate with AO, NAO, and temperature. Daily precipitation and temperature were subtracted from the climatological values (average), and then divided by standard deviation.

Figure 9 shows the extreme rainfall cases listed in the Table 2 over Turkey and the adjacent areas. Each panel of figure display the spatial pattern of the rainfall distributions from APHRODITE for years 1985 and 1997; TRMM3B42 for years 1999 and 2009. The rainfall spots are matching with the geometrical stations locations, we considered in the Figure 1. The spatial pattern of rainfall shows a flow of rainfall from west to east. Most of the extreme rainfalls are occurring over northern part of the Turkey. Daily rainfall in Figure 9 (a) and (c) show the maximum rainfall of 100 mm. The centers of heavy rainfall on a particular day are matching with the location of the station in the Figure 1.

3.2 Surface Evaluation of Case Studies

Here, an attempt is being made to understand the surface conditions and better evaluation, if these events were somehow related to AO and NAO induced low-level jets a number of surface maps were created using the ERA-Interim datasets. The data considered from the 1st of January of each year (1985, 1999 and 2009) to 31st of December. The gridded data involves 2m Temperature, 2m Dew Point Temperature, MSL, SST, precipitation and evaporation parameters.

3.2.1 Surface Temperature Analysis

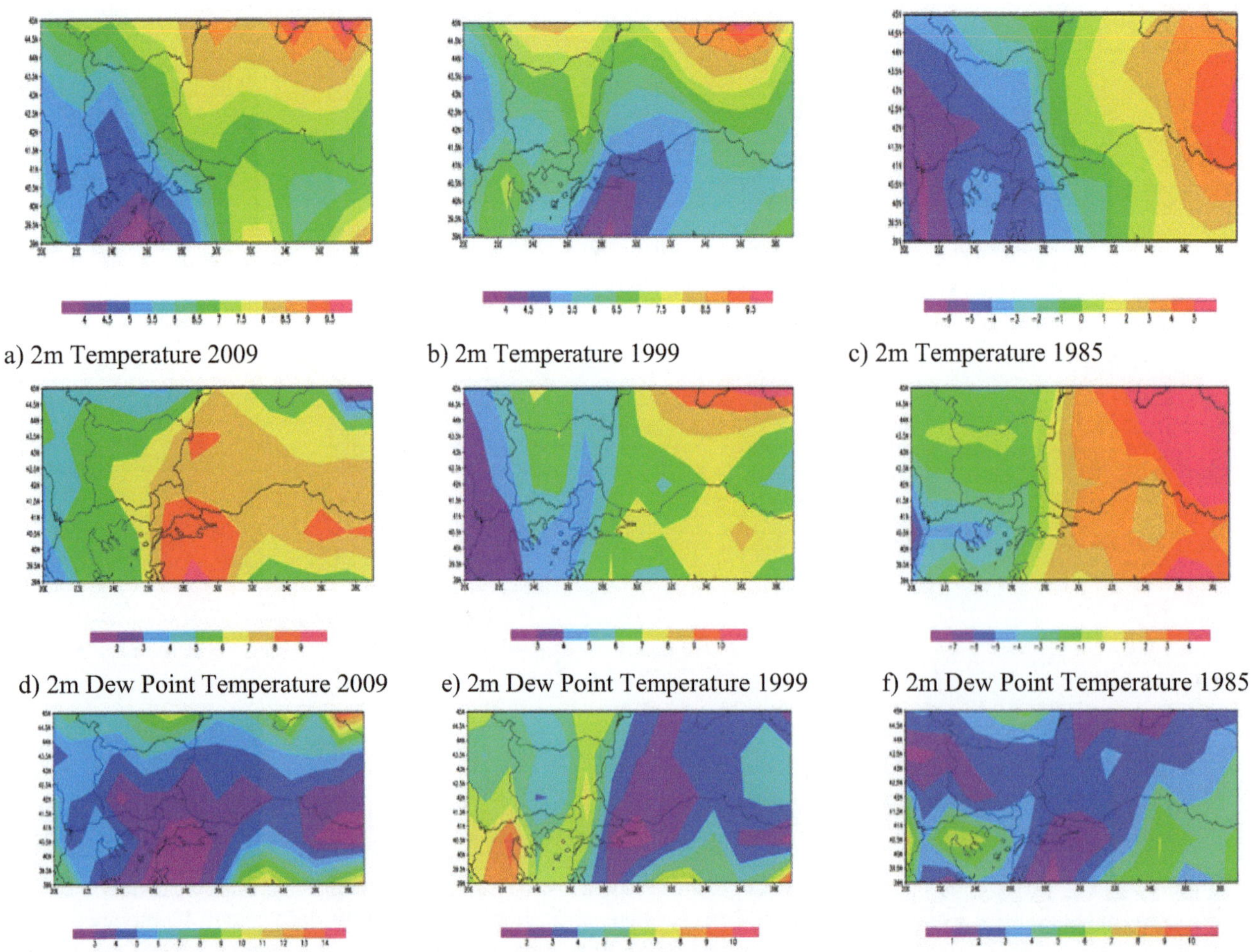

a) 2m Temperature 2009 b) 2m Temperature 1999 c) 2m Temperature 1985

d) 2m Dew Point Temperature 2009 e) 2m Dew Point Temperature 1999 f) 2m Dew Point Temperature 1985

g) 2m Dew Point Depression 2009 h) 2m Dew Point Depression 1999 i) 2m Dew Point Depression 1985

Figure 10. Surface Temperature Maps of Extreme Cases

The three cases show different temperature patterns but each of these can be considered quite low for their relative months in Figure 10. The temperature map of June 26, 1999 (Figure 10b, 2m Temperature) is showing record low temperature of 4 to 5.5°C while the other two cases show temperature of 4 to 7°C (2009) and -5 to 0°C (1985). In each case there is a strong temperature gradient of about 1°C per longitude. The dew point temperature of 2009 case is ranging from 6 to 9°C, 4 to 8 °C (1999) and -4 to 2°C (1985). The dew point depression in each case is around 2°C which is below the average indicator of heavy rain which should be around 4°C.

3.2.2 Evaporation and Precipitation Anomalies

Anomaly maps of evaporation and precipitation are shown in Fig 11. The evaporation charts are exhibiting higher than average evaporation rates for both the cases in June (1999) and September (2009) while the case in October (1985), it is exhibiting a lower than average evaporation rate when compared with the total year. However, the evaporation rates in 1985 are still quite high in the study area when compared with its surrounding area (Figure 11).

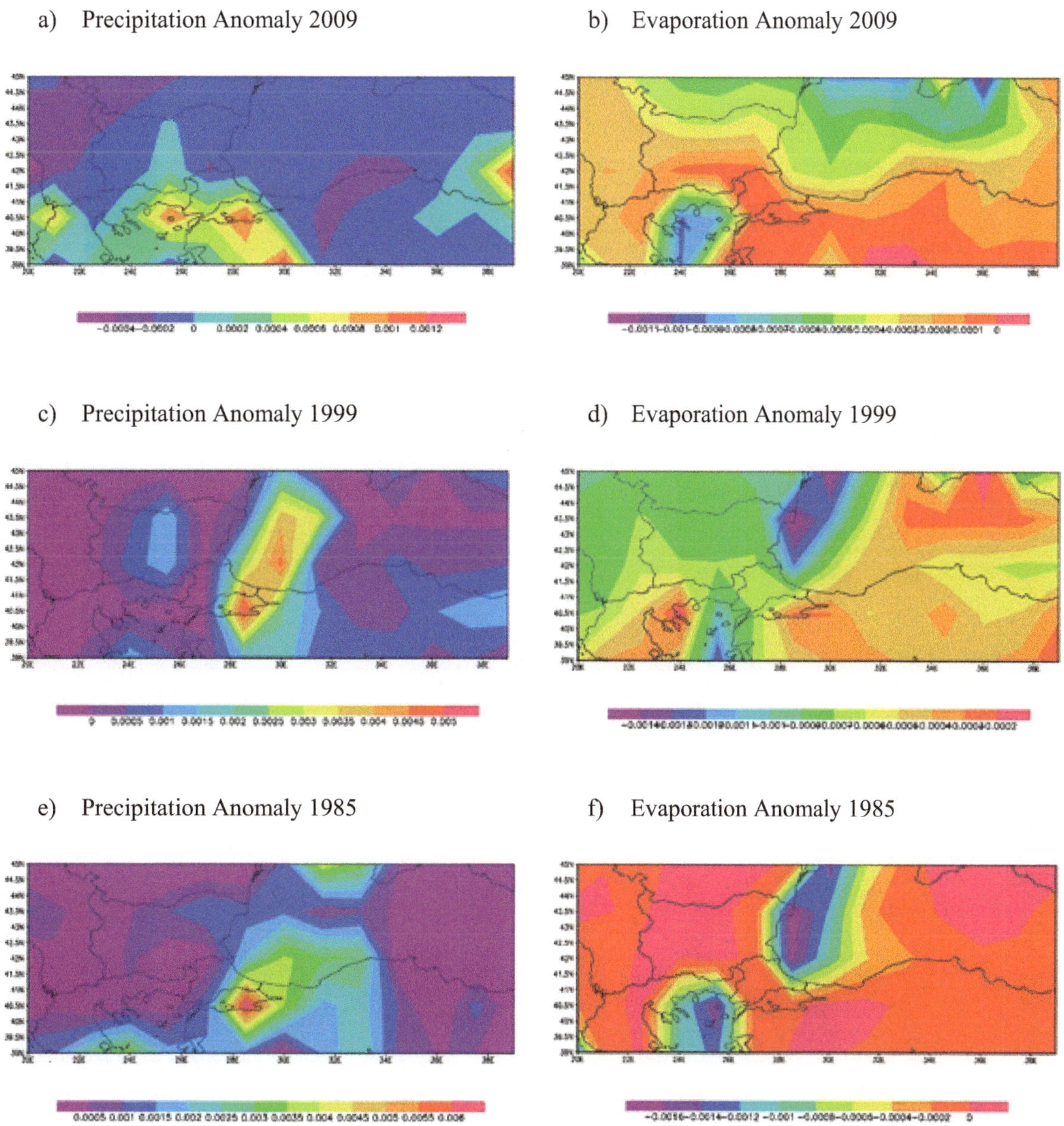

Figure 11. Evaporation and Precipitation Anomalies of the selected extreme cases (2009, 1999, and 1985)

The values shown in the legends of these maps are in terms of dam so a factor of 10000 must be multiplied to get the usual term (mm) of precipitation and evaporation. In Figure 11a, precipitation anomaly chart of 2009 is indicating a relatively low anomaly with a range of 4 to 12 mm whereas both the cases of 1999 and 1985 are indicating a huge precipitation anomaly over the area with ranges of 5 to 50 mm and 5 to 60 mm respectively. However, every one of the charts are showing a condense area of precipitation surrounded by a larger area of low precipitation and no precipitation implying that this might have been a case of convective movement rather than a frontal precipitation.

3.2.3 Mean Sea Level Pressure Charts

a) Mean Sea Level Pressure 2009

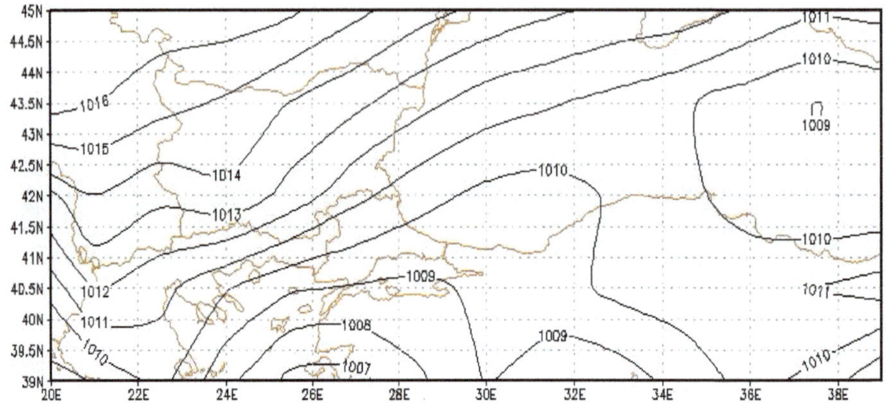

b) Mean Sea Level Pressure 1999

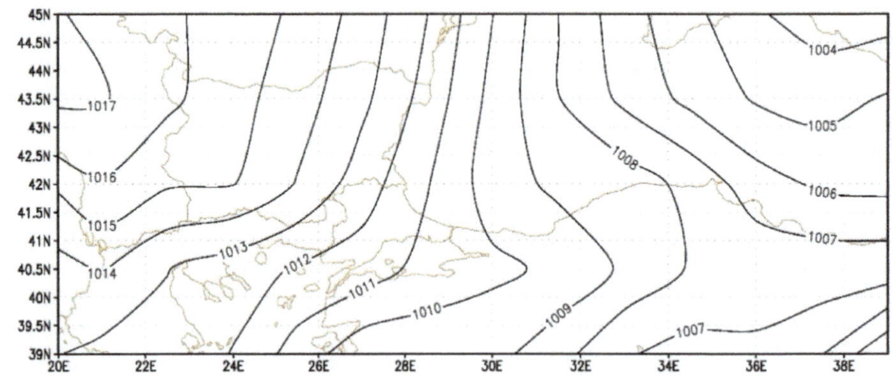

c) Mean Sea Level Pressure 1985

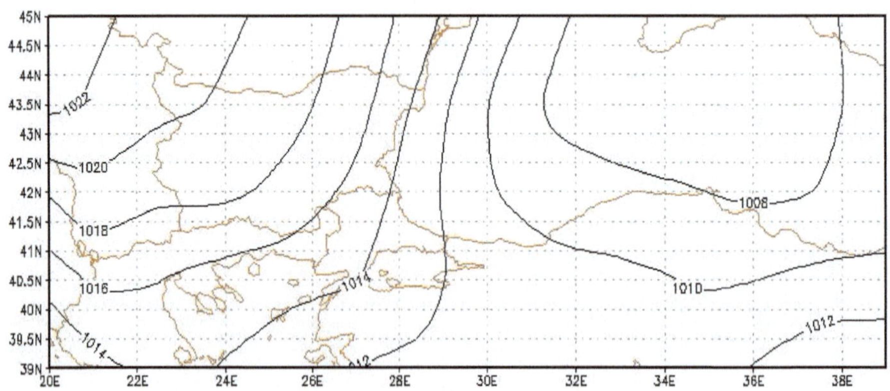

Figure 12. MSL of the 3 cases

The MSL charts are showing standard values for mean sea level pressure over the study area although there seems to be strong gradients over the area especially in the 1999 and 2009 cases. This implies that there are strong jet currents in the upper troposphere in Figure 12.

3.3 Wind Speed Anomalies at Pressure Levels

In the cases of both 1985 and 2009 strong wind speed gradients are present over the study area which indicates a rapid change in the wind velocity vectors (Figure 13).

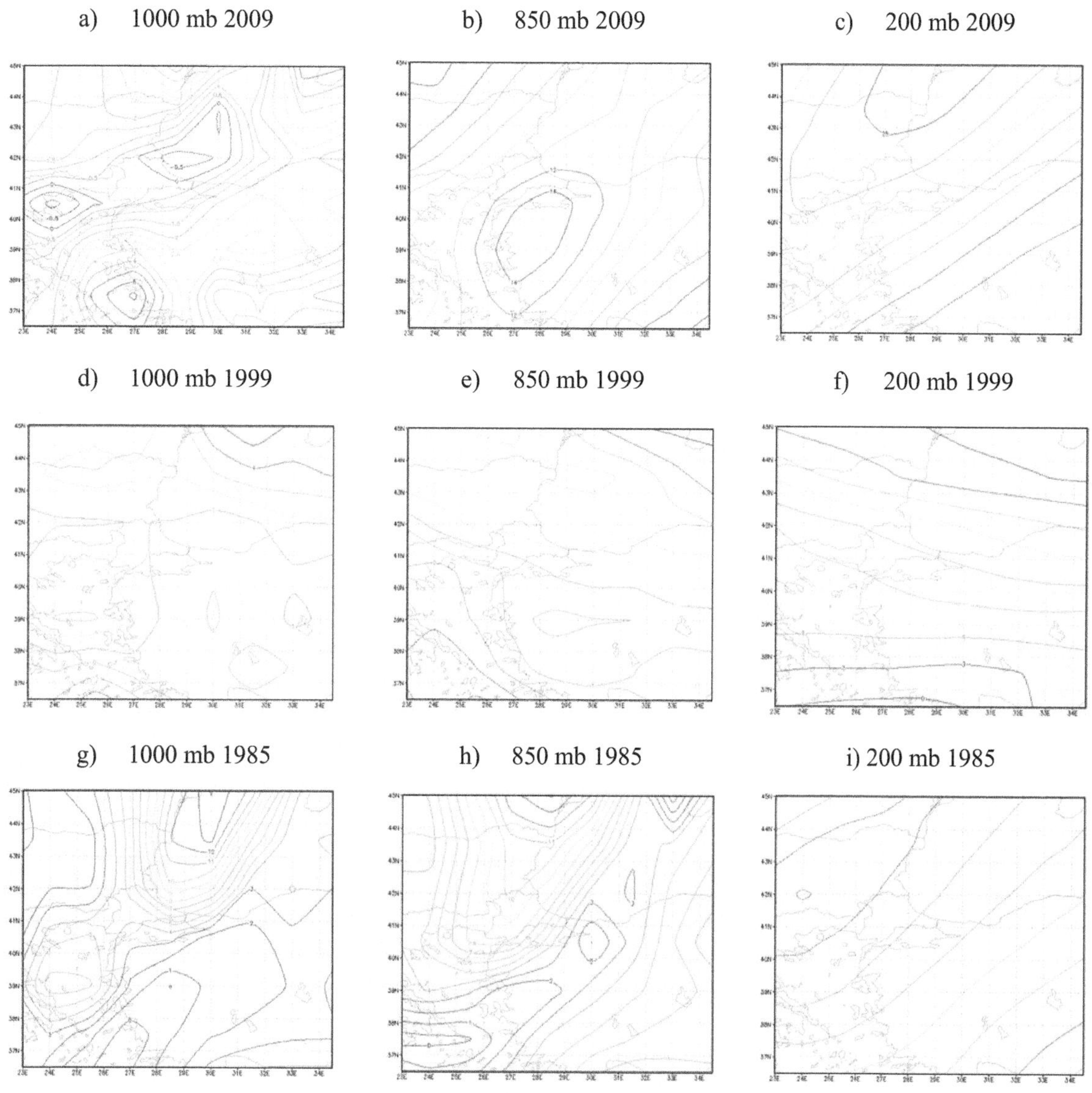

Figure 13. Wind speed anomaly of the cases at 1000, 850 and 200 mb

The general pattern observed in the 1000 mb wind speed anomalies continues to show itself in the 850 mb wind speed anomaly maps as well. A high gradient is persistent at this pressure level as well in the cases of 1985 and 2009 with 1985's gradient being stronger. The 1999 case is showing rapid change in wind speed, which is expected at this pressure level (Figure 13).

The 200 mb wind speed anomalies shows that the highest wind speed gradient was found in the case of 2009 in Figure 13c. Since the winds blowing at this level are already at high speed conditions, but in the 1999 and 1985 cases it is not necessarily indication of a high-level jet pattern with a change rate of 3 and 5 m/h respectively. Although the change rate between the contours in cases of 1985 and 2009 are the same, the density and the high values of the 2009 case indicates a 200mb jet stream (Figure 13).

4. Conclusion and Discussions

Extreme weather events are rare elements of a climate. They are generally not expected, except as an anomaly in the probability density function. They depend on many factors including local effects from meso- to synoptic-scale climates and remote effects from the tele-connection patterns (ENSO, AO, NAO) and air mass movements. In this study the precipitation regime of the local climate has been examined statistically for a 40-year period. The study includes observational data taken from 23 meteorological observation stations in the Marmara region.

The linear regression analysis of the precipitation data from the stations and AO and NAO data do not show any significant sign of linear relationship between these three variables. PCA is been done using 8 meteorological stations (Bandırma, Bursa, Çanakkale, Edirne, Florya, Kırklareli, Kocaeli, Tekirdağ) between the precipitation, temperature and AO, NAO data. The results of the PCA don't show a well-established linear relationship between these variables but the analysis shows a significant correlation coefficient of temperature-precipitation and AO-NAO pairs.

With the help of these analyses 3 cases of extreme weather events have been identified and cross-checked. Since a linear correlation between the precipitation regime and the AO and NAO found to be almost zero; it was a curiosity if low-level jets have any relationship with extreme precipitation events. The examination of these parameters have shown higher than normal wind speeds at 1000 mb and 850 mb for the cases of 2009 and 1985 and extremely higher than normal wind speed at 200 mb for the case of 2009.

Sirdas et al. (2013) have revealed very significant results in their research related with global warming effects on extreme weather events. The study advised, which is quoted here; "The average daily sunning time in July and August is 11-12 hours where evaporation is also visibly higher. In winter times, however, this declines to 3-4 hours and for this reason, it ends without causing evaporation of the water. In general, it has been found out that regions such as the Mediterranean region are more sensitive and vulnerable to climate change. As a country in the aforementioned region, Turkey needs to be prepared for inevitable severe weather and extreme climate conditions".

To sum up, linear regression analysis and/or principle component analysis of the parameters used in this study are too simple to express the linear relationship between teleconnection patterns, low-level jets and extreme precipitation. The examination of the cases shows that there is a physical relationship between low-level jets and extreme precipitation events; the relationship between teleconnection patterns and low-level jets has been established. The possibility exists that if there is a relationship between these three concepts it is more complex than a linear relationship and/or involves more physical parameters.

Acknowledgement

We would like to thank NCEP, ECMWF and Turkish State Meteorological Services (MGM) for letting us use their data.

References

Amiri, A. M., Amerian, Y., & Mesgari, M. S. (2016). Spatial and temporal monthly precipitation forecasting using wavelet transform and neural networks, qara-qum catchment, iran. *Arabian J. Geosci, 9*(5), 1-18.

Amiri, A., M., & Mesgari, M. S. (2016). Spatial variability analysis of precipitation in northwest iran. *Arabian J. Geosci, 9*(11), 1-10.

Carvalho, L., Jones, C., & Liebmann, B. (2002). Extreme precipitation events in Southeastern South America and large-scale convective patterns in the South Atlantic Convergence Zone. *J. Climate, 15*, 2377–2394.

Cassou, C. (2008). Intraseasonal interaction between the Madden-Julian Oscillation and the North Atlantic Oscillation. *Nature, 455*, 523-527.

Cavalcanti, I. F. (2012). Large scale and synoptic features associated with extreme precipitation over South America: A review and case studies for the first decade of the 21st century. *Atmospheric Research*, 27-40.

Diaz, H. F., & Murnane, R. J. (2008). *Climate Extremes and Society.* Cambridge University Press.

Khandekar, M. L. (2013). *The Global Warming - Extreme Weather Link: A Review of the State of Science.* The Global Warming Policy Foundation.

Durkee, J., Mote, T., & Shepherd, M. (2009). The contribution of mesoscale convective complexes to rainfall across subtropical South America. *J. Climate 22*, 4590–4605.

Grimm, A., & Tedeschi, R. (2009). ENSO and extreme rainfall events in South America. *J. Climate 22*, 1589–1609.

Hansen, J., Jones, J., Kiker, C., & Hodges, A. (1999). El Niño-Southern Oscillation Impacts on Winter Vegetable Production in Florida. *J. Climate*, 92-102.

Haylock, M., & Goodess, C. (2004). Interannual Variability of European Extreme Winter Rainfall and Links With Mean Large-Scale Circulation. *Int. J. Climatol. 24*, 759–776.

Khandekar, M., & Neralla, V. (1984). On the relationship between the sea surface temperatures in equatorial Pacific and the Indian monsoon rainfall. *Geophysical Research Letters, 11*, 1137-1140.

Kumar V, & Krishnamurti T.N. (2012). Improved seasonal precipitation forecasts for the Asian Monsoon using a large suite of Atmosphere Ocean coupled models: Climatology. *J. Clim.*, 25, 39-64.

Kumar V, & Krishnamurti T.N. (2015). Mesoscale modeling for the rapid movement of monsoonal isochrones. *Atmospheric Science Letters, 17*, 78-86.

Lazarus, D. S. (2009). Florida's Climate: Past, Present and Future. *AIP Conference Proceedings* (p. 32). AIP Publishing.

Martin, E. R., & Schumacher, C. (2011). The Caribbean Low-Level Jet and Its Relationship with Precipitation in IPCC AR4 Models.

Schmidt, E. L. (2001). ENSO Influences on Seasonal Rainfall and River Discharge in Florida. *J. Climate*, 615-628.

Sensoy, S. (2004). The Mountains Influence on Turkey Climate. *Balwois Conference on Water Observation and Information System for Decision Support.* Ohrid, FY Republic of Macedonia.

Shlens, J. (2003, March 25). A Tutorial on Principle Component Analysis Derivation, Discussion and Singular Value Decomposition.

Sirdas, S, (2003). Meteorological Drought Modeling and Application to Turkey. (Doctoral dissertation), Istanbul Technical University, Meteorological Engineering, Istanbul, The Republic of Turkey.

Sirdas, S. and Sen, Z. (2003). Spatio-temporal drought analysis to the Trakya region, Turkey, Hydrological Science Journal, 28, 809-21.

Sirdas, S., Ross, R.S., Krishnamurti, T.N. and Chakraborty, A. (2007). Evaluation of the FSU Synthetic Superensemble Performance for Seasonal Forecasts over the Euro-Mediterranean Region. *TellusA, 59*(1), 50-70(21).

Sirdas, S., Sen, Z., & Öztopal, A. (2013). Climate Change Expectations in the Next Half Century of Turkey. *Causes, Impacts and Solutions to Global Warming,* 103-127.

Strangeways, I. (2007). *Precipitation: Theory, Measurement and Distribution.* University Press, Cambridge.

Sumner, G. N. (1988). Precipitation: Process and Analysis. John Wiley & Sons Ltd.

Vasconcellos, F., & Cavalcanti, I. (2010). *Serra do Mar. Rev. Bras. Meteorol. 25*, 287-294.

Vasconcellos, F., & Cavalcanti, I. (2010). Extreme precipitation over Southeastern Brazil in the austral summer and relations with the Southern Hemisphere annular mode. *Atmos. Sci. Lett., 11*, 21-26.

Xin, Y., & Xiao, G. S. (2009). *Linear Regression Analysis Theory and Computing.* Singapore: World Scientific Publishing Co. Pte. Ltd.

Türkes, M. 1996. Spatial and temporal analysis of annual rainfall variations in Turkey. *Int. J. Climatol, 16*, 1057–1076.

Yarav, R., Ramu, D., & Dimri, A. (2013). On the relationship between ENSO patterns and winter precipitation over North and Central India. *Global and Planetary Change*, 50-58.

11

The Sustainable Livelihood Challenge of REDD+ Implementation in the Philippines

Rose Jane Peras[1,2], Juan Pulhin[1], Makoto Inoue[2], Abrar Jurar Mohammed[2], Kazuhiro Harada[3] & Masatoshi Sasaoka[4]

[1] Department of Social Forestry and Forest Governance, College of Forestry and Natural Resources, University of the Philippines Los Banos, Philippines

[2] Global Forest Environmental Studies, Graduate Schol of Agricultural and Life Sciences, The University of Tokyo, Japan

[3] Graduate School of Bioagricultural Sciences, Division of Regional Resources Management, Lab. of Forest Resources Utilization, Nagoya University, Japan

[4] Graduate School of Letters, Hokkaido University, Japan

Correspondence: Rose Jane Peras, Department of Social Forestry and Forest Governance, College of Forestry and Natural Resources, University of the Philippines Los Banos, Philippines. E-mail: rjperas@up.edu.ph

Abstract

The forestry sector in the developing world has been continuously challenged by the unsustainability of forest resources and the threat of climate change. Reducing Emissions from Forest Degradation and Deforestation (REDD+) was launched to address the problem, and the Philippines accepted the challenge by undergoing the 10-year phased process. Using the sustainable livelihoods framework, this paper examines the challenges of REDD+ implementation in the Philippines using the case of Southern Leyte REDD+ pilot area and highlights the co-benefits and trade-offs of pilot project activities on the five (5) capital assets. Our findings suggest greater impacts of CBFM on the key indicators of change than REDD+. There is very high association of the natural and financial capital assets with REDD+ pilot project activities, yet financial benefit is short-lived. Local people highly regarded the contribution of assisted natural regeneration and reforestation activities in sequestering carbon, while agroforestry is perceived to sustain agricultural production in the future. The major drawback of REDD+ pilot project activities is that it perpetuates the failures of CBFM initiatives giving little attention to sustainable livelihood objectives. Forest conservation policy like REDD+ as a mechanism for addressing climate change can still be adopted by local communities if livelihood capital assets will be further enhanced.

Keywords: CBFM, co-benefit, perpetuate, project mentality, REDD+, sustainable livelihoods, trade-off

1. Introduction

Pursuing the goal of sustainable development poses great challenge to most developing countries, especially in the context of climate change. Tropical deforestation and forest degradation in combination with rural poverty have likewise added to the continuing threat to forest sustainability and vulnerability to climate change impacts. Solutions to combat the impacts of climate change were made through incentivizing developing countries in managing and protecting their forests through REDD+ (otherwise known as reducing emission from deforestation and forest degradation) primarily to contribute to carbon emission reduction in the forestry sector.

REDD+ is also envisaged to provide multiple benefits to local communities (Gunilla, Olsson, & Quattara, 2013). These benefits are also referred to as "co-benefits" or other benefits derived out of the REDD+ project in addition to reduced emission. Among the REDD+ co-benefits are forest (biodiversity) conservation and the provision of more sustained income source for local communities once carbon trading progresses (Angelsen, 2009). Another benefit is the improvement of forest governance system, especially in developing countries known to have an overall poor governance index (Wertz-Kanounnikoff & Metta Kongphan-apirak, 2009). These potential positive (co-benefits) impacts have been enumerated by the supporters of REDD+ while critics highlight the potential negative impacts (trade-offs) to the marginalized local communities, including the indigenous peoples. Potential co-benefits include increases in employment through entrepreneurship development, diversification of income sources and access to markets, food crop production, and provision of ecosystem services as a result of ecosystem conservation and sustainable management. The adverse side effects, on the other hand, include employment decline as less labor intensive technologies in agriculture intensify, increased income concentration, declining local food production due to large-scale monocultures of non-food crops, and rise of land use competition (IPCC, 2014).

Many scholars believe that REDD+ will have considerable implications on the livelihoods of local communities such as deprivation in resource requirement, displacement of traditional goat herders, risking existence of blacksmiths with tightened wood supply (Poudel, Thwaites, Race, & Dahal, 2015); benefit sharing framework of uneven distribution (Howson & Kindon, 2015); and reducing contribution to poverty alleviation by marginalizing and criminalizing the artisanal and small-scale mining sector (Hirons, 2011). On the contrary, parallel studies involving participatory forest management (PFM) such as community forestry (CF) and community-based forest management (CBFM) saw the direct link between resource conservation incentives and/or benefits to household's livelihoods (Karki, 2013; Nath & Inoue, 2010; Mohammed et al., 2016). Some went to suggest the small and medium forest enterprises (SMFEs) to further promote sustainable use and conservation of forests (Tomaselli & Hajjar, 2011). However, few studies link the different forest conservation activities such as REDD+ pilot demonstration project activities to sustainable livelihoods. In 2014, IPCC assessed the different potential co-benefits and adverse side effects of the different mitigation measures in the AFOLU sector on the economic, social, environmental and institutional objectives. Moreover, Le, Smith and Herbohn (2013) noted sustainable livelihoods as one of the key drivers to a successful reforestation project in the Philippines.

Countries participating in REDD+ are expected to shift from high-carbon emission development pathways to forest carbon stocks enhancement actions for multiple benefits, including biodiversity, livelihood enhancement, water, climate change adaptation, and development (Minang & Noordwijk, 2014). The Philippines accepted the challenge of REDD+ implementation. Even with only 0.32% share in the 2012 global greenhouse gas (GHG) emissions, the country has committed to 70% GHG emissions reduction by 2030 as reflected in its recent submission of Intended Nationally Determined Contributions (INDC) to the UNFCCC (Philippines INDC, 2015).

In 2010, even without support access to multilateral REDD+ mechanism, the Philippine National REDD Plus Strategy (PNRPS) was developed and affirmed. (ADB, 2010). At the heart of PNRPS is the empowerment of forest managers especially indigenous peoples (IPs) in managing forestlands and ancestral domains sustainably and equitably with enhanced carbon stock and reduced green house gas (GHG) emission. REDD+ implementation in the Philippines is a three-phased approach spread in a span of 10 years (2010-2020). This involves readiness, scaling-up and engagement phases. Currently, the country is still at the readiness stage which runs for 3-5 years. This comprises the majority of the strategies and activities identified in the PNRPS. Readiness phase capacitates the country before actual implementation or engagement of REDD+ schemes at the national level. Among the major activities are a) development of the PNRPS; b) capacity-building, consultation, communication, integration and reform; c) national level bodies and carbon accounting establishment; d) pilot/demonstration sites, provinces and regions establishment; e) implementation of the National REDD Plus Strategy (NRPS) Readiness Strategies; f) Scale up from site-level to provincial and regional levels; and g) establishment of new sites.

REDD+ pilot/demonstration establishment is one of the major activities of the readiness phase. At present, the three pilot demonstration sites established in the provinces of Quezon, Palawan and Southern Leyte have been completed. The Deutsche Gesellschaft für Internationale Zusammenarbeit (GIZ) GmbH, the major funding agency for the 3-year (2009-2013) Southern Leyte REDD+ pilot project, is moving towards other REDD+ pilot areas in the provinces of Davao Oriental, Eastern Samar and Albay. Hence, with the completion of the REDD+ pilot demonstration project in Southern Leyte, it is appropriate to determine the major contribution of the Young Innovators for Social and Environmental Development Association(YISEDA), the CBFM-People's Organization (PO), in the achievement of sustainable livelihoods attributed to REDD+ pilot project activities. Specific project activities implemented at the pilot site differ depending on the funding support. The degree of participation of PO members to REDD+ pilot activities is vital to sustain the effort as well as achieve the goal of forest carbon stock enhancement and reducing deforestation and forest degradation.

Sustainable livelihood principle is central in REDD+ and a major objective in the implementation of CBFM in the country. It is promoted by Amatrya Sen and used by many scholars to assess the impacts of forestry development initiatives such as CBFM to the livelihoods of the poor (DFID, 1992). Many REDD+ scholars acknowledged the importance of sustainable livelihoods in assessing the impacts of REDD+ (Jagger, Atiadia, Pattanayak, Sills, & Sunderlain, 2009; Angelsen et al., 2009; Caplow, Jagger, Lawlor, & Sills, 201; Karki, 2013). As development pushes further to cover remote rural areas in the countryside through the promotion of sustainable development and inclusive growth, the forestry sector has long been putting forward all efforts towards sustainable forest management. Forestry initiatives' contribution to the general well-being of the communities is very much sought so that good experiences and lessons can be replicated while challenges and problems can be addressed.

REDD+ intends to learn from the implementation of past participatory forest management (PFM)/CBFM projects (Agrawal & Angelsen, 2009; Peras et al., 2015). In the Philippines, the three (3) decade-long of CBFM implementation is considered ineffective resulting to limited impacts due to unstable policy, complex procedures and requirements, project mentality view of CBFM (rather than as an approach that replaced the traditional commercial large-scale forestry), and weak institutional support system (Pulhin, Inoue, & Enters, 2007). The same pattern leading to the failures of CBFM in most REDD+ sites emerges. Hence, we argue that REDD+ implementation has the tendency to repeat the mistakes of forestry-related development initiatives, such as CBFM, in its pilot demonstration areas. Furthermore, REDD+ may reproduce these failures.

This paper analyzes the sustainable livelihood challenge of REDD+ implementation in the pilot demonstration area based on the perception of CBFM members by addressing the questions: How do CBFM members associate REDD+ pilot project activities to sustainable livelihoods? Can the current REDD+ pilot project activities improve capital assets of the poor? Will the failures of CBFM be reproduced in a REDD+ area?

2. Research Method

2.1 The Study Site

The study selected the REDD+ Readiness Pilot Demonstration Project implemented by the CBFM PO-YISEDA in Bgy. Lunas, Maasin City, Southern Leyte (Figure 1), with a funding support of PhP 2.64 million from GIZ. This project is part of the GIZ Southern Leyte "Climate-Relevant Modernization of Forest Policy and Piloting of REDD in the Philippines" covering 31 848 ha. Focused on forest policy reforms, the REDD Plus pilot activities served as source of information for REDD Plus implementation in the country. YISEDA is one of the CBFM-POs under this REDD+ readiness project from October 2009-March 2013 covering 150 ha intended for reforesting denuded areas and rehabilitating degraded forests, as well as promote conservation, sustainable management of forests and enhancement of carbon stocks. YISEDA was awarded the 2012 Best Performing PO in Region 8 during the 2012 First Community-Based Forest Management People Organizations (CBFM-PO) Congress. It was assessed to have a profound capability to implement forestry development projects based on its timely, effective and efficient accomplishment of previous projects, with dynamic leader and PO members highly committed to greening the environment.

2.2 Data Collection and Analysis

Data was gathered using a combination of qualitative and quantitative approach which includes a household survey, focus group discussions (FGDs) and key informant interview (KII). These were complemented by actual observations in the conduct of boundary/peripheral reconnaissance survey of the CBFM area.

A total of 51.85% of the 108 YISEDA members was randomly interviewed using a semi-structured interview schedule. For purposes of this study, the respondents were selected based on their availability during the duration of the survey and their active membership status (in terms of regular attendance to meetings and activities of the organization). The household survey provides information on the current household livelihood capital assets which are associated with the present socio-demographic and economic characteristics of the respondent's household. Respondent's knowledge and participation to the different REDD+ pilot activities were also gathered. The level of participation to these activities was assessed using a 5- interval Likert scale: where 1 is very low, 2- low, 3- moderate, 4- high, 5- very high. Lastly, respondent's perception on the associated impacts of REDD+ readiness pilot activities, either positive (co-benefits) or negative (trade-offs), on the livelihood capital assets were determined. Descriptive statistics was used to analyze the household livelihood capital asset characteristics and used to support the information obtained from the FGD.

Impacts of CBFM and REDD+ were also gathered using the household survey. The impacts between the three (3) periods -- before CBFM (1999), during CBFM (2000-2010) and with REDD+ (2009-2013) -- were assessed by the respondents using the ladder diagram. Pomeroy, Pollnac, Katon & Predo (1997) describes the ladder diagram as a visual, self-anchoring technique in making finer ordinal judgements. Each period is compared with another period using a two-tailed t-test to determine the degree of variation of scores given by the respondents. Just like a ladder with 10 steps, each indicator per period is scored based on the existing condition of the period, taking careful consideration of the development initiative at each period, i.e. with CBFM and REDD+. Highest score of 10 is given if there was perceived great improvement while lower score means little or no improvement during the period.

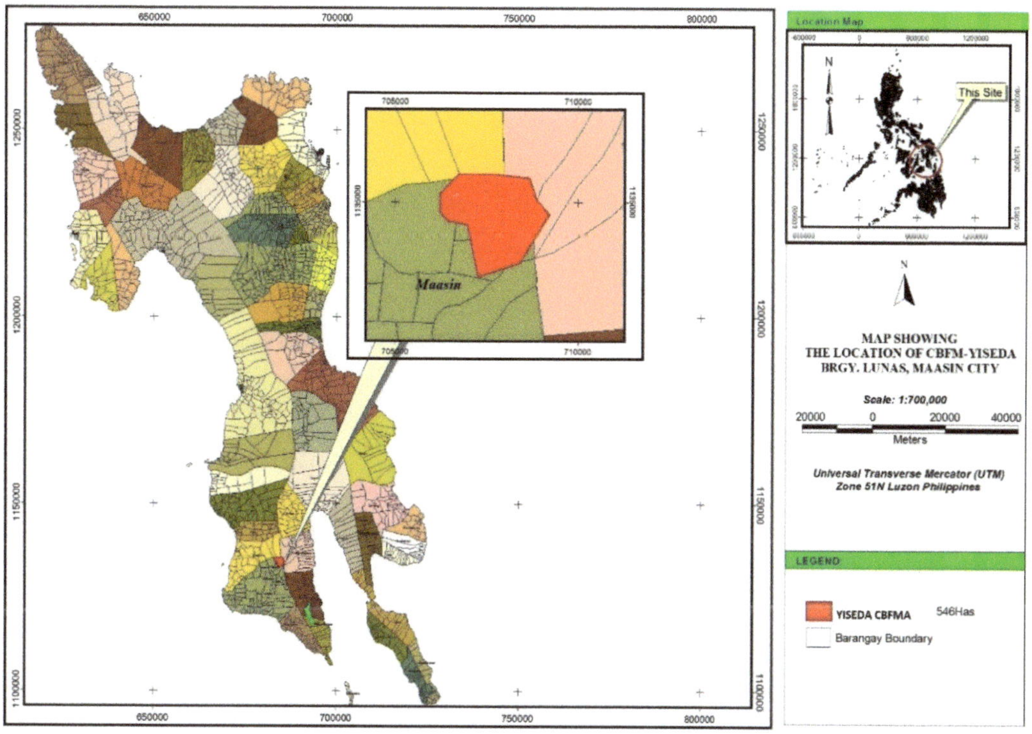

Figure 1. Location map of the REDD+ Pilot Demonstration Activities managed by YISEDA in So. Canlugoc, Bgy. Lunas, Maasin City, Southern Leyte, Philippines

Focus group discussions (FGD) were conducted to have a collective understanding on the impacts of the REDD+ pilot demonstration project on sustainable livelihoods objective as well as on the issues and challenges for future full blown REDD+ implementation. The associated impacts of the REDD+ pilot activities on sustainable livelihoods were derived from the two FGDs conducted: YISEDA (participated by 12 members) and local institutions/organizations working with CBFM and REDD+ (with 10 representatives from the local government units, Department of Environment and Natural Resources (DENR), and GIZ field staff).

Also, a total of nine (9) key informant interviews (KII) were conducted, comprising key officials of YISEDA, GIZ field personnel, DENR and LGU (province, municipal and barangay) field staff. This was undertaken to solicit the perspectives of key individuals knowledgeable on the impacts of CBFM and REDD+ implementation to sustainable livelihoods. This method provided key insights leading to a comprehensive analysis of the future challenges of REDD+ implementation.

Finally, a peripheral reconnaissance survey of the CBFM area was made to supplement the above-mentioned data gathering activities. Actual observations focused on the condition of the CBFM area and in verifying the most recent activities administered by individual members and YISEDA, the current challenges, threats (illegal forestry-related activities), and opportunities for sustainable livelihood generation.

3. Results

3.1 The Pilot Site-Level Context of REDD+ (Readiness and its Pilot Demonstration Project Activities)

The Southern Leyte REDD+ pilot/demonstration site is funded by the GIZ together with the Department of Environment and Natural Resources (DENR) – Community Environment and Natural Resources Office (CENRO) of Maasin City as the lead executing agency. The project intends to reduce GHG emissions by improving climate-related forest policy and conserve biodiversity. More specifically, the project: 1) modernized forest policies and develop specific incentives for reducing emissions from deforestation and forest degradation (REDD); 2) forged conservation/co-management agreements among stakeholders; 3) implemented REDD pilot activities as innovative measures to avoid deforestation and rehabilitate degraded forests in and around selected protected areas in the Philippines; and, 4) enhanced the capacities of DENR, local government units and local population for planning and implementation of climate relevant forest measures, conflict mitigation, securing land use rights and improving local livelihoods.

At the pilot sites, GIZ forged a financing agreement with the CBFM-POs in the implementation of reforestation, agroforestry and conservation activities. The REDD+ pilot site implementing CBFM-POs were the Young Innovators for Social and Environmental Development Association (YISEDA), Maasin, Southern Leyte; Anahaw Multi-Purpose Cooperative (AMPCO), Bontoc, Southern Leyte; Kahupian Upland Farmers' Association (KUFA), Sogod, Southern Leyte; Tomas Oppus Forest Developers Farmers' Association Inc. (TOFDA), Tomas Oppus, Southern Leyte; and Tomas Oppus-Malitbog Farmers' Association Inc. (TOMFA), Tomas Oppus, Southern Leyte (Quitoriano, 2013).

This study only highlights the experience of YISEDA-CBFM in its REDD+ pilot project implementation. Among the activities financed by GIZ are reforestation, agroforestry and assisted natural regeneration (ANR). Reforestation strategy or multi-purpose tree plantation aimed at the establishment of 77 hectares production forest for lumber and fuelwood production as well as for the protection of forest areas needing rehabilitation. This strategy targeted open and denuded areas. The farms of 63 members who availed of the project were planted with mahogany (*Swietenia macrophylla*), narra (*Pterocarpus indicus*), molave (*Vitex parviflora*) and indigenous species. Agroforestry was promoted to provide livelihood and income to YISEDA members and increase forest cover. About 25 hectares of agroforestry production had been established within the farms of 50 members. Assorted grafted fruit trees like jackfruit (*Artocarpus heterophyllus*), lanzones (*Lansium domesticum*), rambutan (*Nephelium lappaceum*), durian (*Durio zibethinus*), banana suckers, or coconut or native coffee were planted while pineapple was encouraged (optional) for areas not suitable for vegetables/corn (*Zea mays*), peanuts (*Arachis hypogaea*), ginger (*Zingiber officianale*), sweet potato (*Ipomoea batatas*), gabi (*Colocasia esculenta*) and other root crops. Areas targeted are those within existing agricultural production areas, with soil and water conservation measures, or in newly established plantation. ANR was established under patches of remnant secondary growth forest and emerging pioneer vegetation in previously forested areas. It is located within the farms of 63 household beneficiaries that participated in the project. While project management is handled by YISEDA, stakeholders were enjoined to undertake collaborative maintenance. Under protection forest category, the ANR species include mixed dipterocarps and fast growing species that can assist naturally growing trees to rehabilitate and expand their vegetative cover (YISEDA REDD+ Project Document, undated).

3.2 Impacts of CBFM and REDD+

The impacts of the three (3) decades of CBFM implementation and the recent completion of the REDD+ pilot project call for an assessment in order to determine whether the challenges and failures of CBFM will persist in REDD+ particularly when carbon trading commences. The impact areas of CBFM and REDD+ were determined using key variables such as over-all condition of the resources (forest, land, water, etc.), general well-being of the household, local income, access to resources, control over resources, participation in community affairs, compliance with resource management and amount of traditionally harvested resources. For change to be attributed to development programs implemented in the area, the use of a ladder diagram in scoring the indicators was done by the respondents. This scoring is made quick and easy and less demanding on respondent's memory such as in tracing back the amount of income derived in the 3 time periods.

Table 1. Result of the two-tailed t-test comparing the three (3) periods with selected key indicators of change

Key variables of change	Before and during CBFM				With CBFM and REDD+				Before CBFM and REDD+			
	T1	T2	Obs.	Sig	T2	T3	Obs.	Sig	T1	T3	Obs.	Sig
over-all condition of the resources	2.35	6.81	52	**	6.81	7.06	52	ns	2.35	7.06	52	**
general well-being of the household	4.76	5.71	51	*	5.71	5.43	49	ns	4.86	5.43	49	ns
local income	5.98	5.37	51	ns	5.37	5.08	51	ns	5.98	5.08	51	*
access to resources	7.00	4.18	49	**	4.18	3.88	49	*	7.00	3.88	49	**
control over resources	5.70	3.50	50	**	3.50	3.40	50	ns	5.70	3.40	50	**
participation in community affairs	4.40	6.29	42	**	6.29	6.60	42	*	4.40	6.60	42	**
compliance with resource management	3.66	6.82	50	**	6.82	6.90	50	ns	3.66	6.90	50	**

Legend: T1 – before CBFM; T2 - CBFM; T3 - REDD+.

** highly significant at 0.01 level, * significant at 0.05 level, n.s.- not significant.

Table 1 presents the result of two-tailed t-test comparing the variation of means on the score per indicator for the 3 periods. The variation is very evident between the period before CBFM (T1) and during CBFM (T2), with high to very high significance level of variation. The same is true between before CBFM (T1) and REDD+ (T3).

However, majority of the scores on indicators slightly declined between CBFM and REDD+, but the variation is not significant.

At a close range, the result indicates that local income is perceived to have declined during CBFM implementation, but the variation is not significant. This decline is attributed to a reduction in the size of farmland, where previously it was all a kaingin area (ISF/CBFM) until it was converted to other purposes such as for reforestation, rehabilitation, assisted natural regeneration (ANR) and agroforestry areas (Mohammed et al., 2016). The decline in income also coincides with the "bunchy top" infestation of abaca farmlands. REDD+

3.2.1 Over-All Condition of the Resources

The condition of the available forest, land, and water resources reflects the type of management implemented in the study site. YISEDA members candidly revealed the past practices in utilizing these resources. Majority of members were into kaingin making, timber poaching, and charcoal making. But the devastating impacts of forest fires during the 1982 El Nino, together with their destructive forest activities, made them realize the ramifications to their income and livelihood sources. The forest areas were denuded before CBFM implementation, hence, given an average rating of 2.35. The awarding of CBFM Agreement (CBFMA) to YISEDA paved the way for the implementation of forest conservation activities such as reforestation and rehabilitation of denuded forest areas. The transition from a denuded state to forest vegetation was evident during the CBFM implementation, hence, on the average it was rated by members at 6.81. Meanwhile, the 2 years implementation of REDD+ somewhat improved the resource's condition, with a rating of 7.06; but the variation is not significant.

3.2.2 General Well-Being of the Household

The household's general well-being refers to overall satisfaction in life which includes food, furnishings and shelter, health, education, employment opportunities, and gender equity. There is a slight improvement in the score between before and during CBFM time periods, from 4.76 to 5.71. Members believed that life was supposed to be in good condition before CBFM due to continuous income they enjoyed from abaca sale. But only few families were able to send their children to school located at the center of the barangay, about 1-2 hours walk, depending on the weather condition, from their homes in Sitio Canlugoc. The same is true for the visit to the health center which is seldom made by members and their families due to the rough road system. When illness strikes, the members relied heavily on "faith healers or "albularyo" and herbal medicines. CBFM implementation contributed some degree of improvement to this indicator. Although indirectly associated with CBFM implementation, the frequent visits of foreign nationals, representatives of national agencies, and LGU officials to the project site triggered the improvement of some infrastructures in the area, such as the road system, education and health facilities. The sorry state of these facilities prompted the LGUs to push for their development, owing to the popularity of the project. LGUs were quick to counter that infrastructure development in the site has long been planned since the priority fell on remote areas of the province where most CBFM projects are located.

However with REDD+ implementation, the score slipped to 5.43. This could be attributed to the difficulty of finding food for the family and lack of employment opportunities, despite the employment of majority of members to the different project activities of REDD+.

3.2.3 Local Income

The local economy is normally based on the amount of available money circulating in an area. The higher the paid employment opportunities, the more income can be generated, and more money will circulate in the area. However, during times of bad weather, local income is greatly affected. Local income before CBFM was perceived to be on the average, at 5.98. Some members claimed that before CBFM was the period when they enjoyed higher income from sale of abaca and illegal logging activities/ poaching. During CBFM, the farmlands of members became the project area which have to be utilized for reforestation and rehabilitation activities. The once cultivated areas have been planted with mahogany (*Swietenia macrophylla)*, narra (*Pterocarpus indicus)*, falcata *(Albizzia falcataria)*, gmelina*(Gmelina arborea)* and others. Although the denuded landscape was transformed into forest (Mohammed et al., 2016) income of members gradually declined, hence, the average score is 5.37. The negative impact of the natural capital transformation on farming and income was later realized when agricultural crops no longer survived under the canopy of growing trees. During CBFM, alternative livelihood sources were introduced to augment the meager income derived from farming. These income sources include livestock production and dispersal (goat, duck, and chicken) and vegetable farming. However, these initiatives were short-lived, not sustained after project termination; hence, CBFM investment was wasted.

Meanwhile, income during REDD+ pilot project implementation somehow declined, although the variation is not significant. Some claimed that the lower income from farming is attributed to the use of farmlands for reforestation and ANR activities. Others doubted this because they claimed that members were compensated for the services rendered during plantation establishment, though admittedly they considered this compensation as temporary or good until the project lasted. In addition, REDD+ implementation coincided with 3 major project/activities, i.e., the first harvesting operation of YISEDA, implementation of Comprehensive Agrarian Reform Program (CARP) and the National Greening Program (NGP). Although YISEDA did not earn much from their timber harvesting operation, they invested in physical assets such as the multi-purpose hall, saw mill and stock yard (Peras et al., 2015). The numerous projects YISEDA ventured in provided the organization ample savings which was used to sustain previous activities especially site monitoring. But with changes in the organizational leadership (Note 1), along with absence of new externally-funded development projects, funds of the organization drained fast.

3.2.4 Access to Resources

This indicator pertains to the freedom to utilize the resources within the CBFM area. This freedom diminished with CBFM and eventually with REDD+ implementation. Before CBFM is rated at 7.00 as resources were generally accessible. Even for illegal logging purposes, timber was easily sourced in the area. With CBFM, the numerous trainings/seminars on forest conservation informed members that access to resources have to be case-specific. Extraction of timber needed for household construction and repair is allowed but with appropriate permission from the organization specifying the number of trees that need to be cut. Extraction of timber without permit from the organization and DENR is considered illegal, especially if intended for commercial purposes. With REDD+, the same restriction on access is carried out as with CBFM implementation. The members are apprehensive of more stringent rules on resource access once REDD+ is in full swing. REDD + entails "forest enclosure," a phenomenon that will constrain future access to forest resources (Peras et al., 2015).

3.2.5 Control over Resources

This indicator pertains to the overall management of the resources found within the CBFMA. The power of the national government agency such as DENR in securing State ownership over forest lands is a good example of control. More specifically, CBFMA enhanced government's control over local communities by limiting devolution to responsibilities on forest development and protection (Pulhin *et. al.*, 2007). Individual farmer's control over their farmlands located inside CBFM areas diminished when they subjected their area under DENR control in the guise of CBFM-PO management. This is consistent with the decline of score for this indicator from 5.70 before CBFM to 3.50 and 3.40 with CBFM and REDD+, respectively. Before CBFM, individual farmers had sole and full control over their farmlands in forest areas. DENR's visibility in forest lands was obscure until the launching of people-oriented forestry projects such as ISFP in 1982 and later on the CBFM in 1995. While CBFM purports to develop people organizations as forest stewards and partners in forest development, over all control over the CBFM area seemingly lies with the DENR. In many instances, CBFM-POs generally remained dependent on DENR-initiated projects.

3.2.6 Participation in Community Affairs

One of the most improved indicator of CBFM implementation is the ability of members to participate in community affairs such as assemblies, community meetings, trainings, seminars, workshops and other activities. The improvement in this indicator is attributed to the continued conduct of trainings, seminars, and IEC (information, education campaign) intended to capacitate members to become forest stewards. Participation in different community endeavors eventually improved with REDD+ as member's knowledge base widened with the inclusion of topics on carbon/ carbon trading, climate change, adaptation and mitigation.

3.2.7 Compliance with Resource Management

The over-all management of the forest resources demands compliance by local stakeholders, most especially forest- dependent communities. Before CBFM, majority did not comply with the existing forest policies. Some attributed this complacence to the non-visibility of DENR officials to monitor their area and ignorance of the law. But these changed with CBFM due to numerous trainings/seminars on environmental protection and reminders on punishment for illegal forest activities. Fear of imprisonment is evident among YISEDA members. YISEDA's compliance to sustainable forest management is reflected in the numerous awards they received such as the Best Performing CBFM-PO in 2012. However, YISEDA's faithfulness was tested with the 3 cases of illegal timber harvesting activities discovered in 2015.

3.3 Participation to the REDD+ pilot project activities

Forestry-related activities that members knew and participated in the CBFM project include forest rehabilitation, reforestation, agroforestry, forest protection, non-forest-based livelihood opportunities, biodiversity conservation, sustainable livelihoods generation and forest policing/ governance (Table 2). The top 3 forestry activities that the members mentioned are reforestation (67.86%), agroforestry (48.21%) and forest rehabilitation (46.43%). Only few respondents mentioned participation in forest protection, non-forest based livelihood opportunities, biodiversity conservation, sustainable livelihood generation and forest governance. Almost all forestry-activities provided financial incentives depending on the degree of services rendered by members. Respondents also associate these activities as part of the REDD+ pilot project in their area.

Table 2. Participation of YISEDA members to forestry-related project activities.

Forestry Activities	Freq.	Percent
Reforestation	38	67.86
Agroforestry	27	48.21
Forest rehabilitation	26	46.43
Forest protection (foot patrol)	14	25.00
Non forest-based livelihood opportunities	14	25.00
Biodiversity conservation	7	12.50
Sustainable livelihoods generation	5	8.93
Forest governance/ Policing	5	8.93

The pilot project activities of REDD+ are the common project activities implemented in the previous projects related to CBFM implementation, i.e. assisted natural regeneration which is associated with forest rehabilitation in strict protection zones within the CBFM Agreement (CBFMA) area, reforestation, agroforestry, and forest protection. Figure 2 provides a graphical presentation on the percentage of member-respondents participating in the REDD+ pilot project activities as well as the degree of their participation. Majority of the respondents participated in reforestation activity (73.08%) followed by agroforestry (59.62%), assisted natural regeneration (53.85%) and forest protection (25.00%). The labor services rendered by members in these activities were paid at P150.00 ($3.26) a day, much lower than the existing wage for hired labor at P200.00 ($4.35) (as the difference goes to YISEDA savings or general fund to be used for other development activities). Even women members were able to participate especially in the production of seedlings that supply the requirement of the GIZ-REDD+ project. Seedling production was very useful for the succeeding DENR-NGP project of YISEDA. Seedlings were also sold to other CBFM organizations in the province. In addition, women also participated in planting in reforestation areas. For ANR, only male members participated in this activity due to the steepness and ruggedness of the site. The same male pattern of participation was also observed in forest protection activity as this entails foot patrol across the entire CBFMA area and along the periphery to locate the boundary marker of the CBFMA.

Participation in the REDD+ readiness pilot project activities was also gauged in terms of determining the degree of participation of members (Figure 2). Majority of the respondents rated their participation to the REDD+ readiness pilot project activity as moderate to high. The implementation of the project coincides with other developmental projects such as the DENR's National Greening Program (NGP), Comprehensive Agrarian Reform Program (CARP), and preparation associated with the harvesting of the 25- year old Mahogany plantation (Block 1). Hence, members were very busy with all project activities of the organization, including the GIZ-REDD+ project. Meanwhile, the highly active male members gave high rating for forest protection. Those who gave very low to low rating were the aged and the sick who were physically deprived to join the activities. Households with sick or aged YISEDA members were replaced by their young members (or youth) to access gainful employment from the project activities.

The findings of Loaiza et al (2015) pointed out that interest to participate in REDD+ pilot project through deforestation avoidance relies on deforestation rates, cash-related forest dependency and off-farm income. In the context of YISEDA, this PO had been pre-selected by the DENR and GIZ as among the 5 CBFM-POs in the province to constitute the REDD+ pilot project site. Also, YISEDA's willingness to undertake the REDD+ pilot demonstration project is in accordance with the goals of CBFM implementation. Hence, all members were expected to participate in the simultaneous implementation of the REDD+ pilot project and the 3 other forestry development projects mentioned. To suffice the needed labor requirement of the different projects, outsiders or non-YISEDA members were also hired to complete the project activities as scheduled. Also, member's active participation in all the projects as well as the organization's activities promises future dividends once income from all economic endeavors materializes. Although the traditional forest-related cash income activities of members such as logging (illegal), charcoal making and kaingin making cannot be totally stopped, REDD+ project activities at the site reduced the intensity of these activities (Carandang et al., 2013).

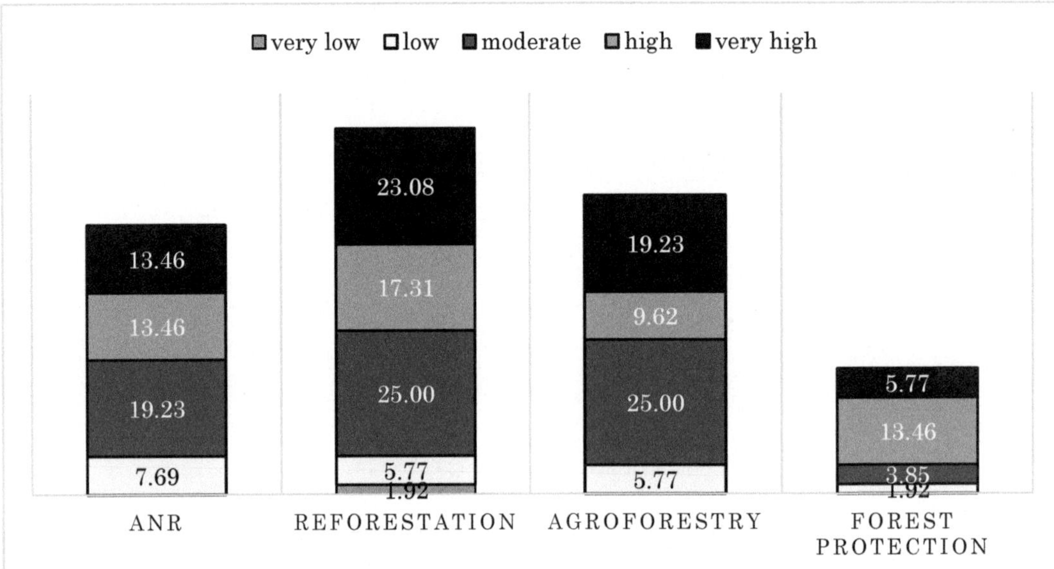

Figure 2. Level of participation of respondents on the REDD+ readiness pilot project activities (%)

3.4 REDD+ Pilot Demonstration Project Activities and Sustainable Livelihoods

Sustainable livelihood is taken by many scholars as a co-benefit of REDD+ (IPCC, 2014; Poudel et al. 2015; Howson & Kindon, 2015; Hirons, 2011; Angelsen, 2009). However, Visseren-Hamakers et al. (2012) interpreted this together with biodiversity conservation and equity as preconditions to the legitimacy and effectiveness of REDD+. The PNRPS Team also promotes sustainable livelihood strategy for successful implementation of REDD+ (PNRPS, 2010). The 3 REDD+ pilot demonstration areas in the country have devised different livelihood activities implemented at the site level. The Quezon and Palawan pilot areas are being managed by indigenous people (IPs) --, the Dumagat-Remontado tribe and Tagbanua, respectively. The REDD+ pilot activities among these IPs are mostly concentrated on enhancing their livelihood strategies such as bee production/farming for Dumagat-Remontado while non-timber forest products (rattan, bamboo) for the Tagbanua tribe. This is quite different from the Southern Leyte site being managed by different CBFM-POs like YISEDA, where activities are focused on forest carbon stock enhancement and protection. All of these REDD+ pilot initiatives have embedded forest protection activities to address the drivers of deforestation and forest degradation.

Table 3 below summarizes the positive (co-benefits) and negative (trade-offs) association of REDD+ pilot activities to livelihood capitals. In general, the different REDD+ pilot activities have both positive (+) and negative (-) impacts on the livelihood capital assets. The positive impacts refer to co-benefits derived out of REDD+ implementation, such as income from pilot activities and during carbon trading. However, negative impacts pertain to trade-offs under REDD+ such as "forest enclosures" that restrict timber utilization.

Only a small number of respondents (58%) answered the question on the impacts of REDD+ readiness pilot project activities on the livelihood capital assets, but their analysis of the impacts provided insights on how to successfully implement of REDD+ at the site level. About more than half of the total respondents associated the activities' impacts on the livelihood capital assets to the natural and financial capitals of reforestation activity. About 20% of the respondents had knowledge on the link of the activities to other capital assets.

The link of the ANR, reforestation, agroforestry and forest protection activities was cited by almost 20% of the respondents to be as high as the human capital asset. This is attributed to the many seminars and workshops brought by GIZ and other NGOs at the local level to capacitate the community and equip them with information related with climate change, global warming and REDD+. Although a smaller number of respondents (5.77%) related human capital with forest protection, it is still worth mentioning that efforts toward stricter forest protection activities reduced the occurrence of illegal forest activities.

The contribution of reforestation to increasing the natural capital asset of the resources in general and the forest in particular is regarded by majority to be high to very high. This means that planting in degraded landscape is key to greening the environment, which is the main lesson learned from most seminars/trainings. Discussions in these trainings focused on the importance of forest and the resources found therein. Some even mentioned the considerable impacts of climate change on people and forests, that is, they saw the urgent need for this kind of activity in their area. Some also narrated the impacts of the 1982 El Nino and the corresponding strong typhoon that followed which left their resource and food grounds highly devastated.

There is a high to very high contribution of the activities, especially reforestation, on the financial capital assets as perceived by the majority of the respondents. This is attributed to the income derived from activities during project implementation. Only a few of the respondents related the activities to be high with physical capital assets. Available cash income allowed households to buy appliances and motorbikes as a means of transportation for the family and for hire. Social capital contribution is high to very high as well due to a more unified and strong organization espoused by GIZ as part of its organizational strengthening.

Some negative impressions were also raised while associating the different REDD+ activities to livelihood capitals. Agroforestry and reforestation activities were rated very low for natural and social capital assets, respectively. The little contribution of agroforestry to improving the natural capital is attributed to smaller diversity and area of species planted that may have little contribution to emissions reduction. However, this is contrary to results of studies showing the higher potentials of agroforestry systems as well as species in reducing carbon emissions (Lasco, Mallari, Pulhin, Florece, Rico, Baliton & Urquiola, 2013). Trade-offs exist in terms of the opportunity cost (income lost) or foregone benefits when farms are converted to reforestation and agroforestry areas, thereby replacing the monocropped agricultural fields.

Table 3. Summary of associated sustainable livelihoods co-benefits and trade-offs of REDD+ pilot project activities

REDD+ pilot activities / Capital assets	ANR	Reforestation	Agroforestry	Forest protection
Human	-/+	-/+	-/+	+/ -
Natural	+	+	+	+
Financial	-	-	+/ -	-
Physical	-	+/-	+	-
Social	+/-	+/-	+/-	+/-

4. Discussion

4.1 Implications to Sustainable Livelihoods

The livelihood capital asset of YISEDA households typifies a family living below the poverty line (83.93% has an annual income that is less than P100 000.00 or \$2 173.91). Members are no longer solely dependent on farming for it cannot provide the needed food and cash requirement of a family. Majority derived income from other sources such as occasional off-farm work as hired labor in the wealthy farmer's lot and in construction works. The only regular income source, aside from employment in the government, is the buy and sell business of agricultural crops derived weekly by female headed households. Annual income averages to P58 492.85 (\$1 271.58) which is way below the annual per capita poverty threshold level of the province at P19 431.00 (\$422.41) (PSA Region 8, 2016) for 5-member households in YISEDA with an average dependency rate of 146%. Despite this condition, respondents still did not seek help aside from the employment and assistance that YISEDA provided to them. Hence, employment in YISEDA projects is vital in obtaining short-term additional income for most of its members.

Future REDD+ is believed by many members to uplift them out of poverty, the same feeling they had with CBFM during its establishment in the area. The current REDD+ pilot activities at the site do not seem to address the issue of sustainable livelihoods. The activities are still concentrated on enhancing the forest carbon stocks, giving more emphasis on the natural capital assets while leaving very little (or no) room for enhancing other capital assets.

REDD+ future success would be dependent on ensuring sustainable financing schemes either from a) donor financing, b) market-based financing, and c) fund-based financing. The inclusion of sustainable financing as a component strategy of PNRPS is considered a cross cutting strategy involving sustainable and long-term financing scheme that will drive REDD+ initiatives in producing its expected outcomes of reduced emissions, poverty alleviation and social justice for forest dependent communities, biodiversity conservation, and improved environmental services (PNRPS, 2010). Part of the long-term plan is investing in self-sustaining local-level programs that will benefit forest-dependent communities. Hence, assured income maybe feasible during REDD+ engagement phase.

At the households, REDD+ pilot activities are evidently enhancing the natural capitals by incorporating the planting of indigenous species in the farms. It has likewise increased the natural capital at the site level. The activities may not have directly improved the present human capitals of the households, but the knowledge system acquired from the different activities helped them to venture into other form of livelihood sources like seedling production. However, seedling production is again short-lived, good until project termination. It was not seen as an enterprise that can be sustained unlike the smallholder nursery of Falcata in the Caraga region.

Other issues emerged. An important concern was raised by some members on training fatigue, with the series of environmental conservation trainings/seminars conducted in the area in the last 10 years as different projects were implemented. YISEDA's diligence and known capability to complete the scheduled tasks even with little financial resources made it a most sought-after PO among funding agencies. This was largely due to high volunteerism exhibited by members. While YISEDA's high dependence on volunteerism paid off, through time, volunteerism subsided and replaced by "project mentality;" that means, members expect financial incentives for every service rendered. This perspective evolved from the last five years of continued paid labor services to members engaged in the different forestry development projects. In recent years, when projects ended and the financial resources of YISEDA dried up, members were no longer willing to do any volunteer work, leaving YISEDA's officials to do the work themselves.

In the long run, pilot activities of ANR, reforestation and forest protection cannot provide the financial asset of the household as demand for labor services is high during the establishment phase only. Moreover, once timber trees mature, agricultural crops can no longer compete for growth. Only agroforestry is financially feasible in both the short and long terms, if farmlands were managed properly.

The different REDD+ pilot activities also limit timber utilization for house construction and repair, where previously members enjoyed the freedom to extract timber for house construction. Another concern is the accessibility of the CBFM area where a road system traverses the plantation area. With the tightening of financial benefits and imposed restrictions on timber access, securing the CBFM area will continue to pose challenges to YISEDA's REDD+ pilot activities.

The REDD+ activities have positive association to the social capital assets. Different institutions/organizations involved in REDD+ activities also serve as the support system of the households. However, these institutions may not continue to enjoy for long the patronage of members, especially the DENR, since the members have become very critical of the continuing illegal activities happening in the area.

4.2 Challenges and Insights on REDD+ Implementation

4.2.1 Too much Focus on Carbon Stock Enhancement

It is very evident that in the promotion of REDD+ pilot projects, the concern is on enhancing the forest carbon stocks (Lasco, Pulhin, Bugayong, & Mendoza, 2011) concentrating on the natural capital asset formation while leaving very little (or no) room for enhancing other capital assets. Also, with the conversion of farmlands/cultivated lands in steep slopes into either reforestation/rehabilitation or ANR sites that led to a reduction in agricultural crop harvest and farm income, majority of the members were wary of the limited or lack of livelihood sources. The meager income has pushed some members to go back to their old illegal forest activities.

Many scholars reiterate the importance of livelihood in participatory forest management (PFM)/CBFM areas (Mohammed & Inoue, 2013). REDD+ also recognizes this as an important element to realize its goal. However, at the pilot project, aside from the project-based income generation activities which are short-lived in nature, there is no other income source alternative that could sustain income generation for households. Additionally, the REDD+ readiness phase improved the human capital asset by instilling forest conservation and protection through the conduct of different trainings/seminars. But for some members, training fatigue set in with the continued infusion of basic concepts of forest ecology, conservation, and climate change.

4.2.2 Importance of Local Implementers

YISEDA, LGU and DENR perspectives on REDD+ pilot demonstration areas are vital to better understand important conditions that necessitate successful REDD+ implementation at the site level. YISEDA believed there was similarity of goals when it comes to the improvement of sustainable livelihood of the community. However, at the site level, CBFM's livelihood options commonly termed "alternative livelihoods" such as livestock (chicken, ducks, goats) raising, sari-sari store operation, etc. are short-lived while REDD+ pilot activities did not directly address this concern.

The continuing poverty and the lack of available sources of income from PO projects led to the declining participation of members to PO activities. Ever since the development projects were promoted in the organization, members looked forward to participate in the activities. Hence, participation is dependent on project funds. Pulhin et al. (2007) refers to this condition as "project mentality" developed from CBFM implementation due to over dependence on income generated from project activities. This is the major drawback of externally funded projects observed in the past, which exists until today. Volunteer work or the "*bayanihan* system" that once defined YISEDA's success is replaced by the "food for work" slogan. This strategy is commonly used by LGUs in soliciting local communities' participation in collective activities such as clean-up drive in the river, coastal area, etc. and other environmental activities. In return, participants to these activities receive 5 kilos of rice.

Likewise, POs with excellent past performance such as YISEDA (recipient of the 2012 Good Performing CBFM-PO Award) is now being challenged with organizational stability and sustainability. The draining financial resources placed the PO's organizational capability uncertain. Monitoring the REDD+ demonstration pilot project site is then constrained by the non-availability of financial resources. Even with only a few members (mostly PO officials) who are willing to do volunteer work, their visibility in forest protection activities should be continued to discourage illegal loggers from entering the project area.

4.2.3 Sustained Forest Protection Activities

Minang & Noordwijk (2014) and Robiglio *et al* (2014) found limited evidence that the national level emission reduction targets will be implemented on the ground. Hence, they recommended the need for REDD+ Readiness to focus on addressing drivers of deforestation and forest degradation. The case of YISEDA reflects the intention to contribute to emission reduction (Note 2). But current REDD+ pilot activities do not directly address drivers of deforestation and forest degradation on the ground, like illegal tree cutting, kaingin making, and charcoal making. Although the reforestation and ANR REDD+ pilot project activities have an embedded forest protection activities, there is no guarantee that the newly rehabilitated forest cover would continue to exist.

4.2.4 The Need for a Committed LGU Partnership

The Philippines cannot deny that forest degradation and deforestation problems still happen in the uplands, most especially in CBFM areas. This is where the local government units (LGUs) and local NGOs can provide the needed partnership towards forest conservation and protection. A corresponding governance system is very important in the implementation of forestry policies, more than engendering participation in tree growing. . The local government units as the frontliner in REDD+ implementation face the challenge on sustaining the Provincial Technical Working Group (PTWG) set up by GIZ for REDD+ purposes. This PTWG is composed of representatives from all stakeholders with environmental concern either NGOs, LGUs, CBFM-POs, academe, etc. working in the province. In the national REDD+ strategy, this PTWG is at the base of the proposed organizational structure (PNRPS, 2011). The group met monthly to quarterly, spearheaded by the GIZ, but in recent years, the meetings became infrequent, from twice a year to once a year. This is due to financial resources needed to conduct a TWG assembly/ meeting which cast burden on the host LGU's budget.

4.2.5 Timber for Illegal Logging is Still more Attractive than the Carbon Market

The ultimate challenge of REDD+ in the future is how to convince local implementers (PO members) to maintain REDD+ areas. Mature harvestable timber attracts opportunistic buyers, illegal loggers as well as CBFM members to harvest even without the appropriate permit issued by the DENR. An opportunistic leader of CBFM may find a way to lead the organization into illegal timber harvesting (without permit to harvest). This is already happening in the CBFM area recently. YISEDA members have been accused by the DENR-CENRO to have allegedly committed 3 cases of illegal timber harvesting. Buyers allegedly transacted business with some members, providing them chainsaws to fast track harvesting of mature timber. Some members attributed their eagerness to do illegal harvesting to poverty and lack of livelihood sources in the community. Also, the presence of speculators and buyers of logs are tempting some members to do illegal activities.

4.2.6 The Core of REDD+ Implementation should be Enhancing Sustainable Livelihoods

Too much attention to enhancing the natural capital assets is opposed by some PO members. Enhancing the 5 capital assets of communities in REDD+ areas would likely provide favorable outcome/success. The case of General Nakar, Quezon REDD+ pilot demonstration site allows indigenous peoples (IPs) to know REDD+ by enhancing community livelihoods like honeybee production from collection, processing and marketing. The main idea is to sustain IP's livelihood source to keep them away from timber cutting and other forms of forest degradation and deforestation in the area. Nearby, the Anahao Movement for Productive Community Organization (AMPCO), a neighboring CBFM of YISEDA also with REDD+ project, devised an enterprise that sustains income of the PO. AMPCO adopts the trading post enterprise locally termed as "*bagsakan*" system, where locally produced copra (coconut) products of PO members are bought by the PO and sold in bulk in the market. This enables AMPCO to generate income for the organization and help members and non-members alike in facilitating the transport and marketing of their products.

4.2.7 REDD+ may Perpetuate the Failures of CBFM

The three decades of CBFM implementation in the country is regarded by many as a failure in the forestry sector (Pulhin et al., 2007). This is because the initiative failed to accomplish the rehabilitation of the targeted 5.97 Million hectares of denuded forest lands (Pulhin et al., 2007), but instead managed to rehabilitate only 600 000+ hectares (or 6.7%). The indicator of CBFM success in the past, from the DENR's perspective, is the physical accomplishments of the program such as the number of trees planted as well as the area (size) of forest plantations. The program failed to include the needs of the local communities in terms of livelihood generation. Although CBFM policies and strategies have been revised to suit the existing condition, somehow its failures can be traced to the absence of sustained income of local communities. Basically, all the challenges mentioned above are also the issues confronting CBFM implementation until now. These challenges and failures may perpetuate with REDD+ if left unheeded.

5. Conclusion

The paper highlights the assessment made by YISEDA members on the impacts of REDD+ readiness activities on pilot demonstration areas using sustainable livelihoods highlighting co-benefits and trade-offs. Findings revealed that the natural capital asset is continuously improving with REDD+ readiness initiative. Local people highly regarded the assisted natural regeneration and reforestation activities' effectiveness in reducing carbon emissions. However, REDD+ activities could only cater to short-lived improvement in the financial capital of members, while indirectly benefiting human, physical and social capital assets.

The 3 year span of REDD+ readiness may not have brought much impacts on maintaining the balance between environmental and economic sustainability. Promises and potentials soon to be achieved with REDD+ may be too overwhelming for local stakeholders which have brought REDD+ at the forefront. However, local participation in REDD+ is project-based, short-lived, hence, livelihood capital assets sustainability is a major concern. This has considerable implications on addressing the illegal forest activities even if there were positive impressions on achieving the REDD+ goals, because income and its sources remained unsustainable. Moreover, the major drawback of REDD+ pilot activities is that it perpetuates the failures of CBFM initiatives, giving little attention to sustainable livelihood objectives. Forest conservation policy like REDD+ may still be embraced fully by local communities if livelihood capital assets were enhanced.

Acknowledgement

The authors would like to give special thanks to the Japan Society for the Promotion of Science (JSPS) Ronpaku (PhD Dissertation) Program Scholarship for providing funding support for the dissertation of the main author from which this paper is taken. This is also a result of a collaborative research between the University of the Philippines Los Baños – College of Forestry and Natural Resources, College, Laguna, Philippines, The University of Tokyo, Japan, Nagoya University, Japan and Hokkaido University, Japan.

References

Agrawal, A., & Angelsen, A. (2009). Using community forest management to achieve REDD+ goals. *Realising REDD+: National strategy and policy options*, *1*, 201-212.

Angelsen, A. (2009). *Chapter 1 Introduction.* In A. Angelsen, M. Brockhaus, M. Kanninen, E. Sills, W. D. Sunderlin, & S. Wertz-Kanounnikoff (Eds.), *Realising REDD+: National strategy and policy options* (pp. 1-12). CIFOR, Bogor, Indonesia.

Asian Development Bank [ADB]. (2010). *National REDD+ strategies in Asia and the Pacific: Progress and challenges.* Mandaluyong City, Philippines: Asian Development Bank.

Caplow, S., Jagger P., Lawlor, K., & Sills, E. (2011). Evaluating land use and livelihood impacts of early forest carbon projects: Lessons for learning about REDD+. *Environmental Science & Policy, 14*(2011), 152-167.

Carandang, A. P., Bugayong, L. A., Dolom, P. C., Garcia, L. N., Villanueva, M. M. B., Espiritu, N. O., FDC-CFNR-UPLB (2013). *Analysis of key drivers of deforestation and forest degradation in the Philippines*. Deutsche Gesellschaft für Internationale Zusammenarbeit (GIZ) GmbH: Manila, Philippines April 2013

Eilenberg, M. (2015). Shades of green and REDD: Local and global contestations over the value of forest versus plantation development on the Indonesian forest frontier. *Asia Pacific Viewpoint, 56*(1), 48–61.

Hirons, M. (2011). Locking - In Carbon, Locking - Out Livelihoods? Artisanal Mining And Redd In Sub - Saharan Africa. *Journal of International Development, 23*(8), 1140-1150. http://dx.doi.org/10.1002/jid.1837

Howson, P., & Kindon, S. (2015). Analysing access to the local REDD+ benefits of Sungai Lamandau, Central Kalimantan, Indonesia. *Asia Pacific Viewpoint, 56*(1), 96-110.

Jagger, P., Atmadja, S., Pattanayak, S. K., Sills, E., & Sunderlin, W. D. (2009). Learning while doing: evaluating impacts of REDD+ projects. In *Realising REDD+: national strategy and policy options*. Center for International Forestry Research (CIFOR), Bogor, Indonesia.

Karki, S T. (2013). Do protected areas and conservation incentives contribute to sustainable livelihoods? A case study of Bardia National Park, Nepal. *Journal of Environmental Management, 128*(2013), 988-999.

Lasco, R. D., Mallari, N. A. D., Pulhin, F. B., Florece, A. M., Rico, E. L. B., Baliton, R. S., & Urquiola, J. P. (2013). Lessons From Early REDD. *International Journal of Forestry Research, 2013.* http://dx.doi.org/10.1155/2013/769575

Lasco, R.D., Pulhin, F.B., Bugayong, L., & Mendoza, M. (2011). An assessment of potential benefits to smallholders of REDD+ components in the Philippines. *Annals of Tropical Research, 33*(1), 31-48.

Le, H. D., Smith, C., & Herbohn, J. (2014). What drives the success of reforestation projects in tropical developing countries? The case of the Philippines. *Global Environmental Change, 24*, 334-348. http://dx.doi.org/10.1016/j.gloenvcha.2013.09.010.

Minang, P. A. & Noordwijk, M Van. (2014). The political economy of readiness for REDD+. *Climate Policy, 14*(6), 677-684. http://dx.doi.org/10.1080/14693062.2014.912979.

Mohammed, A. J., & Inoue, M. (2013). Forest-dependent communities' livelihood in decentralized forest governance policy epoch: case study from West Shoa zone, Ethiopia. *Journal of Natural Resource Policy Research, 5*(1), 49–66.

Mohammed, A. J., Inoue, M., Peras, R. J., Nath, T. K., Jashimuddin, M., & Pulhin, J. M. (2016). Transformation Strategy for Managing Coupled Socio-Ecological Systems: Case Studies from Bangladesh and the Philippines. *Small-scale Forestry, 15*(2), 213-227.

Nath, T. K., & Inoue, M. (2010). Impacts of participatory forestry on livelihoods of ethnic people: experience from Bangladesh. *Society and Natural Resources, 23*, 1093-1107.

Olsson, E. G. A., & Ouattara, S. (2013). Opportunities and challenges to capturing the multiple potential benefits of REDD+ in a traditional transnational Savanna-Woodland region in West Africa. *Ambio, 42*(3), 309-319. http://dx.doi.org/10.1007/s13280-012-0362-6.

Pomeroy, R. S., Pollnac, R. B., Katon, B. M., & Predo, C. D. (1997). Evaluating factors contributing to the success of community-based coastal resource management: the Central Visayas Regional Project-1, Philippines. *Ocean & Coastal Management, 36*(1), 97-120.

Poudel, M., Thwaites, R., Race, D., & Dahal, G. R. (2015). Social equity and livelihood implications of REDD+ in rural communities–a case study from Nepal. *International Journal of the Commons, 9*(1), 177-208. Retrieved from http://www.thecommonsjournal.org

PSA (Philippine Statistics Authority) Region 8. (2016). *Provincial StatWatch.* Regional Statistical Services Office VIII released on January 29, 2016. Retrieved May 18, 2016, from http://www.nscb.gov.ph/ru8/statwatch/2015/Provincial/4th_Qtr/EV_ProvincialStatwatch_Jan2016.pdf.

Pulhin, J. M., Inoue, M., & Enters, T. (2007). Three decades of community-based forest management in the Philippines: emerging lessons for sustainable and equitable forest management. *International Forestry Review, 9*(4), 865-883.

Quitoriano, E. L. (2013). *Supporting Philippine REDD+ readiness. Final evaluation of the Climate-relevant Modernization of the National Forest Policy and Piloting of REDD Measures in the Philippines Project.* Deutsche Gesellschaft für Internationale Zusammenarbeit (GIZ) GmbH: Germany. 98 p.

Wertz-Kanounnikoff, S., & Metta Kongphan-apirak, K. (2009). *Emerging REDD+ A preliminary survey of demonstration and readiness activities.* CIFOR Working Paper No. 46. Center for International Forestry Research: Bogor, Indonesia.

Notes

Note 1. Change in organizational leadership gave equal opportunities for members to lead the organization especially those who criticize the organization the most.

Note 2. Philippines commitment to a 70% emissions reduction by 2030

12

Evaluation and Prediction of Regional Water Resources Carrying Capacity: A Case Study of Shandong Province

Hongyan Liao[1], Yuabiao Zhang[2], Zhifeng Chen[2] & Zexin Meng[2]

[1] Finance Department of International Business School, Jinan University, Zhuhai, China

[2] Packaging Engineering Institute, Jinan University, Zhuhai, China

Correspondence: Hongyan Liao, Finance Department of International Business School, Jinan University, Zhuhai, China. E-mail: 812602882@qq.com

Abstract

Growing pressure on the world's water resources is having major impacts on us. In this paper, we discuss on water resources carrying capacity. We have a case study of Shandong Province which is one of the most arid regions in China. Considering the dynamics of water supply and demand, we combine the Falkenmark indicator and the binary dynamics model to establish an evaluation model of regional water resources carrying capacity. According to the result of our model, Shandong Province is heavily exploited. The per capita water resources in Shandong province were less than 300 m^3 in the past ten years. The increasing destruction and increasing waste make the situation even worse. Then ARIMA model and BP neural network is combined to propose a prediction model. We use it to predict the supply and demand of water resources in Shandong Province in the next 15 years

Keywords: water resources, carrying capacity, Falkenmark indicator: ARIMA model: BP neural network

1. Introduction

Growing pressure on the world's water resources is having major impacts on our life. According to the United Nations, 1.6 billion people (one quarter of the world's population) experience water scarcity. Water scarcity is the lack of sufficient available water resources to meet water needs within a region (Brown, 2011). Water use has been growing at twice the rate of population over the last century. Water is vital for human being. Humans require water resources for industrial, agricultural, and residential purposes. Every community and ecosystem on Earth depends on water for sanitation, hygiene, and daily survival.

There are two primary causes for water scarcity: physical scarcity and economic scarcity. Physical scarcity is where there is inadequate water in a region to meet demand. Indicators of physical water scarcity include: acute environmental degradation, diminishing groundwater, and water allocations that support some sectors over others (Molden, 2007). Economic scarcity is where water exists but poor management and lack of infrastructure limits the availability of clean water. In addition to these two causes, the increasing personal consumption, increasing industrial consumption, increasing pollution and lack of correct management also lead to water shortage.

Since the end of the twentieth century, the calculation methods of carrying capacity have been constantly improved, among which, ecological carrying capacity model and ecological footprint model are the most representative models (Wackernagel & Rees,1998; Wackernagel & Onislo, 1997). These models are widely being used in the field of regional sustainability evaluation in China, and the Water Resources Carrying Capacity has become a topic of great debate since 2001 and represents a new academic frontier at present (Liu & Chen, 2007). And many methods were used to assess the Water Resources Carrying Capacity at present (Yao, Wang & Jiang, 2002), such as usual tendency prediction method, fuzzy comprehensive assessment model (Chen, Ban & Zhang, 2008; Gong & Jin, 2009), main component analysis model (Fu & Ji, 1999), artificial neural networks model (Liu & Chen, 2007; Lu, Xu & Wang, 2009), multi-objective computation model (Fang Hu & Xu, 2006), system dynamics model (Chen Guo & Li, 2000; Feng Zhang & Luo, 2008; Zhu, 2008) matter-element model (Tang & Zhu, 2006), etc.

In this paper, we study the sources of fresh water and water uses, and we also learn about water crisis. Based on BP neural network and the Dynamic Model, we develop a model that provides a measure of water resources carrying capacity. We have a case study of Shandong Province. According to our model, we predict how its water resources carrying capacity will be in 15 years and how does this situation will impact the lives of citizens of this region.

2. Methodology

2.1 A Model for Evaluating Water Resources Carrying Capacity

In this model, the per capita water resources is used as indicator, and a widely-used measure of water stress proposed by Falkenmark (Brown, 2011) is used to assess the water shortage degree of a region at a certain moment. According to the water shortage degree, the regional water resources carrying capacity can be measured. Considering the total amount of water resources in a region affected by the impact of population growth and the development of industry and agriculture, a binary dynamics model for population and freshwater resources is established to analyze the total quantity change trend in a region in the future for a period of time.

2.1.1 The Falkenmark Indicator

From the perspective of the ecological system of the earth, the total amount of water resources is fixed, even becoming less, while the population is increasing with the development of society, resulting less and less water resources per capita in the world, what's worse, some areas have a serious shortage of water.

The Falkenmark indicator (Brown, 2011) is perhaps the most widely used measure of water stress. It is defined as the fraction of the total annual runoff available for human use. Multiple countries were surveyed and the water usage per person in each economy was calculated. Based on the per capita usage, the water conditions in an area can be categorized as: no stress, stress, scarcity, and absolute scarcity (Table 1). The index thresholds 1,700 m^3 and 1000 m^3 per capita per year are used as the thresholds between water stressed and scarce areas, respectively.

Table 1.Water barrier differentiation proposed by Falkenmark(1989)

Index (m^3 per capita)	Category/Condition
>1700	No Stress
1000-1700	Stress
500-1000	Scarcity
<500	Absolute Scarcity

Set x_t as the population in a region in the t year, y_t as the total amount of freshwater resources and O as the per capita water resources. Their relationship is as follow.

$$O_t = \frac{y_t}{x_t} \tag{1}$$

According to (1), when the population in a region is constant, the less the total amount of freshwater resources is, the less the per capita water resources is. Correspondingly, when the total amount of freshwater resources is fixed, the less the population is, the less the per capita water resources is.

Assuming that fertility policy unchanged of the region and its host country in the next 15 years. The population is

$$x_t = rx_{t-1} \tag{2}$$

Where x_t is the population of a region in the t year, r is the average population growth rate.

2.1.2 A Binary Dynamics Model

Based on the system dynamics (Chen, 2006), a nonlinear dynamic model of the evolvement between population and freshwater resources is established. It considers many influence factors, such as population, agriculture and water cycle.

Without considering other resource constraints of the population, population is proportional to population base and the average population growth rate, that is

$$\frac{dx_t}{dt} \propto rx$$

Where x_t is the population of a region in the t year, r is the average population growth rate.

Let β be the population that water resources unit can support. Since y_t is the total amount of freshwater resources, βy the largest population that fresh water resources can support. The population is proportional to the remaining population capacity:

$$\frac{dx_t}{dt} \propto (\beta y_t - x_t)$$

According to Logistic curve, the dynamic model where population changes under the freshwater resource constraint is proposed:

$$\frac{dx_t}{dt} = \alpha x_t (\beta y_t - x_t) = f(x_t, y_t) \tag{3}$$

As the united nations environment program announced in "Global environment outlook", population growth, industrial development and the expansion of irrigated agriculture are three main element lead to the increasing need of water (The UNEP GEO team, 2000). By correlation analysis, It can be concluded that the development of industry and agriculture is closely related to the population growth. Therefore, the consumption capacity of water resources can be defined as

$$U = (a + bI + cA)x_t \tag{4}$$

Where a is per capita water consumption, I is industrial output, b is the correlation coefficient of industries and population, A is farmland scale, c is the correlation coefficient of agriculture and population.

Besides, Slobodan P. believes that the amount of water resource is mainly influenced by five factors: population, agriculture, industry, non-renewable resource and continuing pollution (Simonovic, 2002). Non-renewable resource is an influencing factor for industry. In addition, with the development of science and technology, we take sewage purification rate into consideration and propose a dynamics model of water resources change:

$$\frac{dy_t}{dt} = (R + K - U - P)y_t = g(x_t, y_t) \tag{5}$$

Where R is water cycling speed, K is sewage purification rate, P is pollution degree.

Since the correlation coefficient of industries and population as well as the correlation coefficient of agriculture and population changes in the degree of industrial and agricultural development in the region, we aim at specific region and get the concrete values for analysis later.

2.2 Prediction Model Based on Neural Network

Due to the expansion of industrial scale, the increase of population, the reduction of surface water and other social and environmental factors, the supply and demand of water resources are in a dynamic process of change. In order to better understand the future water situation, the future supply and demand of water resources need to be accurately predicted.

Based on ARIMA model, a prediction model of neural network is established to predict the future supply and demand of water resources. The water demand includes urban living water, industrial water and agricultural water consumption.

2.2.1 Prediction Model Based on Neural Network

ZHAO (2014) applied BP neural network model to forecast the water consumption in Shandong Province. Following his model, a prediction model is proposed.

- *Establishing a Grid Topology*

Firstly, we determine the structure of the grid topology, including input layer, hidden layer, output layer, number of element and so on. The topology is shown in Figure 1.

As shown in Figure 1, there are n nerve cells in input layer. Input vector is $X = (X_1, X_2, \ldots, X_n)^T \in R^n$. There are l nerve cells in hidden layer. Hidden vector is $Z = (Z_1, Z_2, \ldots, Z_l)^T \in R^l$. There is only one nerve cell in output layer and the output value is Y. W_{ij} is the connection weight between the nerve cell i in input layer and the nerve cell j in hidden layer. Its threshold value is $\theta_j (j = 1, 2, \ldots, l)$. W_j is the connection weight between the nerve cell j in hidden layer and the output value Y. Its threshold value is $\theta_j (j = 1, 2, \ldots, l)$.

- *Training the grid*

1) Preprocessing sample data

Before training the grid, the sample data must be preprocessed. One important measure of pretreatment is normalization processing which can be adopted to process the data. The function is:

$$X_i^* = \frac{X_i - X_{\min}}{X_{\max} - X_{\min}} \tag{6}$$

Where X_i is the value before normalization processing and X_i^* is the processed value; X_{min} represents the minimum value; X_{max} represents the maximum value.

2) Training the grid

The linear transfer function is selected as the neuronal activation function. The processed data is used to train the grid with supervision. The training process is composed of forward and backward propagation.

Step 1: Forward propagation

At the beginning of the forward propagation, all the connection weights are randomly set to a smaller initial value. One sample set is selected as input to the input layer, and then turn to the hidden layer unit, and get the output value of the model in the output layer; At this time, there is a big error between the output value and the expected value.

Step 2: Backward propagation

The change of weight and threshold of each layer is calculated by the backward propagation process. Let the change of threshold be $\square W_{ij}$. After correcting the weights and thresholds of each neuron in the network, the network can be output in accordance with the forward propagation. The error between the actual output value and the expected value generates a new round of weights and thresholds.

Step 3: The forward propagation and backward propagation are repeated until the evaluation error of the system comes to a convergence. At this point, the network coefficients of the corresponding model set can be got and the training process is complete.

3) Making predictions

We make use of the nonlinear mapping ability of BP neural network, adopt the Levenberg-Marquardt back propagation algorithm to predict. The algorithm flow is as follows:

Step 1: standardize the initial data;

Step 2: Use raw data to train the grid;

Step 3: Predict the output value of the *t* year;

Step 4: Use the data obtained from the forecast in the t year as a sample to train the grid;

Step 5: Let *t* be *t*+1, return to step 2 until *t*=15.

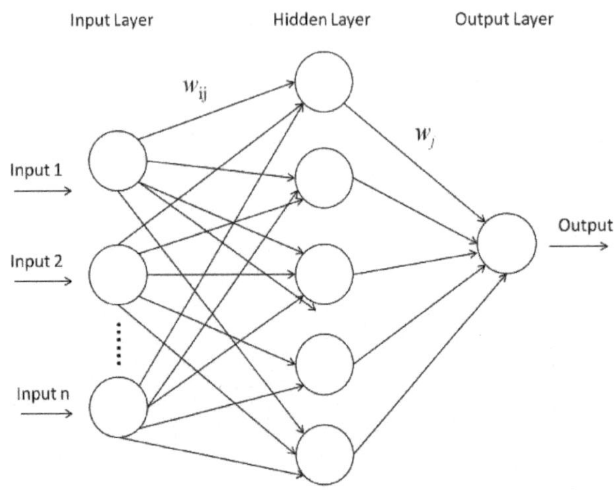

Figure 1. The structure of the BP neural network

2.2.2 Autoregressive Integrated Moving Average Model (ARIMA)

Autoregressive Integrated Moving Average Model is actually the expansion of Autoregressive–moving-average model (ARMA). ARIMA model (Wang, 2012) is proposed by Box and Jenkins in 1970, so it is also named Box-Jenkins model. It is used to find the best fit of a time-series model to past values of a time series. In ARMA model, the time series is required to be stationary. But in practice, the time series has a certain trend or cycle characteristics, and cannot meet the requirements of balance. Therefore, the ARMA model cannot be used directly. However, if the non-stationary sequence Z_t become stationary after the d order by phase difference, the ARMA model can be used to propose a model on the stationary sequence and then the original sequence by inverse transformation can be obtained. The above process is the modeling method of ARIMA. Theoretically, the mathematical description of the ARIMA model is

$$\Delta^d Z_t = \theta_0 + \sum_{i=1}^{p} \phi_i \Delta^d Z_{t-1} + \varepsilon_t + \sum_{j=1}^{q} \theta_j \varepsilon_{t-j} \tag{7}$$

Where $\Delta^d Z_t$ represents the sequence that Z_t turn to after d order by phase difference; ε_t is the random error of t and is an uncorrelated white noise sequence that obeys normal distribution whose mean value is 0 and variance is σ^2; $\phi_i\,(i=1,2,\cdots,p)$ and $\theta_j\,(i=1,2,\cdots,q)$ are the estimated parameters of the model; p and q are the orders of the model. From (7), it can be inferred that the ARIMA model is a linear model in essence and its ability of describing the nonlinear characteristics of time series is greatly restricted.

The establishment of ARIMA model includes four steps as follows:

Step 1: Make sure that the variables are stationary, identify seasonality in the dependent series and seasonally differencing it if necessary.

Step 2: Identify the model, mainly by using the autocorrelation plot and the partial autocorrelation plot to identify p and q.

Step 3: Estimate the model and diagnose the model. We estimate the parameters of the model and go on significance testing on parameters and randomness testing on residual error. Then we can determine whether the model is desirable.

Step 4: Using the model to predict after selecting appropriate parameters.

3. Results and Analysis

Shandong Province in China is selected for case study. Based on the situation of water supply and water demand in Shandong Province in recent ten years, the Falkenmark indicator and a binary dynamics model for population and freshwater resources is used to analyze the ability of Shandong Province to provide clean water to meet the needs of its population. The real data is adopted to further analyze the reason for water shortage.

Considering that the water supply and demand in Shandong Province is in a dynamic process of change, a prediction model of neural network based on ARIMA model is established to predict the supply and demand of water resources in the future 15 years in Shandong Province. The demand of water resources includes urban living water, industrial water and agricultural water consumption. The impact of the water situation on the lives of citizens is also analyzed.

3.1 Evaluation and Analysis of Water Supply Capacity

3.1.1 Evaluate Situation of Water Resources

Table 2. Statistics of water supply and population by year

year	Total water supply (1 billion m^3)	Population (millions)	the per capita water resources(m^3 per capita)
2000	24.947	89.97	277.275
2001	25.161	90.41	278.303
2002	25.239	90.82	277.906
2003	21.934	91.25	240.370
2004	21.488	91.80	234.074
2005	21.102	92.48	228.179
2006	22.553	93.09	242.271
2007	21.955	93.67	234.387
2008	21.989	94.17	233.503
2009	21.999	94.70	232.302
2010	22.247	95.79	232.248
2011	22.405	96.37	232.489
2012	22.179	96.85	229.004
2013	21.794	97.33	223.919
2014	21.452	97.89	219.144

According to water resources supply and population quantity in recent ten years given by Shandong Bureau of statistics (Shandong Bureau of Statistics, 2014), the per capita water resources is calculated as shown in Table 2.

According to water barrier differentiation proposed by Falkenmark, the per capita water resources in Shandong province were less than 500 in the past ten years. It can be known that Shandong Province is absolute scarcity of water resources.

3.1.2 Analyze the Development Trend of Water Resources

The binary dynamics model is used to get the stable solution of the system as follow:

$$\begin{cases} f(x,y)=0 \\ g(x,y)=0 \end{cases} \Rightarrow \begin{cases} A(0,0) \\ B\left(\dfrac{R+K-P}{a+bI+cA}, \dfrac{R+K-P}{\alpha(a+bI+cA)}\right) \end{cases} \tag{10}$$

The stability of the equilibrium state will determine the evolution direction and trend of the system. The future development of ecology and water resources in Shandong Province is discussed by stability analysis theory. The two stable solutions correspond to the respective characteristic root equations. The development trend of the region can be got by analyzing the solution of the characteristic root equation(Liu & Chen, 2007).

Substitute the statistics in 2014 into the binary dynamics model, It makes sense that $R+K>P$, $\alpha>4(a+bI+cA)$. The first inequality shows that the total effect of sewage purification and self-cycling ability of water in Shandong province is greater than the pollution effect. Over time, water pollution problems can be well controlled and will not cause a vicious circle. The benign effect not only avoids the destruction of ecosystem, but also promotes the surrounding areas. However, the second inequality means that the population growth rate is greater than the water consumption rate that the system can afford. If the rapid growth of population is not timely controlled, the water supply will be insufficient.

3.1.3 Reasons for Water Scarcity in Shandong Province

The average amount of groundwater resources in Shandong province for many years is 15.257 billion m^3, the groundwater available yield is 12.593 billion m^3. The average amount of fresh water resources in Shandong province for many years is 30.582 billion m^3 except the repeated calculation of the mutual conversion between surface water and ground water. From the lack of physical analysis of the reasons for water scarcity in Shandong province, there are several points:

- The total amount of the water resources is insufficient and the per capita possession is low.

With 1.09% of the total water resources, Shandong Province raises 7% of the Chinese population and produces 10% of the country's food. The per capita share of water Shandong Province is only 356 m^3, equivalent to 1/24 of the world's per capita share.

- The annual and intra annual drastic changes.

The precipitation and water resources in different parts of Shandong vary from year to year. Wet year and dry year can emerge alternately. It is also common that wet years or dry years emerge continuously.

On the one hand, Shandong Province is a heavily overloaded, on the other hand, increasing destruction and increasing waste lead to the inefficiency of water resources utilization. The primary reasons are as follows:

- Excessively exploit groundwater and damage ecological environment
- The water pollution is serious.
- The utilization rate is low and the waste phenomenon is serious.

Now the process of agricultural irrigation is relatively extensive and the waste of water resources is serious. The recycling rate industrial water is about 50%, while that in the developed countries is 75%~85%.

- The retention ability of surface water is weak and the utilization rate is low.

3.2 Prediction of Future Water Resources in Shandong Province

In this part, firstly the water supply and demand of Shandong Province in the next 15 years is predicted and later the future water resources combining the future population is analyzed.

3.2.1 Prediction of Water Supply

We first analyze the water cycle system of Shandong Province, and explore the source of urban water supply. Then we consider the influence factors of the water supply and predict the supply in the next 15 years.

- Fresh water cycle in Shandong Province

The total amount of water resources in a region is composed of surface water, soil water and ground water. Precipitation is a supplement to the water resources. The process of human activities cannot be separated from water. Industrial water, agricultural water and domestic water are all derived from nature.

Human use fresh water, and then discharge them to nature through the sewage purification. For the total amount of water resources on the earth has not changed, the sewage is increasing, available clean water is decreasing because of the human activities. The water cycle (Gleick, 1993) influenced by human is as shown in Figure 2.

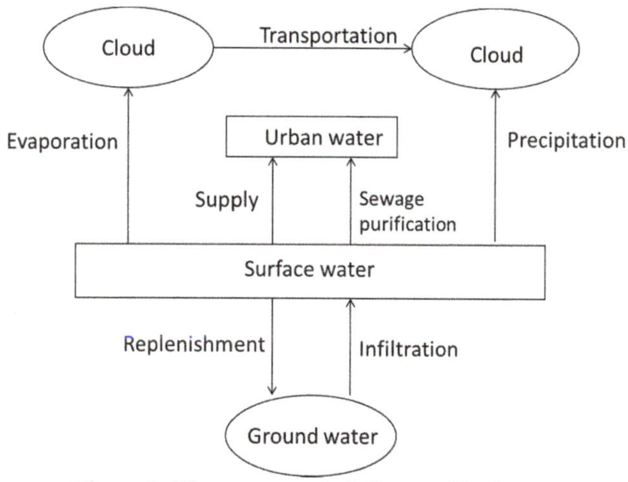

Figure 2. The water cycle influenced by human

- Prediction of fresh water supply in Shandong Province

In order to know the situation of water supply in the future, the prediction model of neural network based on ARIMA model is used to predict it in Shandong Province.

The supply of water resources S_t is influenced by many factors. The surface water and ground water are direct acting factors. The surface water and ground water will increase in regions where the precipitation is large. The pollution degree of water resources is another important factor. The increasing Sewage will decrease the available clean water.

Therefore, the amount of surface water, groundwater, rainfall and sewage can be used as the input layers of neural network and water supply as the output layer.

According to relevant data of Shandong Province in 2000 -2014, the prediction model is used to predict the industrial water consumption of Shandong Province in the next 15 years, and the results are shown in Figure 3.

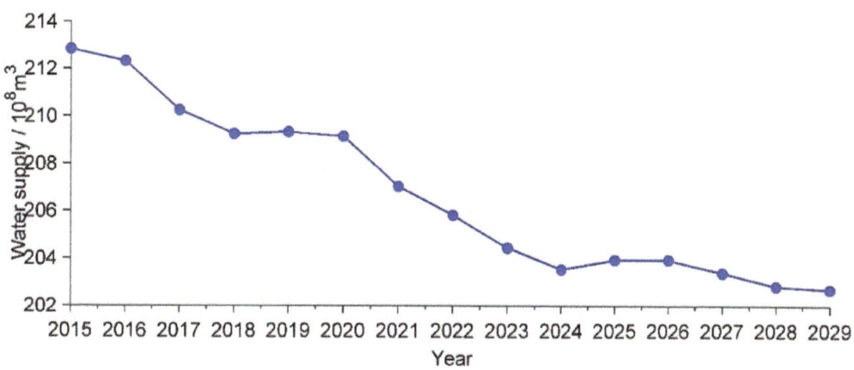

Figure 3. The industrial water consumption of Shandong Province in the next 15 years

The maximum residual error is 1.3, and the average prediction accuracy is 98.71%, which indicates that the model has high accuracy.

From Figure 3, it is shown that the trend of water supply is declining, and there will be a quickly decrease in 2020. By 2029, the amount of water will drop to less than 20.3 billion cubic meters, 1 billion cubic meters less than 2015.

3.2.2 Prediction of the Water Demand in Shandong Province

The situation of water in a region cannot be fully described only from the aspect of water supply. Therefore, the water demand in Shandong Province for the next 15 years should be forecast, and then the result can be properly analyzed. The water demand contains industrial water G_t, agricultural water A_t and domestic water C_t, we will forecast them in order to get a better prediction.

- Prediction of industrial water

We analyze the factors that influence the industrial water demand. Finally, we identified three indicators as the input layer of neural network, including the gross industrial output value, the number of important scientific and technological achievements as well as heavy industrial output value. Output layer is industrial water demand.

By using relevant data of Shandong Province in 2000 -2014, we use the prediction model to predict the industrial water of Shandong Province in the next 15 years, and the results are shown in Figure 4.

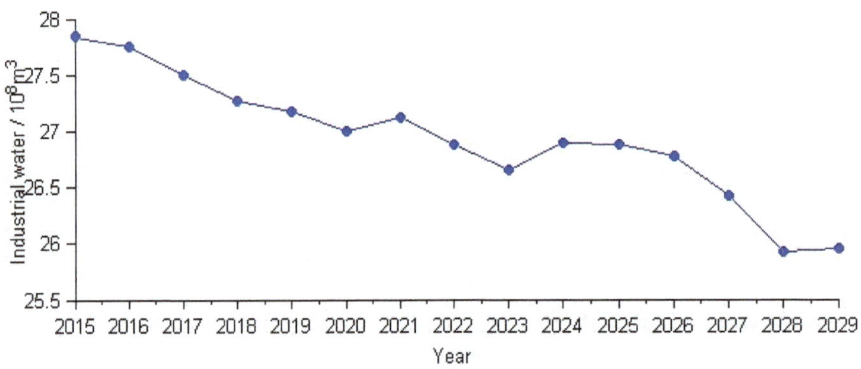

Figure 4. The industrial water of Shandong Province in the next 15 years

The maximum residual error is 1.3, and the average prediction accuracy is 97.50%, which indicates that the model has high accuracy.

It is shown in Figure 4 that the trend of industrial water is declining, and there will be a small increase in 2020 and 2023.

The higher the output value of the industry is, the more water is used. However, with the improvement of science and technology, the utilization of water will increase, and then industry water will reduce.

- Prediction of agriculture water

The main factors of agricultural water consumption are farmland irrigation area, total agricultural output value, crop sown area.

We forecast the farmland irrigation area, total agricultural output value and sown area of crops separately in order to get a better agriculture water prediction. The three factors above will be as the input layer of the neural network, the agriculture water consumption as the output layer.

According to the relevant data of Shandong Province in 2000 -2014, the prediction model is used to predict the per-capita water use of Shandong Province in the next 15 years, and the results are shown in Figure 5.

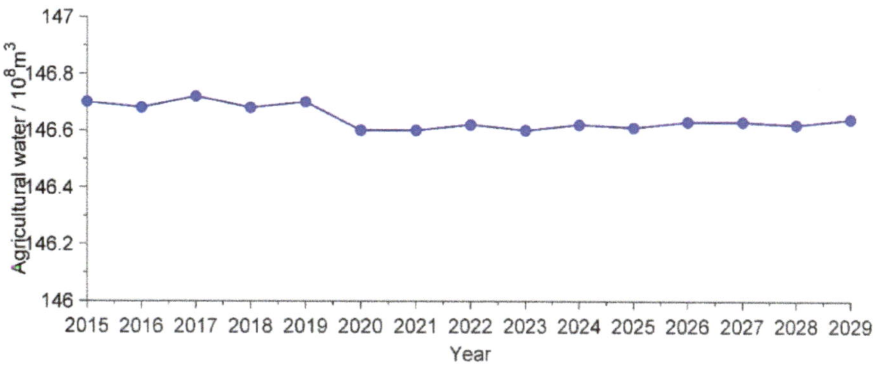

Figure 5. the per capita water use of Shandong Province in the next 15 years

The maximum residual error is 1.3, and the average prediction accuracy is 96.25%, which indicates that the model has high accuracy.

It is shown in Figure 5 that the agricultural water in the next period of time are stable at around 14.67 billion cubic meters.

- Prediction of domestic water

We forecast the population quantity and per capita water consumption separately in order to get a better domestic water prediction.

- Prediction of population

Assuming that the natural population growth rate is constant, we predict the population in the next 15 years. The results are showed in Figure 6.

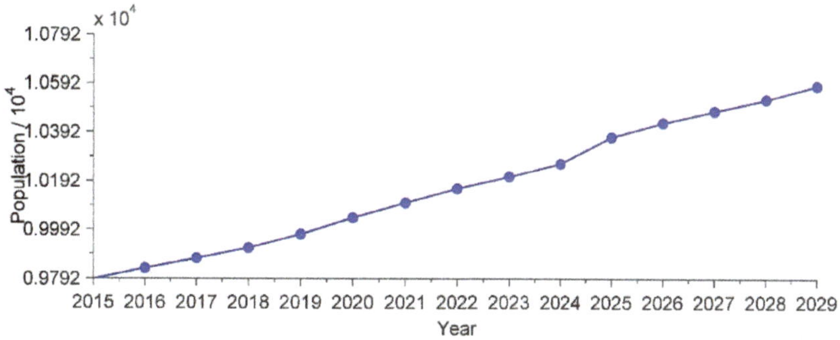

Figure 6. The population in the next 15 years

- Prediction of per capita water use

The factors affecting the per capita water consumption are divided into two categories: economic level and living conditions. The former is measured by per capita industrial and agricultural output and per capita disposable income, and the latter is measured by per capita industrial waste water discharge, per capita construction area and per capita afforestation area. The living conditions will be used for predicting per-capita water use.

Per capita industrial waste water emissions, per capita construction area and per capita afforestation area will be as the input layer of the neural network, the per capita water consumption as the output layer

According to the relevant data of Shandong Province in 2000 -2014, the prediction model is used to predict the per-capita water use of Shandong Province in the next 15 years, and the results are shown in Figure 7.

The maximum residual error is 2.1, and the average prediction accuracy is 97.70%, which indicates that the model has high accuracy.

With the development of economy and society, the per capita water consumption is slowly increasing. However, coupled with the increase in the number of people, the future of living water will be very large.

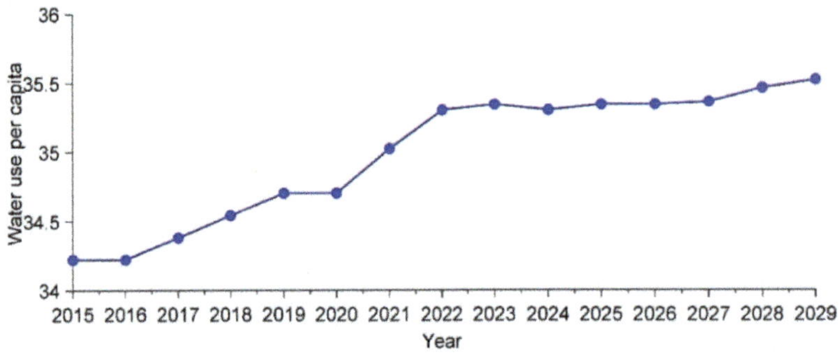

Figure 7. The per-capita water use of Shandong Province in the next 15 years

- Prediction of Total domestic water consumption

Total domestic water consumption C_t can be calculated by the following formula:

$$C_t = per_t * pop_t$$

Where per_t and pop_t is per-capita water use and population in t year.

Total domestic water consumption in the next 15 years are shown in Figure 8.

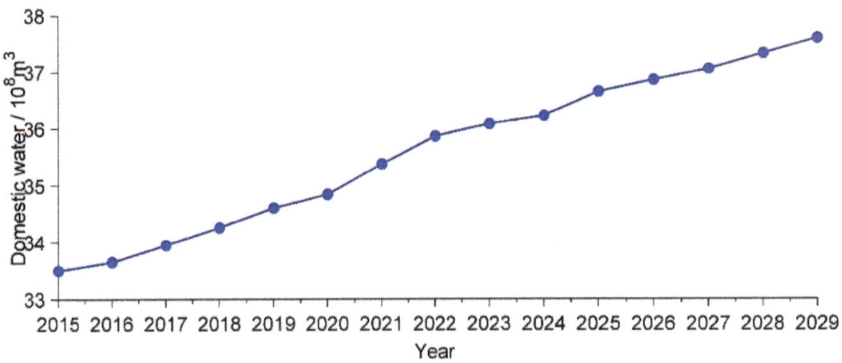

Figure 8. Total domestic water consumption in the next 15 years

3.2.3 Future Water Condition Analysis

Based on the above data and analysis, the supply and demand of water resources in the next 15 years are shown in Figure 9.

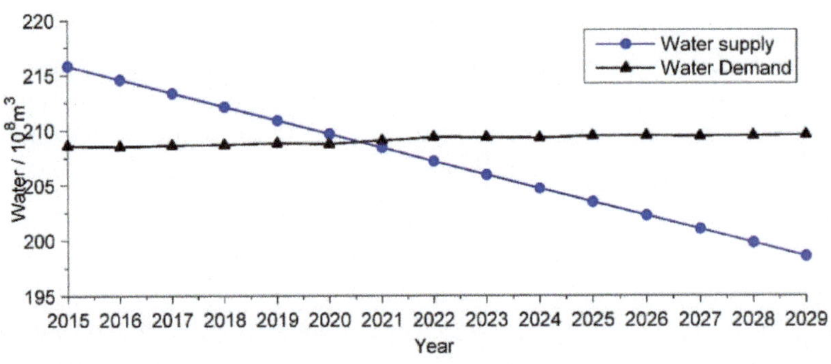

Figure 9. The supply and demand of water resources in the next 15 years

From the image above shows that the region's future demand for water resources is not changing, with an extremely slow rate of increase, while the supply of water resources to a relatively rapid decline. If the relevant measures are not taken, in 2020, the region's demand for water resources will be greater than the supply, so that the city into the water resources crisis.

It is known to all that the most closely related to the life of the residents is the water. In order to thoroughly analyze the future life of the residents of the region, the future of all aspects of water need to be analyzed. Detailed water conditions in Shandong in the next 15 years are shown in Table 3.

Table 3. Water conditions in Shandong in the next 15 years

Year	Water supply (10^8 m^3)	Water demand (10^8 m^3)	Industrial water (10^8 m^3)	Agricultural water (10^8 m^3)	domestic water (10^8 m^3)
2015	215.81	208.62	28.42	146.70	33.50
2016	214.57	208.53	28.20	146.68	33.65
2017	213.34	208.65	27.98	146.72	33.95
2018	212.10	208.69	27.76	146.68	34.25
2019	210.87	208.84	27.54	146.70	34.60
2020	209.64	208.76	27.32	146.60	34.84
2021	208.40	209.07	27.10	146.60	35.37
2022	207.17	209.36	26.88	146.62	35.86
2023	205.93	209.34	26.66	146.60	36.08
2024	204.70	209.28	26.44	146.62	36.22
2025	203.46	209.48	26.22	146.61	36.65
2026	202.23	209.49	26.00	146.63	36.86
2027	200.99	209.46	25.78	146.63	37.05
2028	199.76	209.50	25.56	146.62	37.32
2029	198.53	209.56	25.34	146.64	37.58

It is shown in Table 3 that the total water demand is increasing at a slow speed. After 15 years, the total water demand has only increased by less than 100 million cubic meters. In contrast, the region's water supply to a faster rate of decline, while the population has a certain amount of growth, as shown in Figure 10, in the end to the per capita water share of the rapid decline.

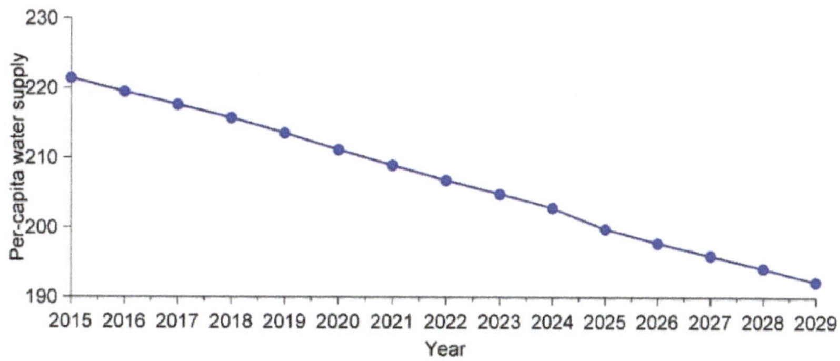

Figure 10. The per capital water supply in the next 15 years

From the part of the water demand, agricultural water will be in a relatively stable state, the industrial water slightly decreased, while the life of water consumption increased significantly. In the case of a sustained reduction in the amount of water supply, water will become increasingly tense, and may even block the supply of water. Therefore, the relevant departments may reduce agriculture water and industrial water be in order to ease the tension of water. However, it will reduce the total output value of agriculture and industry, and hinder the development of the city.

4. Conclusion

With the help of the dynamics model, we establish a model to measure the water resources carrying capacity of a region to provide clean water. Combining ARIMA Model and Neural Network Model, we predict what the water resources carrying capacity will be in Shandong Province. The total effect of sewage purification and self-cycling ability of water in Shandong Province is greater than the pollution effect. Over time, water pollution problems can be well controlled and will not cause a vicious circle. The benign effect not only avoids the destruction of ecosystem, but also promotes the surrounding areas. However, the population growth rate is greater than the water consumption rate that the system can afford. If the rapid growth of population is not timely controlled, the water supply will be insufficient in Shandong Province. The prediction of water resources carrying capacity of Shandong province in the next 15 years shows that the water demand will grow while the water supply will be less. Without taking relevant measures, the water resources in Shandong Province will be unable to meet the needs of the population in 2020. After analyzing the water situation in Shandong Province, we also find the reasons for its water scarcity which include the shortage of total water resources, the large population and the serious pollution, etc.

References

Brown, A. (2011). A Review of Water Scarcity Indices and Methodologies. *The Sustainability Consortium.*

Chen, B., Li, L.-J., & Guo, H. C. (2000). System Analysis on Water Resouces Supporting Alternatives for Chaidamu Basin. *Chinese Journal of Environmental Science, 21*(3), 16-21. Retrieved from http://en.cnki.com.cn/Article_en/CJFDTOTAL-HJKZ200003003.htm

Chen, L.-L. (2006). Study of Human Activities and Freshwater Resources Based on the Dynamic Model. *Research of Soil and Water Conservation*, Aug. http://doi:10.3969/j.issn.1005-3409.2006.04.075

Chen, N.-X., Ban, P.-L., & Zhang, W.-B. (2008). Fuzzy evaluation of the water resources carrying capacity based on the maximum entropy theory. *J Irrig Drain, 27*(2), 57-60. https://doi.org/10.1109/rsete.2011.5964734

Fang, G.-H., Hu, Y.-G., & Xu, Y. (2006). Research on the multi-objective evaluation model of regional water resources carrying capacity and its application. *Water Resour Protect, 22*(6), 9-13. https://doi:10.3969/j.issn.1004-6933.2006.06.003

Feng, L.-H., Zhang, X.-C., Luo, G.-Y. (2008). Application of system dynamics in analyzing the carrying capacity of water resources in Yiwu City, China. *Mathematics and Computers in Simulation, 79*(3), 269-278. http://dx.doi.org/10.1016/j.matcom.2007.11.018

Feng, S.-M. (2010). A forecast method for trip production based on BP neural network. *Journal of Harbin Institute of Technology*. http://doi:10.11918/j.issn.0367-6234.2010.10.023

Fu, X., & Ji, C.-M. (1999). A comprehensive evaluation of the regional water resources carrying capacity—application of main component analysis method. *Resources and environment in the Yangtze basin, 8*(2), 168-173. Retrieved from http://en.cnki.com.cn/article_en/cjfdtotal-cjly199902011.htm

Gleick, P. H. (1993). Water in crisis: a guide to the world's fresh water resources. Oxford University Press, Inc. https://doi.org/10.2307/2623756

Gong, L., & Jin, C. (2009). Fuzzy comprehensive evaluation for carrying capacity of regional water resources. *Water resources management, 23*(12), 2505-2513. https://doi.org/10.1007/s11269-008-9393-y

Huang, J. (2009). Agricultural water consumption trend and its influence factors in Beijing over the past 20 years. *Journal of China Agricultural University*. http://doi:10.3321/j.issn:1007-4333.2009.05.017

Liu, S.-F., & Chen, J.-H. (2007). Water resources carrying capacity based on the theory of ANN. *Res Sci., 29*(1), 99-105. http://doi:10.3321/j.issn:1007-7588.2007.01.015

Lu, F., & Xu, J., & Wang, Z. (2009). Application of GA optimized wavelet neural networks for carrying capacity of water resources prediction[C]//Environmental Science and Information Application Technology, 2009. ESIAT 2009. *International Conference on. IEEE, 2009*(1), 308-311. https://doi.org/10.1109/esiat.2009.59

Molden, D.(2007). A Comprehensive Assessment of Water Management in Agriculture. *Internatinoal Water Management Institute*. Colombo, Sri Lanka: IWMI.

Shandong Bureau of Statistics. (2014).Shandong Statistic Year Book. Beijing: China Statistics Press.

Shandong Provincial Bureau of Statistics. (2016). Retrieved from http://www.stats-sd.gov.cn/

Simonovic, S. P. (2002). World water dynamics: global modeling of water resources. *Journal of Environmental Management, 66*(3), 249-267. https://doi.org/10.1016/s0301-4797(02)90585-2

Song, X., Kong, F., & Zhan, C. (2011). Assessment of water resources carrying capacity in Tianjin City of China. *Water Resources Management, 25*(3), 857-873. https://doi:10.1007/s11269-010-9730-9

Tang, Y.-L., & Zhu, S.-B. (2006). Application of matter-element model to evaluation of water resources carrying capacity in Xinjiang. *Water Resour Protect, 22*(3), 490-494. Retrieved from http://en.cnki.com.cn/Article_en/CJFDTOTAL-SZYB200603011.htm

The UNEP GEO team[R]. (2000). UNEP Global Environment Outlook.

Wackernagel, M. (1998). Rees W. Our ecological footprint: reducing human impact on the earth. *New Society Publishers.*

Wackernagel, M., Onislo, L, & Bello, P. (1997). Ecological footprints of nations. *International Council for Local Environmental Initiatives.*

Wang X. (2012). GDP prediction based on intervention model and BP neural network[A]. Statistics and Decision. http://en.cnki.com.cn/Article_en/CJFDTotal-QZSD200703013.htm

Wang, S.(1991). Discussion on the prediction of urban industrial water consumption. *Water supply and drainage in China*, 1991.

Wang, S.-W. (2006). Analysis on the degree of water shortage in some areas in Zhejiang Province. *Water saving irrigation.*

Yao, Z., Wang, J., & Jiang, D. (2002). Advances in study on regional water resources carrying capacity and research on its theory. *Advances in Water Science, 13*(1), 111-115. http://doi:10.3321/j.issn:1001-6791.2002.01.019

Zhao, L. M. (2014). Multivariate Linear Regression and BP Neural Network Model for Per Capita Water Consumption Forecasting Model Research in Shandong Province. *Journal of Chongqing University of Technology (Natural Science).* https://doi:10.3969/j.issn.1674-8425(z).2014.03.024

Zhu, Y.-Z. (2008). Assessment of water resources carrying capacity: a case study of Zhangye in China. *3rd IASWE/WSEAS international conference on water resources, hydraulic & hydrology, UK* (pp. 145-150). https://doi.org/10.1007/s11269-014-0849-y

13

Social and Economic Impacts of Community Managed Reforestation and Natural Regeneration of Forestry Development, the Case of Humbo District, Ethiopia

Asamere Wolde[1], Tadesse Amsalu[1] & Molla Mekonnen Alemu[1]

[1] 35 Frazer Street Wilkinson Road, Freetown, Sierra Leone

Correspondence: Asamere Wolde, 35 Frazer Street Wilkinson Road, Freetown, Sierra Leone.
E-mail: asamerewol@yahoo.com

Abstract

This study was undertaken in Humbo District, Ethiopia. The objective of this study is to assess the effects of reforestation initiatives on the socio-economy of the rural households brought by the existence of the Community Managed Natural Regeneration (CMNR) project. Four Kebeles (administration units in a district) were selected for the study, based on their geographical location. One Kebele Administration (KAs) from four corners (North, South, East and West) of the closure area was selected. A total of 203 households were involved in the survey. Data were generated through questionnaire, FGDs, KIIs, and physical observation. The study depicted that respondent households were participating in the designation process and subsequent management of the Humbo CMNR project which in fact is considered to be good indicator for its sustainability. They also have good perception on planting trees, reforestation programs and on the ownership of forests as well. It was noted that in some of the surveyed KAs, drought, water constraints and strong wind are the main challenges of the communities. There are available institutional setups and bylaws to manage and protect the forest which was formulated by many stakeholders including local communities. To mention some; bylaws workable for penalizing illegal intrusion into the closure, igniting fire on the forest and benefit sharing.

Keywords: Community, natural regeneration, reforestation, area closure, coping strategy, sustainability, institutional set ups

1. Introduction

Nowadays, forest resources are treated with a good attention because of people's perception on their direct and indirect benefits for the community in specific localities and for a nation in general. The benefits of reforesting a given area are expressed in terms of economic, social and environmental impacts. Economic benefits can be employment opportunities and increased income at household, community, country and global level in general. Institutional set ups emerging at local level for the protection and management of community based forest resources, active involvement of women and the youth can be considered as social benefits as they might lead to strong social capital development. The value of forests to the world's human population is becoming increasingly evident. All these incidents seem to have been taking place largely in the absence of any legal frameworks or by-laws on how to responsibly use and maintain the forest resources. Tree cutting and charcoal making are the livelihoods of last resort. They involve significant labor inputs for very little income and mainly people who have no other options for income generation would go for the activities like this. It is widely accepted that smallholder and community-based forestry development can help alleviate rural poverty (Tipper, 2002). Such interventions have the potential to be in a win– win situation (UNEP, 2002). Community-based reforestation programs though having great potential for local community development, they face many challenges as the poor record of accomplishment of such programs testifies. As Edward Wolf of World Watch Institute wrote in the foreword to Reforestation in Arid Lands (Warner K, 1997), ''reforestation has become a centerpiece of rural development in arid lands, a key to conserving soil and water supplies, securing food production, and reducing the hardships of rural life''. Farmer Managed Natural Regeneration (FMNR), however, is believed to be the systematic regeneration of this pre-existing ''underground forest''.

The Community Managed Natural Regeneration (CMNR) pilot project in Humbo involves the restoration of tree cover and biodiversity in a degraded mountainous part of Humbo Woreda/District of South Western Ethiopia. The CMNR project is intended to contribute to the local community livelihood by improving the quality of water, reducing soil loss, giving an opportunity for fodder and fuel wood collection thereby protecting cropland by controlling flood hazards and improving biodiversity of the area. The move to bring the Humbo forest under rehabilitation through active involvement of the community is envisaged to trap many local socioeconomic benefits and ecological stability. It is evident that improvements in the vegetation cover of Humbo forest are quite visible. However, despite the great interests to rehabilitate the Humbo degraded forest, efforts to generate data indicating what is really happening in Humbo forest in the context of the objectives of its establishment is limited. The knowledge gap in the overall management of Humbo forest makes it difficult to give a clear account of the state of the forest and more importantly, to substantiate the significance of initiating and implementing community based forestry development programs.

It should be noted that FMNR in one form or another is not new. In fact examples that are centuries old can be found in various parts of the world including Europe, Japan and Africa. There is anecdotal evidence from Niger and neighbouring countries that spontaneous "rediscovery" and organic spread of FMNR occurred within the same period as the SIM experience.22 Also in 2004, World Vision Australia and World Vision Ethiopia initiated a forestry-based carbon sequestration project as a potential means of stimulating community development while engaging in environmental restoration. An innovative partnership with the World Bank, the Humbo Communitybased Natural Regeneration Project involved the regeneration of 2,728 hectares of degraded native forests. This brought social, economic and ecological benefits to the participating communities. Within two years, communities were collecting wild fruits, firewood and fodder. They reported that wildlife had begun to return and erosion and flooding had been reduced. In addition, the communities are now receiving payments for the sale of carbon credits through the Clean Development Mechanism (CDM) of the Kyoto Protocl.23 24 Figures 3 and 4: Satellite photos taken in 1975 (left) and 2005 (right) show greatly increased tree cover in southern Niger. Trees show as black dots. Source: US Geological Survey Following the success of the Humbo project, FMNR spread to the Tigray region of northern Ethiopia where 20,000 hectares have been set aside for regeneration, including 10-hectare FMNR model sites for research and demonstration in each of 34 sub-districts (Rinaudo, 2012). In addition, the Government of Ethiopia has committed to reforest 15 million hectares of degraded land using FMNR as part of a climate change and renewable energy plan to become carbon neutral by 2025 (UNDP, 2012). In Talensi, northern Ghana, FMNR has commenced on over 500 hectares and new projects, initiated by World Vision, are introducing FMNR into three new districts (Weston, 2013). In the Kaffrine and Diourbelregions of Senegal, FMNR has spread across 62,000 hectares in five years (Rinaudo, 2012).World Vision is also promoting FMNR in southern African countries. Other examples exist of both independently promoted and spontaneous FMNR movements occurring. In Burkina Faso, for example, an increasing part of the country is being transformed into agroforestry parkland. And in Mali, ageing agroforestry parkland of about six million hectares is showing signs of regeneration (World Agroforestry Center, 2012).

2. Research Problem

This study aims to assess the socioeconomic benefits of reforestation as perceived by households and the community. In addition, government and non-government institutional supports are assessed to develop an insight on the sustainability of CMNR project. Most importantly, as community based forest development involves a cross section of a community with diverse needs, capabilities, interests and attitudes the study is expected to provide plausible ways out for the multitude of problems influencing the effectiveness and sustainability participatory forest development interventions. Assessment of the socioeconomic of the Humbo CMNR project is therefore intended to generate viable knowledge and information that will serve as an important feedback to profitably manage the rehabilitated forest resource. It is evident that there are several abandoned and degraded lands scattered in many parts of the country. Knowledge developed in this study site can serve as an important lesson in efforts to replicate similar forest rehabilitation schemes. Carrying out a critical analysis of the community based forestry development like the Humbo intervention will indeed enrich the knowledge base that is needed to facilitate participatory forest development interventions in general.

3. Methodology

3.1 Description of the Study District

Humbo District is one of the 12 Districts in Wolayita Administrative zone of the Southern Nations, Nationalities and Peoples Regional State (SNNP). The District is bordered by Kindo Kosha District in the east, Soddo Zuria District in the North and Boloso Bombe District in the south. The District is located 25 kms from Zonal capital

Soddo, 195kms from the regional state capital, Hawassa, and about 430 km from Addis Ababa. The geographic location of Humbo (Tebella) township lies between 6° 46"48.47 and 6° 41"04.28 N and between 37° 48"35.44 and 37° 55"14.51 E. The study *site* is about 435 kms South of Addis Ababa, and about 200 kms South-west of Awassa, the capital city for SNNP (World Vision International, 2006).

3.2 Climate and Agro-Ecology of the District

Topographically, the altitude of Humbo District extends from 1300 m.a.s.l. at *Kolla* (hot lowland) to 2400 meters above sea level (m.a.s.l.). Humbo District is part of the Great Rift Valley of Ethiopia and is sub divided in to 2 climatic zones, midland (*Woina Dega*) and lowland (*Kolla*). The *Woina Dega* consists of undulating rugged hills and plateaus while *Kolla* is relatively flat extending to Lake Abaya. The annual rainfall in the District ranges between 700 and 1000 mm with an irregular maximum record of as high as 2500 mm and many areas exhibit the symptoms of erosion and locally severe flooding. Rainfall variability in the study area has been reported as major problem in recent years. Temperature varies according to the season and elevation, and generally it ranges between 12 and 24^{0}C with an average temperature of 18^{0}C (Tino, 2007). The dominant soil types of the district include reddish soil and brown soil and others. The soil between Soddo and the Kenyan border varies between Nitosols, Lithosols, Fluvisols, Luvisols and Xerosols (EMA, 1988).

3.3 Description of the CMNR Project Area

The Humbo Community-Managed Reforestation and Natural Regeneration of forestry development project is located in Humbo district, some 5-10 kms southeast of Tebella town, the capital of the Humbo District, and lies surrounded by seven rural Kebele (Note 1) Administrations namely Abella Longena, Hobicha Badda, Bolla Wanche, Hoibicha Bongota, Abella Gefeta and Abella Shoya and having varying population and mean household size (Table 1).

Table 1. Rural Kebeles involved in the management of the Humbo community managed reforestation and natural regeneration project.

Kebele	Population size	Number HHs	Av. # per HHs
Abella Longena	7006	1367	5.13
Hobicha Badda	4549	2115	2.15
Bolla Wanche	3883	554	7.01
Bossa Wanche	5042	1150	4.38
Hobicha Bongota	13538	2050	6.60
Abella Gefeta	5431	938	5.79
Abella Shoya	3668	456	8.04
	43117	**8630**	***5.6**

Source: CMNR/A/R Project records.

The forest enclosure has a total area of 2,728 ha. The topography of the forest site is characterized by massif mountain and chains of hills interspersed with small valleys, gullies, rocks and flat plains. Part of the forest land (about 35 %) lies in the lowland (*Kolla*) areas and about 65 % in the mid altitude (*Woina Dega*) area (Figure 1).

The forest rehabilitation project is designated as a carbon sequestration project where its day-to-day management is entrusted to the local community living in the Seven Kebeles. The government is in charge for providing technical assistance and financial support from the fund secured from World Bank to facilitate the forest rehabilitation programme. Communities living in the seven Kebeles have established Forest Development and Protection Cooperatives (FDPCs) and membership is on voluntary basis. The cooperatives have formulated their forest management bylaws that should be observed by all members.

Soils in the project area have been identified as vertisols/nitosols, underlain with ancient Precambrian basement rocks in the plains and calcareous soils at the hills. These soil types are common in the Ethiopian Highlands with a nitosol / vertisol gradient often occurring between the higher/steeper areas and the lower flatter areas. Nitosols are highly erodible and often acidic. Over-clearing has led to significant soil erosion problems, with mudslides occurring on the site (Note 2).

The seven kebeles surrounding the forest rehabilitation project land holding size follows similar patterns in the worda and almost all practice settled agriculture producing crops commonly grown in the woreda. They also practice an old age agro-forestry and planting of trees along farm borders. Local communities living close to the forest project site have close attachment to woodland now demarcated as a reforestation project site. Traditionally they have used the woodland/forest as major source of fuel wood, farm tools, construction material, and source of fodder for livestock and there are some segments of the local community whose livelihood is predominantly dependent on sale of wood and charcoal gathered from the woodland (Deribe, 2006). Many springs that emerge from the woodland are the other vital natural resources benefiting the local community though many of them have now become intermittent and small in volume.

The forest rehabilitation site has once been covered with dense mixed broad-leaved tree species, shrubs, herbs and grasses of food, feed, and fodder. Some of the most important trees and shrubs species were *Ficus sur, Terminalia ssp., Acacia ssp. Balanites aegiptica, Croton macrostachys, Grewia flavescens, Erythrina abyssinica, Olea africana, Dodonea viscosa, Oxytenthra abyssinica, Acacia senegal, Acacia abyssinica, Guizotia abyssinica, Urtica simensis, Brassica integrifolia, Pittosporum abyssinica, Asteracantha longifolia, Arundo donax, Commelina benghalensis, Eleusine jaegeri, snowdenia polystachya, Solanum tuberosum, Ricinus communis, Beta vulgaris, and Hyperrhenia rufa.* Currently however due to severe deforestation and continuous illegal encroachment by the local community much of the woody vegetation cover is lost and few remnant trees and shrub species scattered here and there are observed. The severity of the forest degradation is exhibited by the indicator species like *Dodonea viscosa*, and *Acacia dreanolobium* which are already in the process of invading wider spaces in Humbo site (Deribe, 2006). As it is true elsewhere in the district the study kebeles face development constraints such as fragmented and small land holdings, poor soil fertility/soil erosion, dependence on rain fed crop production system, poor infrastructures such as access roads, clean water supply, lack of grazing land, lack of technical skills, etc. Though limited in scale and scope there are some development intervention being undertaken by the local government such as promoting soil and water conservation, small scale irrigation, provision of improved seed, tree seedlings, credit services, veterinary services, health extension support, construction of labor based access roads, opening of schools and heath posts, etc. In general there are efforts made by the government to mobilize the local community to be active partner in local development activities.

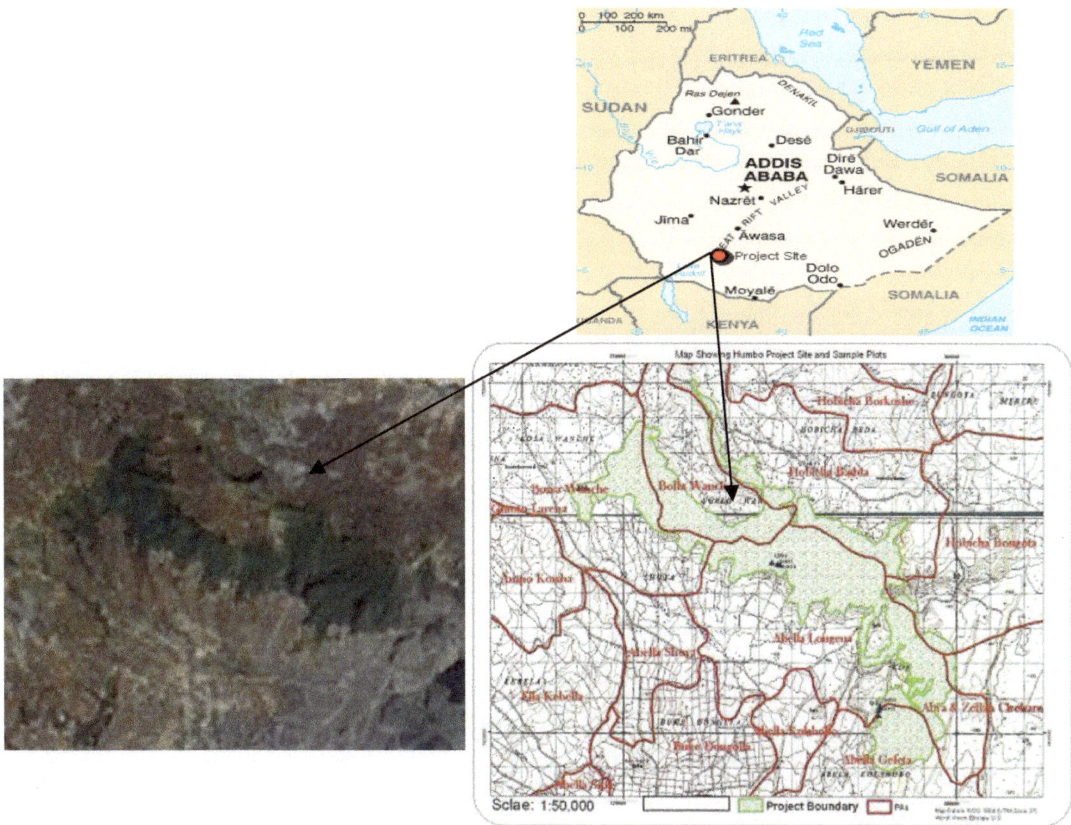

Figure 1. Map of the study area (Source: CMNR/A/R Project records)

3.4 Socio-Economic System of the District

Population data of the 2007 national census shows that the woreda has a total population of 136,679 (CSA, 2007) of which the majority (92 %) are rural inhabitants. With an estimated area of 384,583ha, Humbo Woreda has a population density of 478 per Km^2. The largest ethnic group in the district is reported to be Wolayita tribe (95%). "*Wolayitegna*" language is spoken as a first language by 99.9% of the inhabitants. The majority of the population practice Ethiopian Orthodox Christianity with 99%, while 1% are Muslims (CSA, 2007). Of the total land area of the Woreda, about 24 % is coved with plantation and natural forest, 1 % is grazing land, 66% agricultural land, 8 % settlement, and others occupy 1 % (Deribe, 2006).

Farmers of the district practice a mixed farming system, with crop and livestock husbandry being largely practiced in the traditional management system Information from the district agricultural development office shows that the major crops cultivated include maize, sorghum, sweet potato, Teff (*Eragrostis teff*), coffee, cotton, haricot beans and chickpeas while livestock predominantly include cattle, sheep and goats. Apart from farming some households practice off farm activities such as petty trade, handicrafts, and seasonal migration to supplement their livelihood incomes. Likewise there are also a number of households that are engaged in collection of fuel wood and making of charcoal from remnant woodlands to generate income for the family. Farm land holding in the district is small and it ranges from 0.125 to 2 ha and the average land holding per households is about 0.3 ha. There is high fragmentation of farmland; the average number of plots per household in the district is estimated at 3.56. Many households with small land holdings are reported to be facing food deficit. There are signs where the local farmers are encroaching into fragile ecosystems such as hillsides to cultivate crops and this coupled with overgrazing and complete removal of the woody vegetation seems to have played a big role the degradation of much of the steep slope areas in the district (Deribe, 2006).

3.5 Sampling/Site Selection and Data Collection

3.5.1 Stratification of the Project Site

To enhance the forest rehabilitation interventions the Humbo CMNR site was stratified into four strata based on vegetation cover, terrain, slope and altitude of the sites (Figure 2)

Stratum 1: Southernmost part of the project with few scattered trees. This area is the closest to Lake Abaya, and is at lower in elevation than the rest of the project strata.

Stratum 2: It is in two sites, a small site in the northern part of the project with the major part of the stratum lying between stratum 1 and stratum 3. It has scattered trees and grass cover and is the least undulating of the project strata.

Stratum 3: It has the largest area and accounts for about 40% of the project area. It has shrub growth and encompasses the southern sloping aspects.

Stratum 4: It is the northern most part of the project and has shrub vegetation on rocky hillsides, with a northerly aspect.

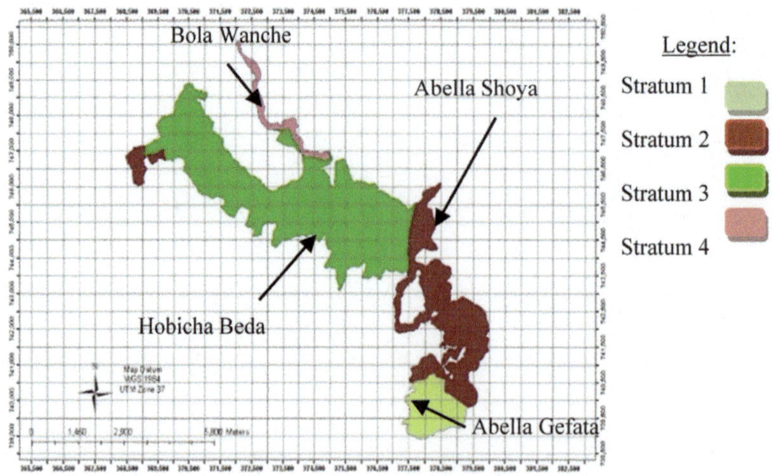

Figure 2. Map of the study area (Source: CMNR/A/R Project records)

3.5.1 Reconnaissance Survey

The fieldwork for this study was carried out from February 2012 to March 2012. The initial field activity was a reconnaissance survey of the study area to establish background information on agro ecological condition, livelihood activities, land use systems, natural resource base, and development activities being implemented. Relevant interactions were also carried out with governmental and non-governmental institutions working in the district and at zonal level and selected individuals having knowledge of their localities were also met to obtain further information about the study area. This exercise in general has helped the researcher to establish a general picture of the proposed study area and prepare a relevant questionnaire and group discussions issues. The preliminary study of the assessment area was followed by study site selection.

3.5.2 Study Site Selection

Humbo CMNR project is aimed to bring socioeconomic development through environmental rehabilitation and the principal stakeholders in the management of the project are the seven Rural Kebeles bordering the reforestation site. In the process of selection of the specific kebeles for the study, consultation was carried out with the local agricultural development agents, and Kebele cooperatives established to facilitate the management of the development activity and criteria for sample kebele was established. Accordingly Kebeles having more households far from the reforestation site were grouped together; others with better access road and market (urban centre) were grouped in another category; and those Kebeles deemed facing more food insecurity were set aside as another category in the site selection process. Taking note of the above variables using purposive sampling method, four Kebele Administrations namely Hobicha Beda, Bola Wanche, Abela Shoya and Abela Gefeta distributed in different parts of the CMNR project were selected.

3.5.3 Selection of Respondent Households

The local communities residing around Humbo closure are used as the primary sources of information. Hence, representative households and community groups were selected from list of households in the four Kebeles that served as a sampling frame. The current Kebele Administration is structured to have many "*gotts*'' or hamlet comprising substantial numbers of households. The number of *gotts* in each Kebele Administration was recorded and their relative position to the location of the Humbo CMNR project site was identified. In this exercise Hobicha, Keranso, Shoya, and Koisha *gotts* were recorded for in Hobicha Beda, Bola Wanche, Abela Shoya and Abela Gefeta kebeles respectively and a total of 12 sample *gotts* were taken proportionally. Households in each *gott* were then stratified in different wealth groups with the help of knowledgeable persons and development agents based on local value judgments accepted by the local community. In this respect livestock ownership, land holding, capability of producing adequate grain for the family were considered as major criteria for classification supported by close knowledge of the economic status of each household.

About 5% of the total beneficiaries out of the four Kebele administration were taken as a sample for analysis (Table 2). Taking into account that this research relies on multiple data sources and considering the ideas promoted by (Fritz Scheuren, 2004) that sample size was fixed at 5% and a total of 203 households were used for the study.

Table 2. Sample Respondent households in the Study Kebeles/ Distribution of Sample Size per Kebele

Kebele Name	Pop.	HH No.	5% of the HH	Ave. HH size
Hobicha Beda	4549	2115	106	2.15
Bola Wanche	3883	554	28	7.01
Abela Shoya	5431	938	46.9	5.79
Abela Gefeta	3668	456	23	8.04
Total	**17531**	**4063**	**203**	**5.7**

After fixing the sample size for each study Kebele, the sample households of each Kebele were proportionally allocated to the three wealth status category (better-off, medium and poor) (Table 3); and sample households for each Kebele were selected by applying simple random sampling statistical technique and using the Table of random number.

Table 3. Distribution of Sample Size by Wealth Status per Kebele

Name of KAs and Total household in sample *Gotts*	Wealth Status Across Gotts Selected in Each kebele			Allocated Sample size		
	Poor	Med	Rich	Poor	Med	Rich
Hobicha Beda (5 gotts)	75	54	40	44	31	29
Bola Wanche (2 gotts)	29	33	41	8	9	13
Abela Shoya (2 gotts)	21	31	30	6	9	8
Abela Gefeta (3 gotts)	35	48	46	11	18	17
Total	**160**	**166**	**157**	**69**	**67**	**67**

3.6 Data Collection Tools

3.6.1 Households Interviews

To realize the objectives of the study, a questionnaire was designed to include a range of issues that could provide an insight into the socio-economic system of the household (demographic characteristics, livelihood activities, access to markets, extension support, etc), and their involvement in the CMNR project, their perception on the importance of the reforestation project, and institutional arrangements to run the project and other related issues. Both close and open-ended questions that address the specific objectives were included to explore the required data from the sampled HHs. The preparation of the questionnaire was preceded by reviewing of literature related to the preparation and analysis of survey questionnaire.

The questionnaire was then pretested on 12 households living close to the project area to make sure that the questions are relevant and easily understood by the interviewee and to improve the content of the questionnaire. Accordingly, some questions on quantifying the benefits were modified in qualitative forms. Redundant and unnecessary questions were also detected and removed, and difficult expressions were made clear.

Four data collectors, one for each sample Kebele, were used to complete the survey questionnaire. They were given training first to thoroughly understand the contentment of the questionnaire and secondly to give them skill on how to approach the respondents during the interview. In general the maximum possible effort was made to gain as much relevant information as possible from respondent households by giving due consideration to their demands on timing and place of the interview as well as carefully recording their responses.

3.6.2 Focus Group Discussions

Group interviews in the form of focus group discussions using semi structured questions gives researchers to have a better opportunity to look into more deeply into issues of concern and develop new lines of inquiry that arise during interviews (Gilham, 2000). Focus group discussion interviews were therefore conducted to supplement and substantiate information that was generated in the household questionnaire and in-depth interviews with key informants. In the group discussions, individuals and institutions familiar with the forestry development project were included. Accordingly, a total of 12 focus group discussions were held in the study Kebeles and the number of people participating in each group varied between 7 and 13. The focus group discussion involved members of the FDPC from the four Kebeles, men and women group of different asset endowments, one in each Kebele, Cabinet Members of the four Kebeles, and informal discussions were also held with school teachers and students.

A checklist of questions was prepared for each group. Issues addressed in the focus group include awareness on the importance of the forestry development, perception matters on ownership, intuitional arrangement, capacity building, sustainability of the forest resource being developed, etc. Likewise, group discussions were held with the staff of the District Agricultural Development Office and the Bilateral Organizations with the aim of generating information that will be useful for future management of the forestry intervention. Hence the discussion gave focus mainly on institutional matters pertaining to the management design of the reforestation project, arrangements of benefit sharing, capacity building made at local level and in the office, sustainability matters and major challenges or issues requiring immediate response, etc. All group discussions were held in a language that everyone speaks well and the researcher was assisted by a note taker who was oriented on how to document and organize the most important issues raised during the discussions.

3.6.3 Interviews with Key Informants

Key informant interviews were carried out by involving selected group of individuals who are likely to provide needed information, ideas, and insights on the proposed researches as argued by (Kumar, 1989). Key informants were selected with the help of the Kebele Administration and agricultural development agents. Three Key informants were used in each Kebele. Checklists of open-ended questions were prepared and used for the interview.

The major topics discussed included information about the process of designating the CMNR project, perceived benefits from the project, ownership and sustainability issues, etc. The interview was conducted at the place and time convenient for the key informants. Key informant interviews were conducted in Amharic and recorded by hand written notes and by digital voice recorder. Digital voice recordings were then transcribed and documents were imported into MS-excel software for coding against the key research questions.

3.6.4 Data Validation Process and Management

Maximum efforts have been exerted to keep the quality and clarity of the household level data. Raw data was well cleaned and verified carefully on the field through supervising the collection process. It was rechecked in the office prior to encoding for its credibility and best results. An effort was also made to reduce defects before the fieldwork is completed if in case some errors and vague records observed. Every questionnaire was checked at the field after being completed by enumerators.

3.6.5 Data Analysis

In conducting this study, both qualitative and quantitative data were collected; hence, a combination of quantitative and qualitative data analysis methods was carried out. Data collected from focused group discussions were immediately summarized through discussion with an assistant note taker/enumerator. Outstanding and prominent issues rose during group discussion and key informant interviews were screened by checking how many of the speakers and which category of households have reiterated the same issue in the process of the discussion. Both diverging and converging issues on particular issues were identified and used for analysis in the context of the research objectives and questions.

Most of the variables collected in the questionnaire and used in the analysis were categorical, nominal or ordinal. In addition, the numeric or measurement variables were often not normally distributed. Hence, non-parametric tests Chi Square (X^2) tests was used in the analysis to see the association between independent variables (sets of household characteristics, villages and institutional variables) that are likely to influence local community involvement in the forestry development efforts and also perception matters on the overall management of the project. An association level of 0.05 was chosen as the minimum significance level. Depending on the relevance of data regression analysis was also carried out. The analytical tests were also in many places supported by descriptive statistics (such as computation of percentages of single variables, the median and average outcomes) and prominent views gained during focus group discussions. The latest SPSS-statistical software was used for analysis.

4. Results and Discussions

This chapter presents the results and discussion of the study. It describes the demographic characteristics and socioeconomic activities of the respondent households in the context of the research. And then discuss about perception and attitudinal matters related to the reforestation project, capacity developed and finally it addresses institutional issues influencing the involvement of beneficiary households and effectiveness and sustainability of the forestry development program.

4.1 Gender Mix and Age of Respondent Households

The sample households interviewed in this assessment are totally 203 of which 148 (72.9%) are male and 55 (27.1%) are female. The households represented in the study encompassed age groups of 30 and above. As explained in Table 4, about 67% of the household heads are between 30 and 50 years of age, while those aged between 51 and 60 years accounted for only 26.6%. The least share (4.4%) is represented by household heads over the age of 61. The predominance of a reproductive population in the study area implies that population density is increasing at a very fast rate and may pose stress on the environment. Hence, it seems logical to strengthen family planning activities already in place in the district albeit in limited extent. On the positive side, mobilizing this potential immense work force for productive activities can bring tremendous development.

The average household size of the study population is 6.5 with a range of 3-10 household members within a family. Although the age distribution is not worked out, the household size is an indicator of labor availability within the household. Literature indicates that households with more family members are most of the time expected to spend part of their time for tree growing and other activities related to forest management. However, there is a fact that the more the family size will result in increased need of the household to feed them all. In pursuit of meeting the family's need, there could be kind of encroachment on the closure particularly by economically weak segments of the community and this was particularly witnessed by group discussion participants stating that poor households still tend to encroach the Humbo community managed natural regeneration site.

The participation on forest management and protection activities differs between male and female respondents. Male household heads do most of the forest protection works though there was some involvement from female household heads. This is one of the areas the institutions that have stake in management of the closure should work further so as to bring about a good gender balance. Based on the assessment result, it can be said that some woman have been participating in trainings conducted by the CMNR project stakeholders ever since the start. All of the respondents stated that women participation in the CMNR project had meaning so as to enable it sustainable. It was noted that there is an association between age of individuals and involvement in the management of forest (X^2=8.275 and df= 6, $p<0.05$).

4.2 Wealth Status and Level of Engagement on Forest Encroachment

All groups of people (poor, medium and better off) have nearly equal contribution to works that should be done by them as a requirement. Sometimes the works of those people categorized under medium weighs a bit more. *Pruning* is one of the main activities undertaken by most of the respondents; about 87 (42.9%) out of the total (n=203). Out of which, about 41 respondent households were medium, about 23 respondents were poor households and the rest 23 of them were found to be better-off. *Hoeing* is also another main activities identified to be practiced by all class of people. Out of the 68 households who were undertaking hoeing, about 26 of them were poor households, about 25 respondents were better-off and the rest 17 respondents were medium class households. The surveyed households according to the findings have found to be less practicing *wild life conservation* and *weeding*. Despite the fact that these activities are less practiced, about 31 respondent households have said that they were involved in wild life conservation works. Out of them, about 13 were poor households, 12 were better-off and about 6 were medium class households. Only 17 of the total respondents (n=203) had stake in *weeding*. Out of the 17 households, about 7 of them were poor households, again 7 were better-off and the rest 3 were found to be medium class households.

Table 4. Kinds of work distribution among female and male headed HHs

Sex of the HH	Kind of work HH undertakes				Total
	Weeding	**Hoeing**	**Pruning**	**Wildlife conservation**	
Male	12	55	57	24	148
	8.1%	37.2%	38.5%	16.2%	100
Female	5	13	30	7	55
	9.1%	23.6%	54.5%	12.7%	100
Total	**17**	**68**	**87**	**31**	**203**
	8.4%	33.5%	42.9%	15.3%	100

Increased involvement of people in forests through community-based management programs has in many cases proven to benefit the environment (soil erosion reduction, water supply and biological diversity) and reducing local poverty (Salim, 1999). In this study, the reasons for growing trees as revealed by respondent households are related to environmental concerns in the first place followed by firewood needs, generating cash income, collection of construction wood, and production of fodder. Tree planting, contribute to conservation of biological diversity, both at the site and landscape level, extensive reforestation with plantation species can help ameliorate long-term environmental degradation in badly eroded landscapes, restoring not only ecological functionality but also site productivity (Maginnis and Jackson, 2003).

4.3 Major Livelihood of the Community

The livelihood of the local community is primarily dependent up on agriculture with some exercises in petty trade for supplementing their household need (Table 5). Crop production is one of the activities commonly practiced in the area for assisting livelihood of local communities. However, it is currently constrained with several shortcomings. The first constraint in crop production according to households' in-group discussion is lack of access to improved agricultural technologies including commercial fertilizers, pesticides and improved seed. Another severe obstacle is small land holding and landlessness. About 90% of the respondents disclosed that there are many young people looking for land but were not able to have access to land holdings due to shortage of arable land.

Table 5. Major Livelihood of local community

Activities	Frequency	Percent
Agriculture and Petty Trade	106	52.2
Agriculture	97	47.8
Total	203	100.0

The area is known for its high population pressure and hence shortage of arable land is the main problem associated with agricultural production. Out of the total respondent households (n=203), about 117 (57.6%) of them have a land size of 0.5 to 1 ha, some 39% hold 1 to 1.5 ha of land, only 2% and 0.5% of the respondent households own more than 1.5 ha and less than 0.5 ha of land respectively. The study also noted that the Humbo CMNR project did not displace households from their land holdings and their land holding sizes remained the same even after the area came under closure. The majority of households (66%) in the study area feel that their land holdings are small and are barely adequate to satisfy their subsistence needs. About 34% of the households claim that they have adequate land and are able to produce enough food for household consumption and for the market as well. During group discussions, many of the participants stressed that there are many young people who are landless and may be a potential danger to the forest under rehabilitation. Unless the project integrates the demands of this segment of the community it eventually become difficult to control encroachment problems into the woodland.

4.4 Benefits Obtained from the Closure

The CMNR project document stated in its base line survey that the community will have a benefit to collect grasses, dried leaves and woods, medicinal plants, spices and wild honey. In addition to this, the local community is privileged to use the employment opportunity created as a result of the CMNR project (World Vision International, 2007). It is quite visible that closures if properly managed have wide range of benefits. In Humbo CMNR project case, the existing income source, which was only focusing on primitive agriculture, is believed to be replaced by the employment opportunity brought by the existence of the project and by trade of carbon offsets as well. The change is believed to come because of the awareness created after the closure is being kept by involving stakeholders thereby conducting several types of trainings for the community as most of the surveyed households reflected in this regard.

In addition, there are four types of benefit which are realized as a result of the existence of the forest resources management project. Harvesting dried leaves, barks and grasses from the enclosure is first ranked benefit as witnessed by 103 (50.7%) respondents out of the total sample households (n=203). According to the household survey, about 64 (31.5%) of respondents have said that Fuel wood (dried ones only) is another type of benefit, reduced flood damage and fruit/NTFPs collection are also the benefits as explained by 36 (17.6%) respondent households and they are ranked as the second, third and fourth ranked benefits respectively. These resources are allowable only through contacting the cooperatives, which have been given main role in managing and protecting the forest. About 62% of the respondent households who are living near the closure have said that they are more accessed for most of the benefits shown on the figure 9 and the remaining 38% of the households who live far from the closure said that they have little access because of the distance between the closure and their settlement. Although there are benefits, there are also problems and challenges faced by the households who are closer to the CMNR site mainly because of wild life attack.

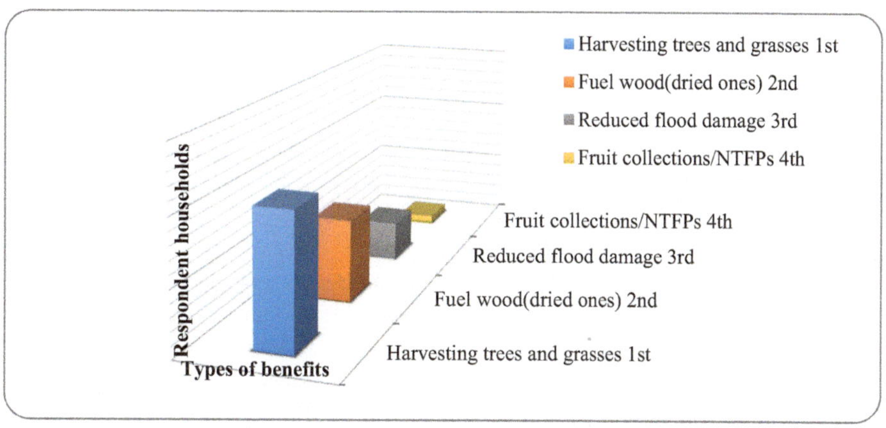

Figure 3. Types of benefits gained by the existence of the project

4.5 Employment Opportunities

The information found with regard to income generation and creating employment opportunity had been triangulated by using the data gathered in FGDs, key informants and household survey. Food insecurity has been a long-standing issue in the Humbo Woreda. A comment by a female community member during a FGD explained as follows:

> *"Keeping children nourished was a priority and in times of shortage of food, meals taken by adults per day were reduced. We use our food to feed the children, and we go to sleep without eating"*

Most households are food secured between September and November. Then the proportion of households that are food secured declines rapidly over the following five months and virtually no households are food secured by April. This is the experience of poor and middle class communities as findings from the household survey depicted. As the trends for household within the CMNR project area shows, during the peak food-shortage months (December to May), households in the project area seem to be working off farm activities in a slightly better manner. This may reflect some change due to the project intervention. This is possibly due to increased access to income from sale of wood and grasses, or to decreased expenditure on forage and firewood at the household level, due to access to wood and grasses in the protected area. Only members of FDPCs have access to the protected area to collect firewood from pruning and harvest dried grasses.

4.6 Income Generated at Household and Community Level

The protection and regeneration of the Humbo forest through the CMNR project has effected many positive changes in the surrounding microclimate, and members of the FDPCs were sustainably harvesting firewood and grasses - thus reducing expenditure on these items at the household level, or earning small amounts of income through sale of these items. However, the scale of savings or income earned was modest and would also provide a limited buffer from food shortages at the household level.

> *"[Humbo] protected area is improving, there is more rain and temperatures is going down, and we use the pruned wood from the protected area to sell at market" [Women FDPC member, Humbo, February 2012].*

> *"There is some evidence that communities were diversifying their livelihoods to include trees for economic return and for improved food security. The main tree types owned (Eucalyptus, Cordia, Grevillea, Mango, Moringa and Coffee) are productive species. Apiculture is forest friendly and it has great potential in the future, communities have asked us to point out where they can buy the materials, it is a forest friendly activity. It has potential in the market. It can even be exported as an organic product" [WVE Project Officer, February 2012].*

Assessing the income created by the project existence was started by interviewing one of the KI from the CMNR project staff at National Office level. It is stated as follows;

> *"Humbo is the first CDM project registered in Ethiopia so the EPA is proud of this achievement. We thank God! We are also very satisfied with the progress of the project, with the objectives being met so far, including the creation of jobs, the restoration of the environment, the flooding has stopped, preventing the stones washing down on the fields. A key achievement expected is that funds will go to the community through sale of carbon Certified Emission Reduction (CER). It means a lot and there is even some money being flow down to the communities"*

A key component of the Humbo CMNR project model is the provision of income from carbon. The potential for communities to receive income from the sequestration of carbon as their communal forest regenerates makes this project a 'first of its kind' for the country. As studies show, clean development mechanism (CDM) projects range from the purely commercial, aimed at generating returns to investors, through to those whose goal is to increase the incomes of local landowners by using carbon markets as a component of a rural development strategy (Jindal et al., 2008).

The carbon-trade registration process with World Bank extended for over the years, understandably testing the patience of communities and other stakeholders in the process. Feedback from community members also highlighted ambitious plans for development outcomes from carbon funds. This is explained by Women FDPC member;

> *"Food, water and wood are the major identified problems in the Woreda. I expect water development works from the carbon project. In future, we want to see a factory, a store for crops to conserve grain. If we get carbon money, we can buy oxen for those who do not have it. Water needs more development. We need a vehicle to serve the community as an ambulance"*

And Male community elder;

> *"We hope we will be benefiting even more from this project, especially from the carbon credit; we have plans to construct a health centre and roads because of this project"*

Community respondents' expectations from carbon revenue included expensive interventions such as provision of better infrastructure and services for health and education, and income generating activities such as eco-tourism. The extent to which these expectations can be met depend on a number of factors, such as the ability of community to meet monitoring and reporting requirements, collectively manage carbon funds, the price of carbon, management and protection of forest from illicit use and natural disasters such as bushfires and prolonged drought.

According to the emissions reduction purchase agreement with the WB, Income from Carbon revenue (1st round payment 34,184USD/ 579,029 Ethiopian Birr (ETB) transferred to community on 15 March 2011). Some Kebeles have more population with large hectare of protected land than others and hence receive a larger share (World Vision International, 2012). Though the share seemed like this, the money will not be directly handed over to the individuals. Because they have agreed to use the income gained by sale of carbon to be invested on development schemes like; building grain store, bridge maintenance, and purchase of mill-grain etc. As a result of which, the communities had shown enthusiasm towards sustaining the CMNR project.

4.7 Energy Sources for Household Consumption

The main energy sources for household consumption were identified to be fuel wood, crop residue and cow dung as explained by the respondent households with a percent proportion of 69%, 18% and 13% respectively (Figure 4). Fuel wood is mostly produced from the trees planted in their homestead and sometimes coop members can collect dried fuel wood from the closure. As the experience of the local communities depicts, they were solely dependent on the fuel wood that is produced from the communal forest over the years. Now, these practices somehow are being changed as a result of the project. The project not only has the physical work components but also it has a full package in order to address the public awareness by conducting various types of trainings with regard to managing and protecting the forest they own in their locality.

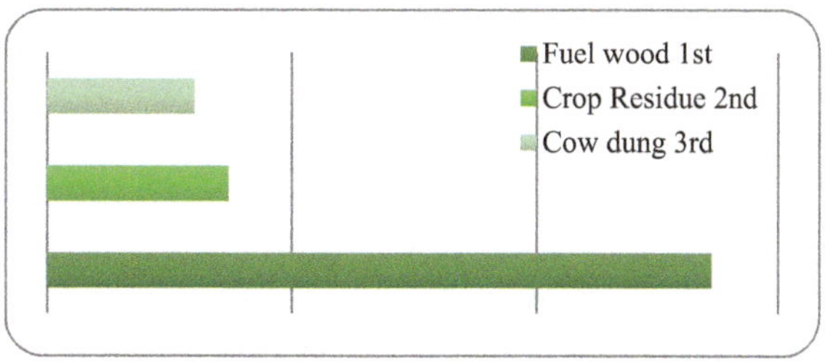

Figure 4. Energy source for HH consumptions

5. Conclusions

The community managed natural regeneration (CMNR) project to date has been implemented largely as planned with high levels of participation and strong ownership demonstrated by community and government partners. Feedback from respondent households and key informants highlighted genuine attitudinal changes in Humbo community towards trees and the forest, now seen as highly valued natural resources worth protecting. The change in the forested area in just three years is dramatic, with previously denuded slopes now blanketed in a cover of young trees, shrubs and grasses. The progress at Humbo demonstrates that communal forest protection combined with natural regeneration techniques (CMNR) and reforestation (tree planting) can produce remarkable, positive outcomes within a very short period. However, these considerable achievements were not without a struggle. Many community members stood to lose by locking up an area of land that had been freely accessible, and advocates of the CMNR project faced opposition and serious risks. Key informants during the study admitted that they were planning to attack those who were promoting the CMNR project. Then witnessing the return of the forest and associated benefits, they now state openly that their earlier judgments were very wrong. Future CMNR project planners need to learn from the Humbo experience; withdrawal of rights to access a communal resource is bound to meet with fierce opposition from those affected. The approach used by government was strategic and empowering; partnering with government to increase the legitimacy of forest protection, and enabling community to identify and determine the compensation for disadvantaged households.

The overall response to the project has been very good, as evidenced by growing memberships in the Forest Development and Protection Cooperatives (FDPCs). According to secondary data, there were over 4200 members in total – and based on a rule of membership, one member per household, this represents coverage of about 50% of the 8630 households across the seven Kebeles. The majority of respondents said they practiced CMNR technique on both communal and on their own farmland. Therefore, at this stage, there is growing evidence that project impact has gone beyond increasing awareness and building capacity, to actual practice change. The study found reliable evidence that the project approach has encompassed key stakeholder knowledge and participation in project design, implementation and management. Community feedback consistently showed genuine ownership of the project and advocacy for continued protection of the forest. Prior to closure of the forest, those households who would be negatively impacted were identified to be directly benefited by the CMNR through support for alternative livelihoods development. Women carry out many agricultural and forestry related activities and are a key target for projects attempting practice change, yet they are usually under-represented in such types of forestry. Under the FDPC membership rules, those who joined up at the start paid a low fee for membership, and then the membership fee increased over time. This would reflect the reduced investment in cooperative and forest management activities by those who joined later compared with the founding members. There was apprehension about the timing of the carbon benefits among community key informants, which suggests that there is work to be done by the CMNR to ensure that the associated processes and decision making includes the community and their partners. The Humbo model has since emerged as a key learning opportunity for Ethiopia and elsewhere in Africa. Moreover, it has also been showcased by the Ethiopian government, the World Bank and WV. Humbo FDPCs, WVE and the Ethiopian government are in a collaborative relationship with a shared goal for sustainable environmental and community development outcomes. It is believed to be effected through the protection of the Humbo forest. The carbon component has underpinned this high level of engagement and commitment, but in addition, the reputation of an organization like the World Bank and the now international profile of the Humbo CMNR has increased community confidence and led to a sense of pride and optimism.

6. Recommendations

- Create an agreed set of indicators, timeframes, and recording templates to enable local stakeholders to collect monitoring data required by WB. Ensure all stakeholders have the capacity and resources required to implement it.
- High profile of the project means there will be many visitors, and they need to plan for this. A protocol is highly recommended to be developed for hosting visitors. This should include agreement between local stakeholders about coordination of visits, visiting fees, role of community members as guides, facilitators required, production of brochures, etc.
- Create a project coordination committee at Humbo Woreda (District) level to coordinate CMNR activities locally and to facilitate communication between external stakeholders, local government and the Union formed by the FDPCs.
- Increased focus on equity of CMNR outcomes for women and poorest households.
- The multiple realities of the various CMNR partner's activities need to be assessed and revisited from time to time. The best way to do this is through face-to-face engagement.
- Take account of financial recording, monitoring environment, identify recurrent problems and set solutions.

Acknowledgement

Most of all, I would like to express my heartfelt gratitude to my supervisor Dr. Tadesse Amsalu, for his continuous and unreserved coaching and follow-up starting from the inception until the end of study period. If there was not his material as well as moral assistance, it would have been very difficult to overcome the challenges I faced during the whole course of the study. Molla Alemu Mekonen (PhD Candidate) has been a vital support and source of encouragement throughout the whole process of putting this article on publishing – my heartfelt gratitude goes to him as well. I am also indebted to Dr. Belayneh Ayele who not only stimulated my interest in this study subject but also assisted me in the planning process, generously sharing his wealth of experience.

I would also like to express my deepest respect to Ato Biyazen for his cooperation in laying out statistical analysis tools so that I could be much comfortable to work on my row data. I would also like to thank Ato Hailu Tefera for his cooperation during proposal development. The input from W/t Tsehay Ataklt is also highly appreciated. I would like to thank Bahir Dar University for giving me this chance. I also need to thank World Vision Ethiopia for assisting me with research finance. The kind cooperation of Elmi Nuri, Kebede, Kibret, Sintayehu Kefato and Sertsebirhan Hailu (Project staff), has been indispensable throughout my study time. I would like to extend my gratitude to Petros Alem, Dereje Mekonen, Ayana, Getahun Asnake and Dereje Mekuria (drivers of WV) for their unreserved collaboration. Finally, I would love to express my deep inside appreciation for my whole family who had always been encouraging me.

Special thanks also go to the farmers of the study area, for sharing their wealth of knowledge and resources for the accomplishment of this research. I also thank you all who have been encouraging me ever since I have started my works in the field until I finish my works; I really thank you once again!

References

Deribe, B. (2006). *Humbo Community Managed Natural Regeneration Project Final Report*, Addis Ababa (pp. 13-29).

FAO (Food and agriculture Organization). (1978). Community Forestry for Local Community Development. *FAO Forestry Paper, Rome, 7*(114).

FAO (Food and agriculture Organization). (2005a). *The state of the world's forests* (pp. 20-95). Food and Agriculture Organization of the United Nations, Rome, Italy.

FAO (Food and Agriculture Organization). (2005b). *The importance of soil organic matter:* Key to drought-resistant soil and sustainable food and production, Soils Bulletin. 80.

FAO (Food and Agriculture Organization). (2007). *State of the world's forests*. Food and Agriculture Organization of the United Nations, Rome, Italy.

Framework Convention on Climate Change – Secretariat. (2011). Executive Board of the Clean Development Mechanism. *Meeting Report. 65.*

Fritz, S. (2004). Qualitative Research Methods: *Qualitative Survey sample size determination, Texas,* 9-10.

Jindal R., Swallow, B., & Kerr, J. (2008). *Forestry-based carbon sequestration projects in Africa: potential benefits and challenges*. Natural Resources Forum.

Kumar, K. (1989). Conducting Key Informant Interviews in developing countries. AID program design and evaluation methodology report. Agency for international development (AID), Washington D.C. 13(37).

Maginnis, S., & Jackson, W. (2003). *The pole of planted forests in forest landscape restoration UNFF inter-sessional experts meeting on the role of planted forests in sustainable forest management.* New Zealand. 88-96.

Reij, C. (2012). *African Re-greening Initiatives Blog*. Retrieved from http:// www.africa-regreening.blogspot.com.au.

Salim, E. (1999). *Our forests our future*: *Summary report of the world commission on forest and sustainable development*. Manitoba, Canada. 6-34.

Tino, R. (2007). *Forest Fire Management Plan*. Humbo Assisted Community Managed Natural Regeneration Project, World Vision Australia.

Tipper, R. (2002). Helping indigenous farmers to participate in the international market for carbon services: the case of Scolel Te. In S. Pagiola, J. Bishop, & N. Landell-Mills (Eds.), *Selling forest environmental services: market-based mechanisms for conservation and development*. Earth scan, London

UNDP. (2012). Programme document for Ethiopia (2012-2015). Retrieved from http://www.et.undp.org/content/dam/ethiopia/docs/Ethiopia%20Country%20Programme%20Document%202012%202015%20-%20English.pdf.

United Nations Framework Convention on Climate Change. (2011). *Subsidiary Body for Scientific and Technological Advice.* Thirty-fifth session Durban.

Warner, K. (1997). The Vision and Role of Community Forestry in Sustainable Development. Proceedings of the XI World Forestry Congress. *Social Dimension of Forestry's Contribution to Sustainable Development, 5*, 57-68.

Weston, P., Hong, R., & Morrison, V. (2013). *Talensi FMNR Project: Endof-Phase Evaluation Report.* unpublished, World Vision Australia/World Vision Ghana.

World Agroforestry Center. (2012). Improving Sustainable Productivity in Farming Systems and Enhanced Livelihoods through Adoption of Evergreen Agriculture in Eastern Africa, Research Proposal to the Australian Centre for International Agricultural Research, World Agroforestry Center, Nairobi

World Vision International. (2007). *Natural disasters: World Vision responds to regional crises across the globe*. Retrieved from http://www.worldvision.org/news.nsf/news/200709_natural_disasters?OpenDocument

World Vision International. (2012). *Ethiopia*. Retrieved from http://wvafrica.org/index.php, Option=com_content & view=article&id=137&Itemid= 68. Accessed Dec July 2011

Notes

Note 1. Kebele is an Amharic word that has a meaning of Chiefdom

Note 2. See fao.org/Wairdocs/ILRI/x5493E/x5493e19.htm

14

Characterization of Water Source Types and Uses in Kirisia Forest Watershed, Samburu County, Kenya

John Warui Kiringe[1], Francis Mwaura[2] & Fiesta Warinwa[3]

[1] Habitat Planners, P.O. Box 10982-00100 Nairobi, Kenya

[2] Department of Geography & Environmental Studies, University of Nairobi, Box 30197-00100, Nairobi, Kenya

[3] African Wildlife Foundation (AWF), P.O. Box 310-00502, Nairobi, Kenya

Correspondence: John Warui Kiringe, Habitat Planners, P.O. Box 10982-00100 Nairobi, Kenya. E-mail: jwkiringe@gmail.com

Abstract

Kenya's dry-land water catchments are valued for their water provision services but their conservation is given little attention. This study was carried out between October and December 2015 and documented water source types and uses by humans and livestock in Kirisia Forest watershed. Different water source types exist in the watershed including earth dams, water pans, shallow wells, boreholes, springs and streams. The estimated population of livestock and locals using these sources was 180,645 and 147, 060 respectively. Earth dams and water pans provided water to the highest population of community members estimated at 11,564 people followed by boreholes at 9,886 people while streams, springs and shallow wells were used by the least number of people. They also provided water to the highest number of livestock estimated at an average of 15,422 animals. The highest amount of water was abstracted from boreholes at nearly 197,720Litres/day (197.72m^3/day) followed by earth dams and water pans at 91,960Litres/day (91.96m^3/day), and the least was from shallow wells, springs and streams at about 38,000Litres/day (38m^3/day). Daily water abstraction from all the water source types by humans and livestock was nearly 366,540Litres/day or 366.54m^3/day. Twenty three sub-locations rely on water from the watershed and based on the 2009 population census, their projected water demand was approximately 182,238,520Litres/day (182,238.52 m^3/day). Water demand by livestock was estimated at 12,172,600Litres/day (12,172.60m^3/day) based on the 2013 population data. Overall water demand by humans and livestock in all the sub-locations was estimated at 194,411,120Litres/day (194,411.12 m^3/day); and these findings demonstrate the critical role played by the watershed in sustaining locals livelihoods and pastoralism.

Keywords: Kirisia Forest watershed, local livelihoods, Samburu County, water demand, water source types

1. Introduction

Societies the world over rely entirely on the earth's ecosystems and the different services they provide including food production, water provisioning, climate regulation and amelioration, disease management and spiritual fulfillment (MEA, 2005). Thus, Ecosystem Services (ES) can be viewed as benefits that humans get from ecosystems either directly or indirectly (Daily, 1997; de Groot, Matthew & Roelof, 2002). Nonetheless, in the last century, humans have changed ecosystems primarily to meet a growing demand for fuel, fresh water, food, fiber and timber. While exploitation of ecosystems and their transformations have substantially improved livelihoods and contributed to rapid socio-economic development, its ramifications on environmental health has created a lot of concern (MEA, 2005). In recognition of the value of Ecosystem Services to humans, a lot of attention has focused on preserving the earth's ecosystems and natural resources in general. Nonetheless the importance of services provided by ecosystems for human welfare is not new, and dates back to the time of Plato, and economic conceptualization of ecosystem values (Costanza et al., 1997).

Water is central to survival of humans and environmental integrity but it also supports food and energy production, recreation, manufacturing and industrial processing. It's therefore one of the Ecosystem Services that currently receives a lot of attention because of its role in sustaining human life and the environment (Gleick, 2003; MEA, 2005). Most hydrological services occur in highlands of forest watersheds, which work as natural systems that; collects, manufactures, and distributes water and provides hydrological services to lowlands (GWP,

2000; MEA, 2005; Perez-Verdin, Navar-Chaidez, Kim, & Silva-Flores, 2012). Thus, key aspects of the water cycle, mainly; water infiltration, water surface run-off, evaporation is partly dependent on the state of forest cover. If forest cover decreases substantially, it ultimately lowers the water quantity and quality that is available to downstream users (Wu, Kim & Hurteau, 2011). Since watershed services benefit nature and people, they are the most appropriate units for water management because they represent landscapes that harness rainfall and channel it to rivers, streams, springs, wetlands and aquifers (MEA, 2005).

In Kenya, water is the foundation of livelihoods in rural and urban areas but most parts of the country are faced by exceptionally high water demand amidst a decline in its availability (Mogaka, Gichere, Davis, & Hirji, 2006). This situation is worse in dry areas which make up nearly 80% of country's land mass, and where water resources are naturally scarce (Mogaka et al., 2006). Accordingly, the United Nations (UN) has classified Kenya as a chronically water-scarce country (RoK, 1991), and freshwater availability is very limited with an annual renewable supply of about 647m^3 per capita, which is way below the 1,000 m^3 per capita set as the benchmark for water scarcity. The country's economy is predominantly rural-based and mainly dependent on natural resources (Mogaka et al., 2006; GoK, 2007); and water is a key backbone for sectors such as energy, agriculture, fisheries, livestock and tourism (RoK. 1991; Mogaka et al., 2006; GoK, 2007). In this regard, the National Water Policy Sessional Paper No. 1 of 1999 on Water Resources Management and Development, and the National Development Plans laid great emphasis on the importance of water in achievement of the country's development goals and Sustainable Development Goals (SDGs) (GoK, 2007). According to GoK (2007), Kenya's vision of poverty reduction and promotion of strong and sustainable socio-economic development is partly pegged to provision of water resources in all parts of the country. Hence, degradation, general mismanagement and wastage of water present a big setback in the country's socio-economic development agenda. Based on scanty data, Mogaka et al. (2006) estimated that degradation of water resources costs the country nearly 0.5% of the Gross Domestic Product (GDP) or an equivalent of Ksh. 3.3million each year.

The World Bank proposes that; pollution, lack of climate variability preparations, rampant degradation and mismanagement of water resources have increasingly slowed down economic growth and perpetuated poverty not only in Kenya but across the entire African continent (Grey & Sadoff, 2002). Over the years most parts of Kenya have been characterized by wide-spread misuse and poor conservation of water resources, water theft and corruption, and its allocation is mostly made using inadequate hydrological data and information (RoK, 1991; Mogaka et al., 2006; Kiringe, Okello, Tome, & Seno, 2009). The effects of this situation are felt more in water-stressed areas particularly dry-lands or the country's rangelands. Groundwater resources have not been spared either and these have been over-exploited way beyond the natural capacity of the hydrological system to replenish them. Decline in water quality has also been rampant through agricultural, urban and industrial wastes and massive deposits of sediments emanating from degraded landscapes.

Past studies have shown Kirisia Forest is increasingly getting deforested and degraded due to; frequent forest fires, wood fuel harvesting, illegal human settlements and logging of high quality trees, livestock grazing and fodder harvesting (Watai & Gachathi, 2003; Lambrechts, Woodley & Gachanja, 2005; Anne, 2009; KFS, 2012; Nyaligu, 2013). This is a big threat to the ecological integrity and forest cover of the watershed, and its natural ability to replenish water resources and provision of other environmental goods and services. Forests play an important role in water and soil conservation, mainly; regulation of water flow and control of watershed erosion and floods (LVBC, 2011). But when their vegetation cover is interfered with substantially, these functions are usually reduced leading to decline in quantity and quality of water flowing downstream from the watershed (Honeck & Smith 1992; KWS, 2002; Akotsi, Gachanja & Ndirangu, 2006; Olang & Kundu, 2011; Gichuhi, 2013). In this regard, Kirisia Forest watershed will only continue to provide valuable goods and services to locals on a sustainable basis if it's secured and the threats facing it are effectively mitigated. Given its role in sustaining the socio-economic well-being of local communities, it's imperative that it's conserved and managed as a local and national livelihood and economic treasure. Failure to do so will trigger a cascade of negative impacts such as prevalence of water conflicts and decline of; local livelihoods, pastoralism and wildlife based tourism. In the long-term it will impede and slow down the socio-economic aspirations of local communities.

Availability of adequate water is a key bottleneck to sustainable socio-economic development, improved livelihoods and poverty reduction in Samburu County, and other dry regions of Kenya (RoK, 1991, 2010, 2013; Mogaka et al., 2006). So, high altitude forests in the country like Kirisia are critical water catchments and supply water which supports local livelihoods, pastoralism and wildlife conservation (RoK, 1991). Nyaligu (2013) carried out an inventory on the status of water in Kirisia Forest and his findings revealed a worrisome situation on the ability of the forest to continue supplying adequate water to meet the needs of different users. Thus, our study focused on the following objectives; i) delineation of the spatial coverage of Kirisia Forest watershed and its drainage system, ii) document water source types, and, iii) assess water use and demand levels by humans and livestock.

2. Area Description

Kirisia Forest also locally known as Leroghi is one of the forested high altitude landscapes in Samburu County in the northern part of Kenya (Figure 1); and lies at an altitude of 2,000 to 2,200m ASL (Watai & Gachathi, 2003; Hitimana et al.2005). It's one of the oldest state protected areas in the country and was gazetted by the British administration in 1933 (Watai & Gichohi, 2003; Ngaligu, 2013). At time of gazettement, the forest covered nearly 92,000ha but thereafter its acreage was systematically reduced, and currently covers less than 780Km^2 (Watai & Gachathi, 2003; Hitimana et al., 2005; Anne, 2009). The mean annual rainfall in the forest and its environs ranges between 600 – 750mm, and is received thrice a year with January and February being the driest (Watai & Gachathi, 2003; Anne, 2009; Nyaligu 2013). According to Kiringe, Mwaura and Kimeu (2015b) the northern zone of the forest like Porro area typically receive a slightly higher mean annual rainfall of about 575mm compared to the central region around Mararal town which receives a mean annual rainfall of 563mm while the southern zone in Baawa area has the lowest at 552mm.

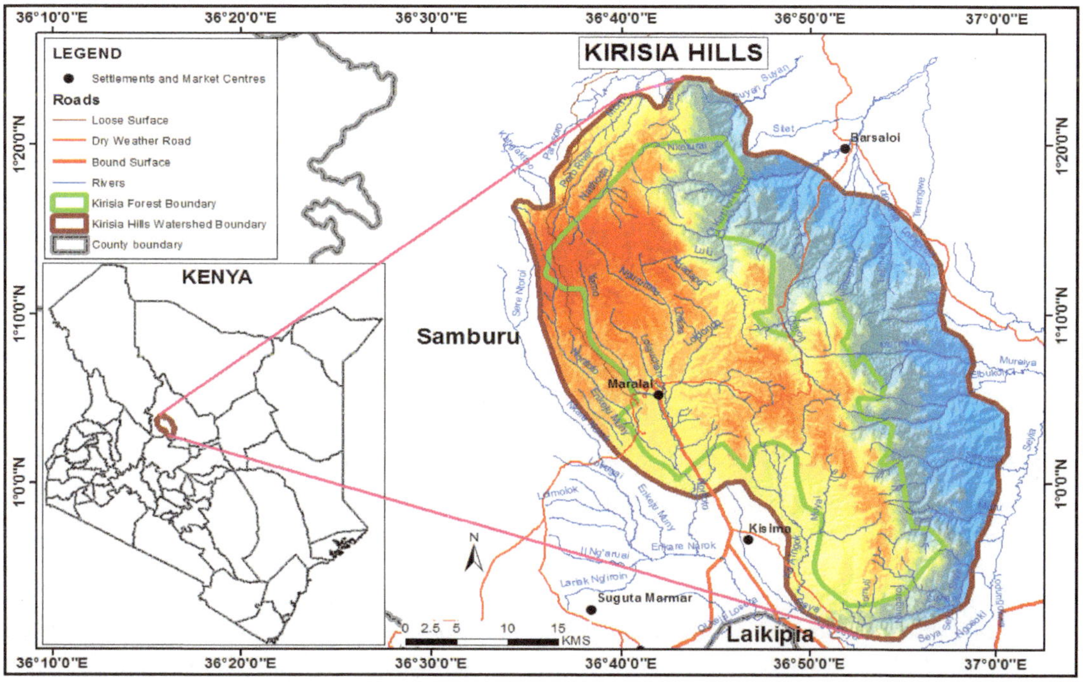

Figure 1. Location of Kirisia Forest and its watershed in Samburu County

The region where the forest is situated is classified as a semi-arid environment, and lies in agro-ecological zone IV-VI (Pratt & Gwynne, 1977). But due to its tri-modal rainfall and high altitude, the forest acts as an important dry-land water tower (Nyaligu, 2013), and is a critical water source for the local communities especially the pastoral Samburu who are dominant inhabitants in the region (Watai & Gachathi, 2003; Anne, 2009; Nyagilu, 2013). Pastoralism is the main land use although mixed crop-livestock farming is done in wetter areas like Porro where maize, wheat and other crops are grown (Hitimana et al., 2005; Anne, 2009). Kirisia Forest is the foundation of local livelihoods and provides various goods and services including dry season grazing grounds and watering sites, livestock fodder, construction materials, herbal medicine, honey and wood-fuel (Watai & Gachathi, 2003; Anne, 2009; KFS, 2012). Communities in the immediate environs of the forest live in thirteen group ranches which were created under the Kenya Livestock Development Project of 1968-1980 (Anne, 2009; Hitimana et al., 2011). But the Dorobo people who are mainly hunter-gatherers live within the forest which forms the backbone of their livelihoods.

According to Beentje (1990) Kirisia Forest has diverse vegetation communities, and four woody plant species tend to dominate the forest top canopy; *Olea europaea* spp *africana* (up to 34 %), *Juniperus procera* (up to 25%), *Podocarpus falcatus* (up to 26 %) and *Croton megalocarpus* (15 %). Those species dominating the middle canopy are; *P. falcatus* (12-45 %), *Olea. europaea* spp *africana* (21-28%), *Juniperus procera* (20 %), *Teclea simplicifolia* (13-15 %) and *Croton. megalocarpus* (12 %). The disturbed and rocky areas of the forest are characterized by small sized trees and shrubs such as; *Euclea divinorum*, *Carissa edulis*, *Rhus natalensis* and *Croton dichogamus*. Apart from woody flora biodiversity, the forest also supports a variety of large mammalian wildlife species including; the African elephant (*Loxondonta africana*), olive baboon (*Papio anubis*), giant forest hog (*Hylochoerus meinertzhageni*), cape buffalo (*Syncerus cafer*), common zebra *(Equus burchelli*), bush-buck (*Tragelaphus scriptus*), eland (*Taurotagus oryx*), common warthog (*Phacochoerus aethiops*), Maasai giraffe *Giraffa camoleopardis tippelskirchi)*, lion (*Panthera leo*) and spotted hyena *(Crocuta crocuta)* (Watai & Gachathi, 2003; Hitimana et al., 2005; Anne, 2009).

3. Methods

3.1 Watershed and Drainage Network Delineation

Kirisia Forest watershed layout was demarcated using 30m ASTER Digital Elevation Model (DEM) satellite data based on the drainage network. ASTER satellite data was presented on ArcGIS software window and overlapped on the watershed drainage layer. The layout of the watershed boundary was then obtained by digitizing a polygon using DEM and surface water flow direction on the landscape; and the polygon then depicted the forest watershed layout in which the study was then focused. The spatial extent of the drainage network within the Northern, Southern, Eastern and Western zones of the watershed was determined using satellite imageries. This data was used to establish the total length (in kilometers) and drainage density of rivers and streams in each zone.

3.2 Documentation of Water Source Type and Usage

An inventory of water source types was done using drive-through transects traversing different zones of the watershed, and local Samburu guides acted as translators. This was supplemented by walking through non-motorable landscapes. When a water source was encountered the following information was noted; i) type of surface water source (e.g. spring, stream, water pan or earth dam) and type of ground water source (borehole and shallow well), ii) geographic location (GPS coordinates) using a hand held Garmin GPS unit), iii) ownership, and, iv) state of the water source i.e. whether it was dry or active and functional. The GPS coordinates were geo-referenced on a topographic map of the watershed using GIS software to create a water source type map. Reports and relevant water documents for the Samburu County where Kirisia Forest is situated were sourced from; Samburu County water offices at Maralal town, African Wildlife Foundation (AWF) Nairobi office and the Water Resources Management Authority (WRMA) office at Rumuruti town. Some of the key information obtained from these reports included the water source types found in the Kirisia Forest watershed and the rest of the county.

Water use at various source types was assessed by interviewing community members with the help of local Samburu translators. Some of the key information gathered from the respondents included; i) number of households, community members and livestock using a given water source type, ii) water uses, and, iii) amount of water used by livestock and abstracted daily (in jerricans) by each household. This information was supplemented with secondary water use data based on reports and records from the Water Resources Management Authority (WRMA) office at Rumuruti Township, AWF office in Nairobi, Samburu County water and works department in Mararal and Samburu Water and Sanitation Company (SAWASCO) office. The key information obtained included; i) typology of water uses, ii) water beneficiaries and their numbers, and, iii) water abstraction (use) levels.

Understanding water demand is important in planning its development and supply. In this regard, we projected human and livestock populations and their associated water demand levels in Samburu Central sub-county which depends on water from Kirisia Forest watershed. Given the dry nature of the region and scarcity of large quantities of water, irrigated agriculture is not possible, and livestock and households (rural and urban populations) are the main water users. In Kenya, human population census is normally done every ten years. Therefore, the projected human population growth in the sub-county for the year 2020 and 2030 was based on the 2009 national population census records (KNBS, 2009). This was done using the equation below and the annual population growth rate for Samburu County which is estimated at 2.10% (KNBS, 2009).

$$P_2=P_1(r/100+1)^t$$

Where:-

- P_1=Population at start of a census period
- P_2=Population at end of the census period in consideration
- t=Time (Number of years of census period in consideration)
- r=Percent annual population growth rate

Estimates of water demand levels in the sub-county by humans for the year 2009 was based on secondary data (KNBS, 2009) while that of 2020 and 2030 were projected based on Samburu County 2009 national households census data. Livestock water demand was based on Samburu Central sub-county 2013 livestock data (RoK, 2013) but it was not possible to account for water use and associated demand level by different species of wildlife found in the watershed.

4. Results

4.1 Watershed and Drainage Network

The drainage system in Kirisia Forest watershed comprises of a variety of seasonal and permanent rivers and springs but due to the state of the environment, they tend to be more seasonal most of the year. Some of the rivers include; Baawa, Baringo, Saanata, Lalmargwet, Nchangalo, Longek, Yaamo, Nashuda, Rapa and Mwata though they are relatively small in size and their permanency and water discharge vary on a seasonal basis. Laggas are also common in the watershed but are usually seasonal in their water provision mainly during the wet season. At the time of this study only three of the many valleys in the watershed had some little water flowing through at; Nachuda stream to the north of Porro, Nangaro springs and Ngonyeki springs in the Baawa area. The drainage network density varied among different zones of the watershed (Figure 2) ranging from 0.461Km to 0.518Km per Km (Table 1).The Northern and Eastern zones had the longest and most dense drainage network followed by the Western and Sothern zones in that order (Table 1 & Figure 2).

Table 1. Drainage network length and density in watershed zones of Kirisia Forest

Watershed zone		Approximate area of watershed zone (Km^2)	Total length of laagas (Km)	Drainage density (length of streams in Km per Km^2)
1	Northern zone	1,096	568	0.518
2	Southern zone	666	307	0.461
3	Eastern zone	1,000	503	0.502
4	Western zone	762	372	0.488

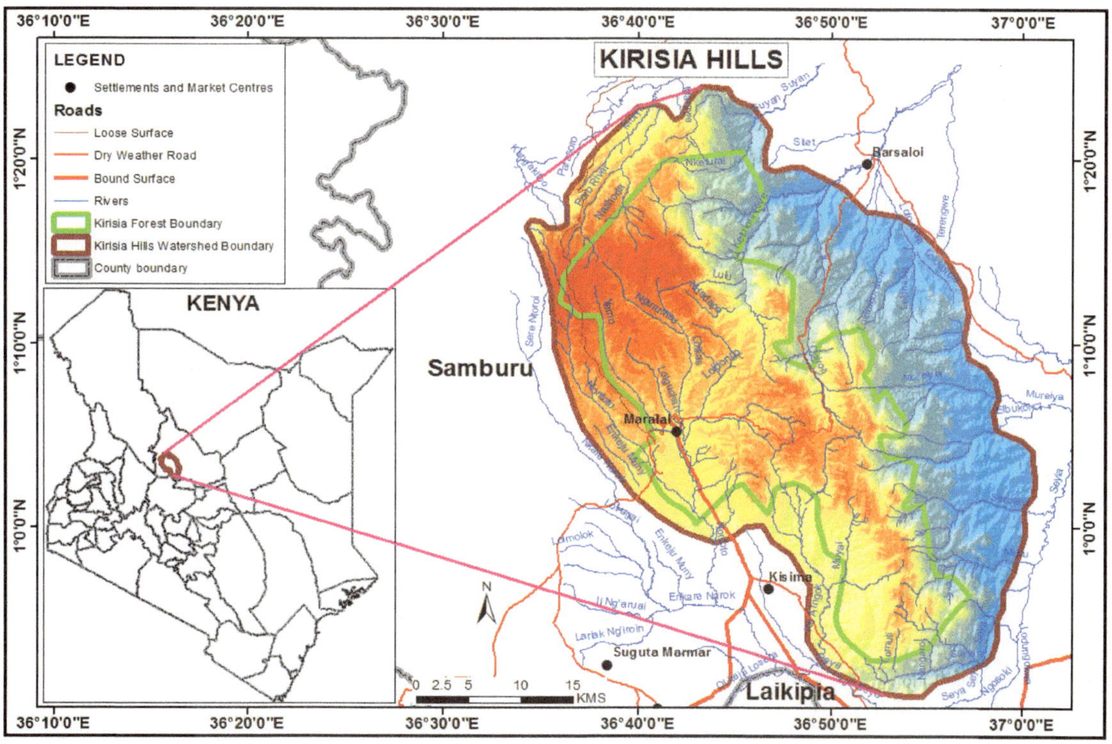

Figure 2. Spatial layout of the Kirisia Forest watershed drainage network

4.2 Water Source Types and Usage

Samburu County is characterized by a dry environment but its climate varies spatially based on variation in landscape altitude. The dry nature and low rainfall characteristics of the county implies that surface water resources are scarce in space and time with ground water in form of boreholes being the most reliable water sources. In this regard, numerous boreholes have been drilled during the colonial days and after independence in an attempt to ensure households have access to reliable water supply. Approximately 644 water sources types exist in Samburu County and are broadly grouped into seven categories (Table 2), and most of them are concentrated in Kirisia Forest watershed and its environs (Figure 3). However, not all of these water sources are operational, and this presents a challenge in regards to ensuring that locals and their livestock have access to water throughout the year. Earth dam are the most common (213) followed by shallows (141), water pans(112) and boreholes(104) in that order (Table 2) with permanents surface water sources in form of streams and rivers being the least common.

Table 2. Typology of water source types in Samburu County

Type of water source		Estimated total number of water source type
1	Permanent rivers & streams	2
2	Protected springs	35
3	Unprotected springs	37
4	Earth dams	213
5	Water pans	112
6	Boreholes	104
7	Shallow wells	141
Estimated total number of all water source types		644

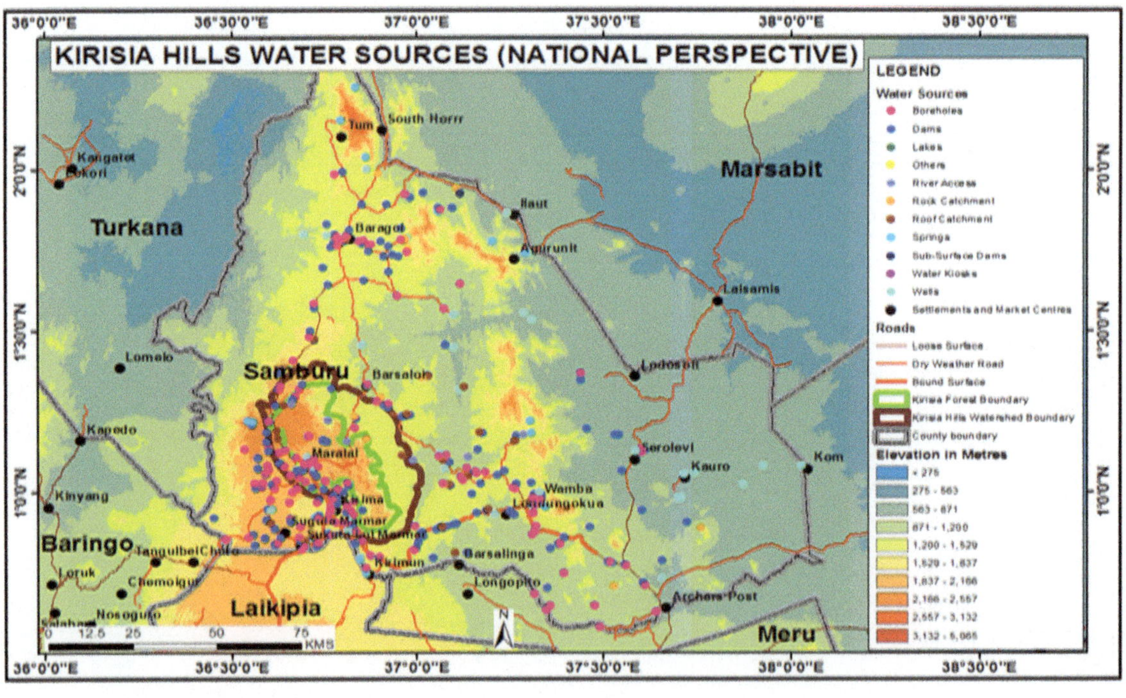

Figure 3. Spatial distribution of water source types in Samburu County (Source: Samburu CIDP 2013-2017)

Our study established that surface and ground water source types existed in Kirisia Forest watershed comprising of; springs, streams, earth dams, water pans, boreholes and shallow wells (Table 3). Some of the key springs included; Ngonyeki, Nang'aro, Soit-Ngablo and Lechoro-Lenanguya while streams were; Naschuda, Naadapo-Elkileku, Lonyunyi and Lorok. Most of these water sources were found in the Northern, North-western and Southern zones of the watershed but no streams and springs were found in the Central zone. Earth dams and water pans were also found in the watershed, and the former were mainly confined in the Southern and Central sectors (Table 3). Only two dry water pans were encountered; Ngabai and Ndikir-Elgwesi Elgwesi, with dams being the most prevalent, and included; Ntontol dam, Su-en dam, Prison dam, Nontotol dam and Suen-Lkidoroto dam. Dams provided water for domestic uses and watering livestock and wildlife. Boreholes and shallow wells were the main ground water sources and were mostly confined to the Northern and Southern zones (Table 3). The key boreholes were; Ngorika, Baawa, Leirr, Lchoro and Partuk 1 owned by SAWASCO and supplied water to Maralal town. Lonkutukie, Ntonlol, Lorok (1, 2 & 3) and Ngablo were some of the shallow wells found in the watershed. The wells and boreholes were critical water sources for households, livestock and wildlife particularly during the dry season when water resources were not readily available in springs, streams, earth dams and water pans.

Table 3. Inventory of water source types found in Kirisia Forest watershed

	Water source types	Location in the watershed	GPS Coordinates
	SPRINGS		
1	Ngonyeki	Northern watershed zone	N01.17356^0;E036.70635^0
2	Nang'aro	Southern watershed zone	N 01.02376^0; E 036.81657^0
3	Soit Ngablo	Southern watershed zone	N 01.00474^0; E 036.85241^0
4	Lechoro Lenanguya	Southern watershed zone	N 01.06251^0; E 036.73471^0
#	**STREAMS**		
1	Nashunda	Northern western zone	N 01.27412^0; E 036.65909^0
2	Naadapo Elkileku	Northern western zone	N 01.25484^0; E 036.62896^0
3	Lonyonyi	Northern western zone	N 01.12215^0; E 036.62797^0
4	Lorok	Central watershed zone	N 01. 16813^0, E 036.77607^0
5	Kirisia	Central watershed zone	N 01.08643^0; E 036.71154^0
#	**EARTH DAMS**		
1	Ntontol	Northern watershed zone	N01.28597^0;E036.66153^0
2	Su-en	Central watershed zone	N01.16161^0;E036.74428^0
3	Suen-Lkidoroto	Central watershed zone	N01.15981^0;E036.74540^0
4	Nontotol	Central watershed zone	N01.05713^0;E036.67558^0
5	Loikurukul	Central watershed zone	N01.05293^0;E036.67929^0
6	Baawa	Southern watershed zone	N01.01952^0;E036.81092^0
7	Prison	Southern watershed zone	N01.06392^0;E036.71875^0
8	Morijoi	Southern watershed zone	N01.05020^0;E036.72626^0
#	**WATER PANS**		
1	Ngabai	Northern watershed zone	N 01.23212^0;E 036.65409^0
2	Ndikir Elgwesi	Central watershed zone	N01.15660^0;E036.75555^0
#	**BOREHOLES**		
1	Ngorika	Northern watershed zone	N01.27412^0;E036.65909^0
2	Partuk 1	Central watershed zone	N01.12249^0;E036.62801^0
3	Partuk 2	Central watershed zone	N01.12212^0;E036.62801^0
4	Baawa	Southern watershed zone	N01.02780^0;E036.81181^0
5	Leirr	Southern watershed zone	N01.01065^0;E036.78466^0
6	Lchoro	Southern watershed zone	N01.03459^0;E036.74981^0
#	**SHALLOW WELLS**		
1	Lonkutukie	Northern watershed zone	N01.34551^0;E036.71298^0
2	Ntonlol	Northern watershed zone	N01.28639^0;E036.66094^0
3	Lorok (1, 2 & 3)	Central watershed zone	N01.13754^0;E036.70920^0
4	Loikas	Central watershed zone	N01.13754^0;E036.70920^0
5	Ngablo	Southern watershed zone	N01.00109^0;E036.85105^0
6	Naingolie	Southern watershed zone	N 01.05102^0;036.74687^0
7	Naingolie	Southern watershed zone	N01.04232^0;E036.74277^0

Table 4 summarizes the findings of the estimated number of community members and livestock using the surveyed water source types as well as the approximate amount of water used per day. The approximate total number of livestock and locals using the sources was about 180,645 and 147, 060 respectively. Earth dams and water pans provided water to the highest population of community members estimated at an average of 11,564 people followed by boreholes at 9,886 members (Table 4). Shallow wells, springs and streams were used by the least number of people. In regards to usage by livestock, earth dams and water pans served the highest population of livestock estimated at an average of 15,422 and the other water source types provided water for an almost equal number of livestock at about 7,000 (Table 4). The highest amount of water was abstracted from boreholes at nearly 197,720Litres/day (197.72m^3/day) followed by earth dams and water pans at 91,960Litres/day (91.96m^3/day), and the least was from shallow wells, springs and streams at about 38,000Litres/day (38m^3/day). Overall, the daily water abstracted from all the water source types for use by humans and livestock was nearly 366,540Litres/day or 366.54m^3/day.

Table 4. Estimated number of community members and livestock using the survyed water sources in Kirisia Forest watershed

Type of water source	Name of water source	Approximate No. of beneficiaries	Approximate water abstraction (jerry cans* per day)	Approximate water abstraction (Liters/day)	Approximate water abstraction (m^3/day)	Approximate No. of livestock using water source
Shallow wells	Loikas shallow wells	9800	4200	**38000**	**38**	12200
	Ntontol shallow well	350	300			9320
	Lorok wells 1, 2 & 3)	1000	1200			2610
	Mean	**3717**	**1900**			**8043**
Earth dams and water pans	Ntontol dam	350	150	**91960**	**91.96**	9320
	Suen dam 1 & 2	6000	7200			12500
	Baawa dam	4000	10800			28020
	Prison dam	14000	6000			7055
	Nondoto dam	1800	1040			40000
	Morijoi dam	4800	2400			6000
	Ngorika water pan	50000	-			5060
	Mean	**11,564**	**4,598**			**15,422**
Boreholes	Partuk 1	40000	25000	**197720**	**197.72**	-
	Leirr borehole	3000	750			20000
	Ngorika borehole	420	420			5060
	Lchoro borehole	948	1476			1160
	Baawa borehole	5062	3311			-
	Mean	**9886**	**6191**			**6555**
Springs and streams	Ngorika springs	420	-	**38860**	**38.86**	5060
	Ngonyeki springs	4900	2100			11180
	Nashuda stream	210	120			6100
	Mean	**1843**	**1110**			**7417**
Total		**147060**		**366540**	**366.54**	**180645**

*A jerry can is equivalent to 20 Litres.

Kirisia Forest watershed is the only source of domestic and livestock water supply for 23 sub-locations in Kirisia and Lorroki Sub Counties (Table 5). According to the 2009 national population census, the total human population in the two sub-counties which includes Mararal town was 92,695 people who can be considered as the watershed beneficiaries. This number is expected to increase to approximately 116,427 people by 2020 and 142,954 people in 2030 (Table 5). Based on the 2009 population census, the human water demand varied among the sub-locations, and ranged from 127,800Litres/day (127.80m^3/day) at Barsaloi to 81,880,600L/day (81,880.60m^3/day) for Maralal town (Table 6). The projected total water demand for all the sub-locations was about 182,238,520Litres/day (182,238.52 m^3/day), and the highest water demand areas were; Mararal town, Ledero, Lpartuk, Loosuk and Opiroi (Table 6). Using the 2013 livestock population data for all the sub-locations, the total volume of water demand by livestock was estimated at 12,172,600Litres/day (12,172.60m^3/day) (Table 6) The highest livestock water demand areas were; Sukuta Marmar, Loosuk, Malaso and Opiroi; and the total water demand by humans and livestock approximately 194,411,120Litres/day (194,411.12 m^3/day.

Table 5. Human population estimate and projections for sub-locations in Samburu Central Sub-County which depend on water from Kirisia Forest watershed

	Sub location	2009 population estimate	2020 population projection	2030 population projection
1.	Angata Nanyukie	2388	2999	3694
2.	Barsaloi	1525	1917	2359
3.	Loibashai	1239	1556	1916
4.	Loosuk	2756	3462	4263
5.	Malaso	2214	2781	3425
6.	Opiroi	3115	3912	4818
7.	Sukuta Marmar	4173	5241	6455
8.	Mabati	1042	1309	1611
9.	Mbukoi	1112	1397	1720
10.	Nkejemueny	924	1161	1429
11.	Seketet	2560	3215	3960
12.	Siambu	745	936	1152
13.	Sirata Oirobi	1346	1691	2082
14.	Ledero	1822	2288	2818
15.	Lpartuk	3452	4336	5340
16.	Lolmolok	3758	4720	5381
17.	Langatolia	1729	2172	2674
18.	Mugur	990	1243	1531
19.	Milimani	4468	5612	6911
20.	Nkuroto	3407	4279	5270
21.	Ngari	5027	6314	7776
22.	Shabaa	6148	7722	9510
23.	Maralal Town	36755	46164	56859
Total		92,695	116,427	142,954

Table 6. Estimated human and livestock population and projected water demand in Kirisia Forest Watershed-Samburu Central sub-county

Sub-location	2009 human population census	Domestic water demand (liters/day)	Domestic water demand (m^3/day)	Estimated livestock population (2013)	Water demand by livestock (Liters/day)
1. Angata Nanyukie	2388	954300	954.30	2800	28000
2. Barsaloi	1525	127800	127.80	1500	15000
3. Loibashai	1239	381700	381.70	10750	107500
4. Loosuk	2756	3075300	3075.30	62100	621000
5. Malaso	2214	1558300	1558.30	59950	599500
6. Opiroi	3115	2138700	2138.70	55330	553300
7. Sukuta Marmar	4173	1133100	1133.10	98700	987000
8. Mabati	1042	183400	183.40	55330	553300
9. Mbukoi	1112	350000	350	1100	11000
10. Nkejemueny	924	604300	604.30	21050	210500
11. Seketet	2560	976800	976.8	33850	338500
12. Siambu	745	1186200	1186.20	55330	553300
13. Sirata Oirobi	1346	1645000	1645	261500	2615000
14. Ledero	1822	49339000	49339	55330	553300
15. Lpartuk	3452	3632600	3632.6	55330	553300
16. Lolmolok	3758	2647100	2647.10	55330	553300
17. Langatolia	1729	5070720	5070.72	55330	553300
18. Mugur	990	5070720	5070.72	55330	553300
19. Milimani	4468	5070720	5070.72	55330	553300
20. Nkuroto	3407	5070720	5070.72	55330	553300
21. Ngari	5027	5070720	5070.72	55330	553300
22. Shabaa	6148	5070720	5070.72	55330	553300
23. Marala town	36755	81880600	81880.60		
Total	92,695	182,238,520	182,238	1,217,260	12,172,600

5. Discussion

A variety of water source types exist in the Kirisia Forest watershed but most of them especially springs, streams, water pans and shallow wells tend to be seasonal. Boreholes, earth dams and some of the shallow wells are more reliable in water provision especially during the dry season due to their permanent nature. Nyaligu (2013) documented a higher variety of water source types (48) in the watershed compared with our study which included; marshes, swamps, springs, shallow wells, earth dams and boreholes. Our survey was done during a prolonged dry spell, and locals mentioned the area had not received adequate rainfall for the last two years or so while that of Nyaligu (2013) was done during the wet season; this may account for the diverse water types reported in his study. The endowment of the watershed with diverse water source types may be attributed to a variety of factors such as; expansive nature of the catchment and its diverse topography and altitude, land cover types like forests, woodlands and bushlands (Hitimana et al., 2011; KFS, 2012; Nyaligu, 2013; Kiringe et al., 2015b). Further, it has an expansive drainage network with the Northern and Eastern zones having the longest and most dense drainage network followed by the Western and Sothern Zones in that order. These attributes coupled with a tri-modal rainfall pattern provide environmental conditions which enhance the hydrological functions of the watershed in terms of rainfall interception and channeling it into the soil, underground aquifer and associated water systems. Consequently, preservation of this environmental heterogeneity is important in sustaining the natural ability of the watershed to harness and replenish water. In a study in the Chyulu Hills watershed which is a dry-land water catchment like Kirisia Forest, Kiringe, Mwaura, Wandera, Kimeu & Gachuga (2015a) documented different types of water sources (springs, shallow wells, rivers and boreholes) and attributed this to the expansive and heterogeneous nature of the watershed.

Kirisia Forest watershed is located in a semi-arid region characterized by water deficit due to low and unreliable rainfall (Watai & Gachathi, 2003; Lambrechts et al., 2005; Anne, 2009; KFS, 2012). Therefore, all the water sources associated with the watershed within it and beyond are key focal points for water use by humans, wildlife and livestock, and are normally subjected to heavy daily water abstraction especially during the dry season. In particular, earth dams and water pans like those found in the watershed are very important water sources in arid and semi-arid areas where rain water is harnessed from surface runoff (RoK, 1991). Normally, pans store water for a short period of up to 3 months after the rains unless they have underground recharge system. On the other hand, dams are designed to store large volumes of water for a long period of time, and in this regard the numerous dams found in the Kirisia Forest watershed are critical and more reliable water supply assets in the region.

Although water dams and pans are supposed to provide water for various uses, their water quality is not guaranteed since most of them are open and easily accessible by humans, livestock and wildlife leading to water contamination. Therefore, other water sources like springs, boreholes and shallow wells are critical in supplying water for household uses since their water is less likely to be contaminated to the extent of making the water unsuitable for domestic purposes. The boreholes found in the watershed also offer a more reliable source of water in the absence of surface water especially during dry periods, and in most cases, their water quality is good and assured unless for cases of salinity and high fluoride levels(RoK, 1991; Nyaligu, 2013;Kiringe et al., 2015b). Generally, boreholes with high water discharge can serve more people and cover larger areas, and are essential water sources especially in dry areas like Kirisia Forest where water resources are scarce in space and time. In their study in the dry-land watershed of the Chyulu Hills, Kiringe et al. (2015a) also found that boreholes were not only prevalent but they provided water to a large population of beneficiaries.

The key beneficiaries of the water provision services associated with Kirisia Forest watershed were pastoral communities living within and in the immediate environs of the watershed mainly for domestic uses and watering their large herds of livestock. Urban populations in centers like Maralal town, Kisima and Sukuta Marmar also depended on water associated with the watershed. However, the diversity of water users in the watershed was much lower compared to that of the Chyulu Hills (Pringle & Quayle, 2014; Kiringe et al. 2015a) included; domestic water users, livestock keepers, small scale irrigators, wildlife conservationists, large scale irrigators and tourism operators. This difference may be attributed to the high water provision capacity of the the Chyulu Hills watershed compared to Kirisia Forest. Documentation of water source types in other dry-land catchments in the country such as Mt, Kulal Forest watershed in Marsabit (Watkins & Imbuni, 2007) and Taita Hills watershed (Eduardo et al., 2011; Johanna, Emmah, Paola & Pellikkaa, 2015) have shown they are also characterized by a low variety of source types like Kirisia Forest.

Most of the water in the Kirisia Forest watershed was mainly used for domestic chores (cooking, drinking, bathing and washing clothes) followed by livestock. Thus, the projected water demand by humans is almost fifteen times more that of livestock although the population of the former is far much less that the later. Humans are physiologically less enduring to low water intake compared to livestock, and also use water for other domestic needs, and as such tend to use more water compared with livestock. In general, the water requirements of livestock is a function of; feed and salt ingested, lactation, size of the animal, animal's genetic adaptation to its environment and ambient temperature (Peden et al.,2003). It's worth noting that water use and demand by wildlife in the Kirisia Forest watershed was not estimated and as such, the estimated water use and projected demand is less than the actual use. Water use and demand for most of Kenya's dry-land water catchments has not been documented but Kiringe et al. (2015a) approximated human and livestock water use in the Chyulu Hills watershed to be 244,238.25m^3/day. In the Taiita Hills watershed (Eduardo et al.,2011; Johanna et al.,2015) documented high water use levels by humans compared to livestock use due to irrigation activities. Watkins & Imbuni (2007) also documented low water use level by humans and livestock for Mt. Kulal watershed in Northern Kenya

The 2009 national population census estimated the human population served by the Kirisia Foest watershed to be 92,695 (KNBS, 2009), and this is projected to increase to about 116,427 and 142,954 by 2020 and 2030 respectively (Kiringe et al., 2015b). This, situation together with a high livestock population will continue to use and demand more water from the watershed which may overwhelm the hydrological capacity of the watershed to sustain such a high abstraction level.This study noted numerous shallow wells and springs which had dried up in the recent past, and which may indicate the hydrological processes in the watershed had declined perhaps due to deforestation and environmental degradation. Kiringe et al. (2015b) found a 55% increase in the built environment in the forest watershed between 1973 and 2015 which might partly be due to increase in human settlements and urbanization. Other factors that have been noted to cause massive degradation of the forest are; illegal settlements in the forest, trampling and vegetation destruction by livestock, wood fuel harvesting, forest fires and illegal logging (Watai & Gachathi, 2003; Lambrechts et al.,2005; Anne, 2009; KFS, 2012). Such changes can lead to decline in ability of the vegetation and the general landscape to intercept rainfall and ultimately lower water recharge in the watershed.

6. Conclusions and Recommendations

Kirisia Forest watershed has diverse water source types which are the oases of life for wildlife, communities living in urban centres and a large population of local communities and their livestock. The surface water sources in the forest are concentrated close to the forest edge while communities living further tend to rely mostly on underground water resources drawn from bore holes and shallow wells and whose water recharge is linked to the forest water catchment. Underground water from the forest is harvested beneath sandy seasonal lagga's, which cautions wildlife, livestock and communities from the adverse effects of water shortage during times of droughts. Communities living contiguous to the forest and locals, domestic and wildlife animals living between Kirisia Forest and the Mathews Range depend almost entirely on water supplied by the watershed. Earth dams, shallow wells and boreholes were the most reliable water sources but wells and dams were more vulnerable to anthropogenic contaminants and runoff during the wet season. Dams were faced by another problem of eutrophication due to defecation by livestock particularly during the dry season when they were heavily used by large herds of livestock. Permanent rivers and springs were very few and most of the drainage system comprised of laggas which had water mostly on a seasonal basis. A notable aspect of the watershed was prevalence of dry water pans, shallow wells and springs in most of the drainage system of the catchment zone. This is a worrisome situation which requires further investigation to understand the possible underlying cause(s).

The watershed was the main water source for many households in 23 sub-locations including a rapidly growing urban population in urban centres like Maralal town and a large population of livestock. Human used the highest amount of water followed by livestock and this is projected to escalate in future. Based on the 2009 census data, the estimated population of the locations was 92,695, and it's projected to increase to 116,427 and 142,954 people by 2020 and 2030 respectively, and this will potentially lead to a surge in water use and demand. Unless effective conservation measures for Kirisia Forest and water management interventions are put in place, trends in water use and demand in the watershed will exceed the available water resources and in the long-term stagnate socio-economic development. Water conflicts among locals and with wildlife are also likely to become prevalence and severe.

Based on the findings of this study, we recommend various strategies that can be used to address the water situation in the watershed. There's an urgent need to enhance protection and conservation of the forest which is the main water replenishment power house in the region. This is not an easy undertaking given the political implications of removing illegal settlers and regulating livestock grazing by local communities but it holds the key to ensuring long-term water provision in the entire watershed. In the same vein it's important to educate and create awareness among different water users and stakeholders on the linkage between water availability and the environmental state of Kirisia Forest, and the importance of embracing sustainable water use ethics. It's also necessary to create an independent and inclusive oversight taskforce to oversee sustainable protection of the forest. Currently, the management of all water towers in the country including Kirisia Forest is under different government lead agencies (i.e. Kenya Water Towers Agency, Kenya Forest Service, Kenya Wildlife Service and more recently Community Forest Associations-CFAs). But this institutional arrangement creates confusion and duplication of oversight responsibility at the expense of ensuring that water towers are effectively protected since they are the main water power houses which drives the local and national economies. Creating awareness among communities living within the watershed on the impacts of climate change on water resources, and training them on how to cope with this change so as to reduce adverse effects on their livelihoods is also essential. There's a need to document and monitor water yield and abstraction in the watershed, and use the data to make informed decisions on effective and sustainable water management. Communities dependent on earth dams, shallow wells and water pans should be trained on how to secure their waters to ensure year round availability of suitable water for domestic uses especially for drinking and cooking purposes. It's also important to have designated watering points for livestock to reduce water contamination by dung.

Acknowledgements

This study couldn't have been accomplished without the financial support provided by African Wildlife Foundation (AWF) in Nairobi-Kenya, as well as the professional support provided by their point person, Ms. Fiesta Warinwa, the County Director. The technical and logistical support provided by; John Kinyua, Inoti Mburugu of WRMA office in Rumuruti, Antony Ndubi (Regional Centre for Mapping of Resources for Development (RCMRD)-Nairobi) also greatly contributed to successful completion of this survey. Many thanks for the technical inputs provided by the County Director of Environment, Water and Natural resources Mr. Benson Lengalen. Mr. Mark Lcharuni, Managing Director of SAWASCO and Mr. David Mumo of WRMA office in Rumuruti. Many thanks to Wilson Lekaaso, Vice-secretary of the Kirisia Forest Community Forest Association for being an excellent field guide and liaison person with local communities and other stakeholders. The excellent field assistance given by Bosco Lekaaso (Mararal) and Teacher Kimani (Baawa Catholic Primary School) and the Chief of Ngorika Location, Mr. Moses Leseiya is appreciated. Mr. William Leleshep, the Chairman of the Kirisia Hills Community Forest Association (CFA) was very helpful. Ms. Tiffany Mwake of Habitat Planners provided incredible support during the study and other aspects of preparation of this manuscript.

References

Akotsi, E. F. N., Gachanja, M., & Ndirangu, J. K. (2006). *Changes in forest cover in Kenya's five water towers (2003-2005)*. UNEP,KFWG and Royal Netherlands Embassy

Anne, P. (2009). *Preliminary ecological survey of Kirisia Forest Reserve, Samburu District*. Report to Laikipia Wildlife Forum for Conservation Enterprise Development Program

Beentje, H. J. (1990). The forests of Kenya. *Mitteilungen aus dem Institut für 14 Allgemeine Botanik in Hamburg, 23a*, 265-286.

Costanza, R., d'Arge, R., de Groot, R., Farber, S., Grasso, M., Hannon, B., … van den B. (1997). The value of the world's ecosystem services and natural capital. *Nature, 387*(6630), 253-258.

Daily, G. C. (ed.) (1997). *Nature's Services: societal dependence on natural ecosystems*. Island Press, Washington, DC.

de Groot, R, S., Matthew A. W., & Roelof, M. J. B. (2002). A typology for the classification, description and valuation of ecosystem functions, goods and services. *Ecological Economics, 41*, 393–408,

Eduardo, E. M., Petri, K.E. Pellikka, P., Barnaby J.F. Clark, H., & Mika, S. (2011). Prospective changes in irrigation water requirements caused by agricultural expansion and climate changes in the eastern arc mountains of Kenya. *Journal of Environmental Management, 92*, 982-993.

Gichuhi, M. (2013). Ecological management of the Mau catchment and its impact on Lake Nakuru National Park. *Jomo Kenyatta University of Agriculture and Technology, 15*(1), 81-101.

Gleick, P. H. (2003). Global freshwater resources: soft-path solutions for the 21st century. *Science, 302*, 1524–1528.

Global Water Partnership (GWP). (2000). *Integrated water resources management: technical advisory committee background paper no. 4*. Global Water Partnership Technical Advisory Committee, Stockholm, Sweden.

Government of Kenya (GoK). (2007). *Vision 2030: a competitive and prosperous Kenya.*

Hitimana, J., Kiyiapi, J. L., Hamphrey, K., Warinwa, F., Lenaiyasa, P., Kibugi, P., … Sumba, D. (2005). *Linking socio-economic significance and conservation for Kirisia Forest, Samburu, Kenya*. Technical Paper to African Wildlife Foundation (AWF), Nairobi, Kenya.

Hitimana, J., Kiyiapi, J. L., Wambui, P., Kibugi, P., Kisio, H., Mayienda, R., Warinwa, R., Lenaiyasa, P., & Sumba, D. (2011). *Challenges of linking socio-economic significance and conservation value of forests in dry-lands of Kenya: case study of Kirisia Forest-Samburu pastoralists coexistence, biological diversity and sustainable resources use*.

Honeck, J. H., & Smith, W. T. (1992). *Watershed ecosystem analysis as basis for multiple-use management of Eastern Forests*. USDA Forest Service, North Carolina.

Johanna, H., Emmah, O., Paola, M., & Pellikkaa, P. (2015). Local assessment of changes in water-related ecosystem services and their management: DPASER conceptual model and its application in Taita Hills, Kenya. *International Journal of Biodiversity Science, Ecosystem Services & Management*. http://dx.doi.org/10.1080/21513732.2014.985256

Kenya Forest Service (KFS). (2012). *Leroghi /Kirisia forest management plan (2012 – 2016)*. The Kenya Forest Service, Nairobi – Kenya.

Kenya National Bureau of Statistics (KNBS). (2009). *2009 population census*. Kenya National Bureau of Statistics, Nairobi, Kenya.

Kiringe, J. W., Mwaura, F., & Kimeu, M. M. (2015b). *Watershed ecosystem services and water management tools for the Kirisia Forest ecosystem in Samburu Landscape*. Report prepared for the African Wildlife Foundation (AWF) by Habitat Planners, Nairobi, Kenya.

Kiringe, J. W., Mwaura, F., Wandera, P., Kimeu, M., & Gachuga, F. (2015a). *Water management tools in the Chyulu Hills watershed*. Report prepared for the African Wildlife Foundation (AWF) by Habitat Planners, Nairobi, Kenya.

Kiringe, J. W., Okello, M. M., Tome, S. R., & Seno, S. K. (2009). *The water situation analysis in Kimana area: causes and consequences of water quality, quantity, and distribution dynamics. Final water situation analysis report:* Report prepared for the project implementation committee (PIC), Kimana water resources conservation project

Lake Victoria Basin Commission (LVBC). (2011). *The total economic value of Maasai Mau, Trans Mara and Eastern Mau Forest Blocks, of the Mau Forest, Kenya.* Report prepared by the Environmental Research and Policy Analysis (K); P. O Box 112-00100 Nairobi, Kenya (Environment.analysis@yahoo.com); for Lake Victoria Basin Commission Secretariat, 6th Floor, Re-Insurance Plaza, P.O. Box 1510 – 40100, Kisumu, Kenya.

Lambrechts, C., Woodley, B., & Gachanja, M. (2005). *Aerial survey of the of the threats to Leroghi (Kirisia) Forest Reserve*

Millennium Ecosystem Assessment (MEA). (2005). *Ecosystems and human well-being: synthesis*. Island Press, Washington, DC.

Mogaka, H., Gichere, S. Davis, R., & Hirji, R. (2006). Climate variability and water resources degradation in Kenya: improving water resources development and management. *World Bank Working Papers Number 69*, 1-105.

Nyaligu, M. O. (2013). *Water resources inventory and assessment, Kirisia Forest National Reserve*. Final report to AWF, Kenya

Olang, O. L., & Kundu, P. (2011). *Land degradation of the Mau Forest complex in Eastern Africa: a review for management and restoration planning*. http://dx.doi.org/10.5772/28532

Perez-Verdin, G., Navar-Chaidez, J. J., Kim, Y. S., & Silva-Flores, R. (2012). Economic valuation of watershed services for sustainable forest management: Insights from Mexico, sustainable forest management. *Current Research*.

Pratt, D. J., & Gwynne, M. D. (1977). *Rangeland management and ecology in East Africa*. Hodder and Stoughton, London, England.

Pringle, C., & Quayle, L. (2014). *The role of the Chyulu Hills in the delivery of ecosystem services in south-eastern Kenya. INR Report No 495/14*. Report Prepared for the African Wildlife Foundation: Nairobi, Kenya.

Republic of Kenya (RoK). (2013). *Samburu County integrated development plan (2013-2017).* Samburu County Government.

Republic of Kenya (RoK). (1991). *Samburu District water development study (1993-2013); Part I-Water demand and water resources*. Ministry of Water Development, Nairobi, Kenya.

Republic of Kenya (RoK). (2010). *National climate change response strategy (NCCRS)*. Ministry of Environment and Mineral Resources, Nairobi

Watai, M. K., & Gachathi, F. (2003). *Conservation for sustainable utilization of biodiversity through enhanced access and benefit sharing with forest adjacent communities in Kirisia forest Samburu District*. Report by Kenya Forestry Research Institute, Nairobi – Kenya.

Watkins, T. Y., & Imbumi, M. (2007). Forests of Mt. Kulal, Kenya: source of water and support to local livelihoods. *Unasylva, 58*, 33-27.

Wilk, J., & Wittgren, H. B. (eds). (2009). *Adapting water management to climate change*. Swedish Water House Policy Brief Nr. 7. SIWI, 2009.

Wu, T., Kim, S. Y., & Hurteau, M. D. (2011). Investing in natural capital: using economic incentives to overcome barriers to forest restoration. *Restoration Ecology, 18*(4), 441–445.

15

Antifouling Effectiveness and Potential Toxicological Risk of an Elastomer-Based Coating against Zebra Mussels

Yves de Lafontaine[1] & Julie Veillette[2]

[1] Department of Fisheries and Oceans Canada, Institut Maurice-Lamontagne, Canada

[2] Environment and Climate Change Canada, Centre Saint-Laurent, Canada

Correspondence: Yves de Lafontaine, Department of Fisheries and Oceans Canada, Institut Maurice-Lamontagne, Canada. E-mail: Yves.deLafontaine@dfo-mpo.gc.ca

Abstract

The fouling of hard substrates by zebra mussels (*Dreissena* sp.) in freshwater ecosystems is a persistent problem which calls for antifouling treatments being fully efficient, long-lasting, and environmentally safe. The present study assessed the potential toxicological impact and the effectiveness of an elastomer-based coating containing salts that would make surfaces repulsive to zebra mussel attachment. Laboratory testing using standard analytical methods for water and wastewater, and a battery of six bioassays confirmed that the leachates from the coating exhibited no toxic response suggesting its safety to the receiving environment. In situ experiments using multi-plate collectors indicated that biofouling by zebra mussels and sponges on coated surfaces was effectively reduced by up to 97% over one growing season. Effectiveness of the coating was slightly less (82%) during the second growing season. Results suggested that the repulsive effect would be due to the presence of salts within the elastomer-based coating affecting both zebra mussels and other freshwater organisms. Saltwater experiments indicated that the coating was totally ineffective to prevent biofouling in marine environments. In addition of being environmentally safe to use, results suggested that the coating can be an effective antifouling product for submerged structures in freshwater environments.

Keywords: antifouling, coating, freshwater, invasive species, toxicity, zebra mussels

1. Introduction

The introduction of zebra and quagga mussels (*Dreissena polymorpha* and *Dreissena rostriformis bugensis*) in North America has resulted in the biofouling of underwater hard substrates which led to important socio-economical and environmental consequences (Pimentel et al., 2005) as well as major impacts to freshwater biodiversity, including the decline of native benthic species in the Great Lakes (Van Appledorn & Bach, 2007), the loss of water column primary production (Claudi & Mackie, 1994; Brines et al., 2007) and the alteration of food web structure in lacustrine environments (Karatayev et al., 2002). Zebra mussels adhere tightly to virtually any underwater hard substrate by their tenacious proteinaceous byssal threads with adhesive plaques at the tips (Farsad & Sone, 2012). In addition to local environmental effects caused by high densities (> 10000 ind. m^{-2}) of zebra mussels, heavy fouling can further increase the risk of transfer and spread of these invasive species.

The biofouling by zebra mussels and other undesirable aquatic organisms can be controlled to variable degree in using various water treatment technologies based upon physical processes or toxic chemical materials (Martin, Mackie, & Baker, 1993; Claudi & Mackie, 1994; Morse, 2009; Bernabeu et al., 2011; Bueley et al., 2014). An alternative approach is to cover underwater surfaces with repulsive products that deter organisms fouling to levels that can be considered acceptable and of low risk of proliferation. Ideally, an antifouling coating should be non-toxic to non-target organisms, produce no toxic by-products in the environment, resist erosion over time, adhere to a variety of substrates, be easy to apply, anticorrosive, long-lasting (3-5 years), resistant to damage and cost competitive (Chambers et al., 2006; Maréchal & Hellio, 2009). The various antifouling coatings being developed, tested and used can be divided into two treatment approaches: 1) the chemically active antifouling compounds and, 2) the surface characteristics of the substrate.

Antifouling paints that continuously release chemicals through the painted surface are the most common chemically active antifouling technologies used for zebra mussel mitigation. In spite of its versatility, this control method poses concerns because of potential harmful impacts on non-target organisms and their environment, and of high economic costs (Cope et al., 1997). Hence, the use of many chemicals have been banned by several countries as a result of their environmental toxicity, a classical example being the use of tributyltin (TBT) in paints applied to ship hulls and pleasure crafts (Regoli et al., 2001; Dafforn et al., 2011; Howell & Behrends, 2010). Similarly, copper-based coatings that do not degrade rapidly were shown to produce residues that accumulate in the sediments and the biota posing a threat to aquatic environments (Dafforn et al., 2011). Environmental concerns about the toxicity of chemicals used in antifouling coatings have prompted the development of alternative chemically active antifouling technologies which would be more selective and not toxic to the environment. These include the use of naturally occurring compounds that repel aquatic organisms (Qian et al., 2010; Dafforn et al., 2011). For example, strains of *Pseudomonas* bacteria and their natural metabolic products were shown to selectively kill zebra mussels by intoxication (Molloy et al., 2013). The derivatives of capsaicin, a natural component of chili pepper, and other structurally similar products were shown to inhibit zebra mussel attachment (Angarano et al., 2007; Angarano et al., 2009). Marine sponges are typically unfouled by macroorganisms, and the potential antifouling capacity of natural products isolated from sponges was found to be effective against zebra mussel (Diers et al., 2004, 2006; Bowling et al., 2010). On the hand, chitin, an extract from the exosqueletton of crustaceans, was not useful to significantly reduce biofouling by zebra mussels (de Lafontaine et al., 2002). Overall, the broad effectiveness of these naturally occurring antifouling products has not yet been reported (Chambers et al., 2006; Webster & Chisholm, 2010). Finally, other approaches have attempted to design innovative application strategies for existing biocides (Costa et al., 2008a), such as encapsulation of toxins in microscopic particles of edible material (Aldridge et al., 2006; Costa et al., 2011a), combination of toxins (Costa et al., 2011b; Costa et al., 2012), and investigation of the seasonal variation of the species' tolerance to toxins (Costa et al., 2008b).

Antifouling solutions based on the surface properties of the substrate to control aquatic fouling are usually considered to be benign to the environment because no chemical or material is released into the environment. These coatings rely on surface physico-chemical and material characteristics to deter organisms to attach or to reduce their attachment strength. These solutions include nanostructured surfaces, surface modification and foul release technologies (Chambers et al., 2006; Hellio & Yebra, 2009). Recent advances in nanotechnology and polymer science, and the development of novel surface designs bioinspired by nature are expected to have a significant impact on the development of these new antifoulants (Callow & Callow, 2011). Foul release technology aims to create surface characteristics that reduce the attachment strength of organisms (Chambers et al., 2006). Hence, attached organisms detach under their own weight as they grow or they are removed by water movement, as when a ship moves through the water. Among the products assessed to date, silicone was the polymer having the lowest adhesion and the best foul release capacity (Townsin & Anderson, 2009; Lewis, 2009; Wells et al., 1997). Foul release coatings provide best results for vessels operating regularly at speeds greater than 15 knots, or they will necessitate regular cleaning (Townsin & Anderson, 2009). Although antifouling solutions based on surface properties have proven to inhibit biofouling, they are still not widely applied to vessels due to practical shortcomings (Webster & Chisholm, 2010; Lewis, 2009).

Another alternative technology are the non-leaching active coatings, where the chemicals are bounded to a surface and exert a repulsive effect to organisms that get in contact with the surface, but without the chemical being released into the surrounding environment (Lewis, 2009). An elastomer-based compound was recently developed to be used as an antifouling coating against zebra mussels (MD Technologies, 2008). This coating (named *ZebraOut*©) which contains no biocide or persistent toxic compounds is an heterogeneous system made of two liquid phases. The first liquid phase is dispersed by forming small polymeric droplets and the second liquid phase is ionically charged with saline solutions to allow the attraction and solidification of the elastomer. Given that survival of zebra mussel larval stages is greatly reduced in salty environment (Kilgour et al., 1994), it was hypothesized that the presence of salts within the product would render surfaces repulsive to zebra mussel attachment. The coating can be easily applied as a spray, dries quickly and adheres to most surfaces, including metal, steel, aluminum, wood, concrete and plastic. Thickness application varies from 0.25 to 6 mm, or more if necessary. This soft coating has an impressive extension factor of >1000, which makes it useful for surfaces that may suffer physical stress (MD Technologies, 2008).

The present study assessed the effectiveness and the potential environmental risk of the *ZebraOut*© elastomer-based coating as an antifoulant against zebra mussels. The specific objectives were two-fold. First, the toxicological potential of leachates from the coating was determined in laboratory experiments to assess whether the use of the coating can pose some risk to aquatic environments. This was viewed as a pre-requisite before conducting any experiment under in situ conditions. Second, the antifouling response of the coating against zebra mussels under different in situ conditions over short (4-5 months) and longer period of time (11 months) was evaluated. Specifically, we quantified the percent reduction in zebra mussel fouling on coated surfaces. We also examined the effect of the coating on mussel size distribution to verify if the coating effectiveness was size-selective. The efficacy of the coating in freshwater and saltwater environments was assessed. Finally, the effects of substrate colour and the potential competitive interaction of zebra mussels with sponges were also analyzed.

2. Method

2.1 Toxicological Analyses

Toxicological testing analyses of leachates from polypropylene 15-cm X 15-cm plates coated with *ZebraOut©* were performed in the laboratory. Leachates were obtained by immersing newly coated plates in glass containers filled with 3.5 L of deionized water for 24 h and the control consisted of the leachates from uncoated plates. Temperature was stable between 23.2 and 23.9 °C during the leaching period. First, contaminant analyses and water quality testing on leachates were made using standard analytical methods for water and wastewater (Greenberg et al., 1992). The results were then compared to the standard "BNQ (*Bureau de Normalisation du Québec*) 3660-950 - Safety of Products and Materials in Contact with Drinking Water" in order to evaluate the potential toxicity of leachates. Second, following the approach used in previous assessment studies (de Lafontaine et al., 2008; de Lafontaine et al., 2013), the potent toxicity of coating leachates on various components of the food web was evaluated by a battery of six bioassays from bacteria to fish.

The six bioassays were the Microtox (*Vibrio fischeri*) test, the algal (*Pseudokirchneriella capricornatum*) microplate assay, the Cnidaria (*Hydra attenuata*) test, the cladoceran (*Daphnia magna*) test, the microcrustacean (*Thamnocephalus platyurus*) test, and the trout (*Oncorhynchus mykiss*) hepatocyte test. The Microtox and the algal tests are two sublethal tests measuring the inhibitory effect of toxic compounds on cell physiology and growth. The four others are lethal tests measuring the proportion of dead organisms resulting from toxic exposure over different time periods. Although their experimental protocols vary slightly, all assays are based on a common procedure that consists of exposing, under controlled laboratory conditions, living organisms to a dilution series of water samples and to note and quantify the observed effects. The Microtox test was conducted using a bacterial concentration of $1x10^6$ cells ml^{-1} in four replicates for each of the 10 diluted concentrations with an exposure time of 15 min. The algal test consisted of 10 diluted concentrations with five replicates, each one with 10 000 cells ml^{-1} for a 72 h exposure time at 24 °C. The Cnidaria test was conducted for 96 h using seven diluted concentrations with three replicates and three organisms per replicate. The cladoceran test used five newly hatched (< 24 h old) individuals in three replicates for the six dilutions tested and the contact time was 48 h at 20 °C. For the microcrustacean test, three replicates of 10 larvae were used for each of the six diluted samples at dark at 25 °C for 24 h. Data were analysed to compute various endpoint values ($LC_{50/25}$: 50/25% lethal concentration; EC_{50}: 50% effect concentration; IC_{50}: 50% inhibitory concentration; NOEC: no observed effect concentration) depending on the assay. The results are expressed in toxic units (TU), which correspond to the inverse of the dilution factor required to reduce potent toxicity at the selected endpoint concentration. All the toxicity tests were conducted according to standard procedures and protocols currently used by Environment Canada (see references in de Lafontaine et al. (2008) and in Gagné (2006) for the rainbow trout hepatocyte test). All bioassays using experimental animals during this study were conducted following approved protocols in accordance to guidelines provided by the GLLFAS/NWRI animal care committee of Environment Canada.

2.2 Coating Effectiveness

2.2.1 Experimental Design

Antifouling effectiveness was determined in 4 in situ experiments conducted in 4 different years using multi-plate collectors as previously described and used for zebra mussel colonization experiments by de Lafontaine et al. (2002). Each experiment consisted in a series of multi-plate collectors submerged for a period of 4-5 months during the summer and autumn months (June-October), at sites in the St. Lawrence River (Canada) where zebra mussels are established (Table 1). During experiment #2, five collectors at Bassin Louise (Quebec City) remained submerged for an 11-month period to assess the long-term adherence of mussels to coatings. Multi-plate collectors were made of three white polystyrene (Experiments #1, 2 and 3) or grey polyvinyl chloride (PVC, Experiment #4) 15 cm x 15 cm plates attached at 3 cm intervals to a central stainless steel rod with an eyebolt (Figure 1). Since different plastic types were used in different years, it was assumed that it did not introduce a confounding factor to the study design. The collectors were deployed horizontally with the upper plate being identified as Plate A, the middle, as Plate B and the lower plate, as Plate C. In experiments #1, 2 and 3, each collector was attached to a polypropylene rope to a sub-surface float, which was suspended at a depth of 4 m and attached to a floating dock at the surface. The collector was maintained straight in the water column by a cement brick that was attached to the bottom of the collector. For experiment #4, eight collectors were attached in pairs of coated and uncoated units inside a rigid steel frame (1 m X 2 m) with four 45 cm-long legs welded perpendicular at each corner of the frame. The frame was then suspended along the side of a dock or pier with the legs keeping the frame away from the dock wall. This design was adopted to eliminate the risk of loosing individual collectors.

Experiment #1 conducted at Bassin Louise (Quebec City) was a preliminary trial for comparing the antifouling effectiveness of *ZebraOut*© against other surfaces having different degree of roughness. Plates coated with *ZebraOut*© were black-coloured and characterized by a surface texture similar to an orange skin. Sand and chitin coated plates consisted of sand or chitin particles sprayed on an epoxy polymer yielding a beige-coloured rough surface. Plates coated with epoxy only and uncoated plates were all slick and served as control plates. In all other experiments, coated plates were black-coloured while control (uncoated) plates were either white or black to test the influence of substrate colour on colonization. Experiment #2 was conducted at four different sites having different mussel densities. Experiment #3 conducted at Bassin Louise was designed to assess long-term durability of the coating, by re-utilizing plates that had been used in the previous year (Experiment #2), cleaned and kept frozen at -20 °C during the winter months prior to be deployed again for another season. This was done to mimic the potential effect of wintertime freezing conditions exposure on the effectiveness of the coated plates. Finally, experiment #4 was designed to test the effectiveness of the coating against saltwater organisms at piers and docks of three sites in the Estuary and the Gulf of St. Lawrence : at the St-Joseph-de-la-Rive marina in the middle St.Lawrence estuary (47.449°N; 70.365°W), at the Maurice-Lamontagne Institute dock in Mont-Joli along the lower St.Lawrence estuary (48.643°N; 68.166°W), and at the Carleton marina in Baie-des-Chaleurs (48.101°N; 66.129°W) in the Gulf of St.Lawrence. Collectors were also deployed at Bassin Louise (Quebec City) and served as a freshwater control. Overall, the Bassin Louise site was selected in all experiments which allowed comparison between years at that site.

Usually 8 collectors per site were used in the different experiments. The number of collectors retrieved at a given site and year varied between 5 and 8, due to some lost or vandalism. Upon their retrieval, collectors were brought back to the laboratory and stored in -20 °C freezers. Prior to laboratory analysis, digital photographs were taken of each multi-plate collector from 6 different angles (4 sides, 1 top, 1 bottom). Once dismantled, digital photographs of each plate side were also taken. These photographs were used to qualitatively describe the spatial distribution of mussels on plates and to estimate surface area by sponges on each plate side.

Table 1. Summary of experimental design for each in situ experiment.

Experiment	Site	Substrate coating	Substrate colour	Number of collectors	Exposure time (days)
		no coating	white	1	102
		Sand	light beige	2	94
#1	Bassin Louise (Quebec City)	Epoxy	light beige	1	94
		15% chitin	light beige	2	102
		ZebraOut	black	1	102
		ZebraOut	black	2	120
	Bassin Louise	no coating	white/black	2/2	120
	BLO2000	ZebraOut	black	2	337
		no coating	white/black	1/2	337
	Montreal marina	ZebraOut	black	2	120
#2	MON2000	no coating	white/black	1/1	120
	Longueuil marina	ZebraOut	black	2	120
	LON2000	no coating	white/black	1/1	120
	Ile Perrot marina	ZebraOut	black	2	120
	IPE2000	no coating	white/black	1/1	120
#3	Bassin Louise	ZebraOut	black	2	148
	BLO2001	no coating	white/black	1/1	148
#4	Bassin Louise	ZebraOut	black	4	145
	BLO2003	no coating	black	4	145

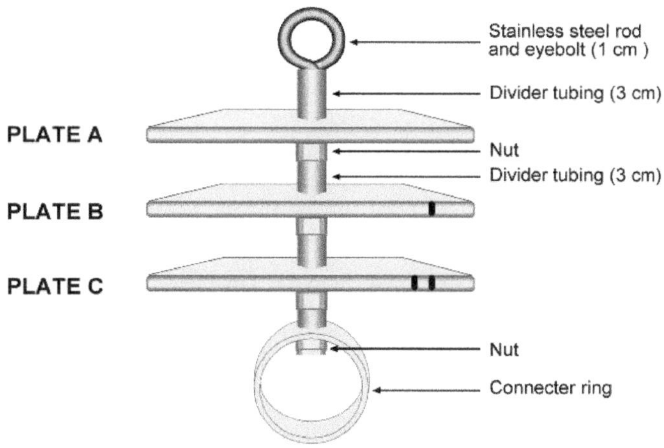

Figure 1. Schematic view of a multi-plate collector

2.2.2 Laboratory Analyses

Zebra mussel abundance was obtained by counting the number of zebra mussels that were scraped off from each plate side of every collector. Due to the small size of some individuals, counting was performed under a stereoscopic microscope. From each plate side, 50 or 100 mussels were randomly selected for shell measurement using a micrometer in the stereomicroscope eyepiece. Shell length was measured (± 0.01 mm) as the distance from the umbo to the opposite shell edge along the animal's frontal plane. For experiment #2 (plates immersed during 11 months at Bassin Louise only), #3 and #4, the percentage of the total surface area of each plate side covered by sponges was determined digitally by means of an image-analyzing system using MapInfo (version 4.5) software. In experiment #4, the dry biomass of the fouling material from each plate side of every collector was also determined by scraping it off with the help of stereomicroscopic magnifying glass or binoculars, and the resulting material was categorized by organisms. When organisms were embedded into algal material, they were categorized as 'algae'. Zebra mussels were enumerated prior to weighing. In all experiments, all mussels were assumed to be alive at the time of collectors retrieval. All samples were returned to the freezer after analysis.

2.2.3 Data Analyses

Differences in mussel abundance, dry weight and sponge percent coverage were fitted to generalized linear mixed models (GLIMMs) with overdispersed Poisson, lognormal and beta distributions, respectively. Treatment effects were tested against whole collector variability to avoid the possibility of pseudo-replication among collector plates. Plates and sides were ranked within collectors to assess their homogeneity with Fisher's exact tests. Variability in settlement densities was compared between plate sides with Conover tests. Collector-sampled mussel length distributions were compared with Wilcoxon (location), Conover (dispersion) and Kolmogorov-Smirnov (shape) tests. The results on mussel abundance variability between plate sides within a collector (see the Results section) confirmed those previously reported by de Lafontaine et al. (2002) who suggested using counts from plate undersides in order to minimize variability. Therefore, statistical analyses in this study were performed on zebra mussel abundance, dry biomass or sponge percent coverage per plate undersides only.

Antifouling coating efficiency was expressed as the difference between the quantity of biological material (expressed as abundance, dry biomass or percent coverage) attached on the coated and the uncoated plates in each experiment and was calculated as:

$$Efficiency(\%) = 100*(uncoated\text{-}coated)/uncoated \qquad (1)$$

Relationships between sponge percent coverage and mussel abundance were assessed using Spearman correlations. All tests were done using SAS/STAT® 12.1 for Windows, at a 0.05 probability significance level.

3. Results

3.1 Toxicological Response

Results from all water quality and contaminant analyses realized on coating leachates were consistently well below the guideline criteria of the *Bureau de Normalisation du Québec* (BNQ) for products and materials in contact with drinking water (Table 2). Results from the six ecotoxicological bioassays revealed no potential toxicity of leachates from the coating (Table 3), as indicated by the toxic response of the different endpoints being all equal or below 2 toxic units. The significant increase in specific conductivity in leachates from the coated plates (0.5 to 93.1 µS cm^{-1}) was probably caused by the salting out from the coating. If trace metals would become bioavailable by this process, they do not however induce any toxic response as suggested by the ecotoxicological bioassays results. Percentage of dissolved oxygen measured at the beginning and at the end of the 24 h leaching period indicated no difference between control and coated plates leachates. Leaching of coating surface may be considered as completed after 24 hours. An additional Microtox bioassay with leachates from coated plates during 72 hours revealed no additional toxicity. All the above laboratory tests revealing no toxic response suggested the innocuity of the coating for the receiving environment.

Table 2. Results of water quality analyses of the coating leachates in comparison to the BNQ (*Bureau de Normalisation du Québec*) criteria for the Safety of Products and Materials in Contact with Drinking Water (protocol 3660-950). Blanks were substracted from the results

Water quality parameters	Analytical methods	BNQ guidelines (mg L^{-1})	Results (mg L^{-1})
Arsenic	3114-B	0.05	<0.001
Baryum	3120-B	1	<0.01
Cadmium	3120-B	5	<0.02
Chromium	3500-C,D	0.05	<0.001
Lead	3120-B	0.05	<0.001
Mercury	3112-B	0.001	<0.0002
Selenium	3114-B	0.01	<0.001
Silver	3120-B	0.05	<0.0003
Tin	3120-B	0.05	<0.01
Cyanides	4500-CN,E	0.2	0.09
Fluoride	4500-F,C	1.5	0
Phenolic compounds	5530-B,C	0.002	<0.002
Vinyl chloride	EPA-524	0.05	<0.001
Nitrites-nitrates	4500-NO3,H	10	<0.05
Odor	2150-B	< 16	1
Taste	2160-C	< 4/5	1
Color	2120-B	<16	<1

3.2 Colonization on Multi-Plate Collectors

All uncoated plates were colonized by zebra mussels with densities on underside plates ranging from 133 to 33 778 individuals per m^2 (mean of 14 452 individuals per m^2). Densities on control plates at the Bassin Louise site were always higher than at other locations, and were 3 to 4 times higher in 2001 and 2003 than in 1999 and 2000 at that site. Both dreissenid mussels species were found but Zebra mussel largely dominated while Quagga mussel accounted for less than 2% of all mussels on uncoated plates in the four experiments at Bassin Louise. Zebra mussels accounted for most (> 99%) of the fouling biomass on the plates, the remainder consisting of algae and sponges. Sponges coverage varied between 0 to 50% of the surface of uncoated plates (mean = 15 ± 17%) and was more important in 2000 and 2001 than in 2003.

Among the six plate sides within every collector, the topside of plate A was significantly the least colonized surface by zebra mussels (Fisher's exact test, $P < 0.0001$). The topsides of plates B and C were significantly more colonized than their undersides for uncoated plates (GLIMMs, $P = 0.05$) but not for the coated plates (GLIMMs, $P = 0.56$). A monolayer of zebra mussels was always observed on the coated plates while very few mussels were occasionally attached to conspecifics on the uncoated plates. As previously described in de Lafontaine et al. (2002), zebra mussels tend to aggregate more at the center than near the edges of the coated plates. Probably as a result of higher densities, mussels were distributed more homogeneously on the uncoated plates. Mussel abundance on plate undersides within every collector showed less variability (CVs of coated plates: 49%, CVs of uncoated plates: 20%) than that of plate topsides (excluding plate A; CVs of coated plates: 72%; CVs of uncoated plates: 29%). These differences in variability were significant for uncoated plates (Conover test, $P = 0.01$) but not for the coated plates (Conover test, $P = 0.17$). These results justified the selection of the underside of plates for statistical analyses. Difference between white- and black-coloured plates was not statistically significant for zebra mussel abundance (P = 0.10) and sponge colonization (P = 0.16) on uncoated plates. Therefore, it was decided to pool results from both coloured plates for statistical analyses.

Table 3. Results of the six ecotoxicological bioassays (expressed in toxic unit - TU) from coating leachates analyses and water specific conductivity measured at the beginning (0h) and the end (24h) of the leaching period. Results for uncoated and coated plates are means of triplicates. Various endpoints for bioassays are: $LC_{50/25}$ = 50/25% lethal concentration; EC_{50} = 50% effect concentration ; IC_{50} = 50% inhibitory concentration; NOEC = no observed effect concentration

Bioassay	Endpoint	Control	Uncoated plates	Coated plates
Microtox	IC_{50}	<2.0	<2.0	<2.0
	NOEC	2.0	2.0	2.0
Algae	IC_{50}	<1.0	<1.0	<1.0 - 1.1
	NOEC	1.0	1.0	1.4
Cnidaria Hydra	LC_{50}	<1.0	<1.0	<1.0
	EC_{50}	<1.0	<1.0	<1.0
Daphnia	EC_{50} - 24 h	1.2	<1.0 - 1.1	<1.0
	EC_{50} - 48 h	1.3	>1.0 - 1.1	<1.0
Microcrustacean	LC_{50}	<1.0	<1.0	<1.0
	LC_{25}	<1.0	<1.0	<1.0
Trout hepatocyte	LC_{50} - 48 h	<2.0	<2.0	<2.0
Specific conductivity	0 h	0.4	0.5	0.5
($\mu S\ cm^{-1}$)	24 h	1.0	1.4	93.1

3.3 Effectiveness of Coating in Freshwater

3.3.1 Differences in Colonization

Results from the preliminary trial study (Experiment #1) showed that *ZebraOut*© was the only really effective coating to significantly reduce zebra mussel abundance on plates (Figure 2; P = 0.0004). The antifouling efficiency of *ZebraOut*© was 92% while that of other coatings was less than 20% relative to the uncoated plates. The sand coated plates yielded the highest density of zebra mussels, being 8% higher than that on the smoother control plates. The efficiency of the *ZebraOut*© coating was confirmed in subsequent experiments and ranged between 82 and 97% (Figure 3a). The abundance of zebra mussels was always significantly greater on uncoated plates than on coated plates in every experiment after 5 months of immersion (BLO2000: P < 0.0001; MON2000: P = 0.0123; LON2000: P = 0.0454; BLO2001: P = 0.0004; BLO2003: P < 0.0001). In experiment #2, the coating efficiency was 95 and 97% at the Longueuil (LON2000) and Bassin Louise (BLO2000) sites respectively, both characterized by relatively high zebra mussel abundance. The efficiency was 82% at Montreal marina (MON2000) but the very low abundance of zebra mussels (<15 individuals per control plate) at that site yielded less precise quantitative estimation of coating efficiency. The absence of any zebra mussel colonization at the Ile Perrot site did not allow the calculation of efficiency at that site. *ZebraOut*© was still effective in reducing zebra mussels abundance on plates submerged during 11 months (P = 0.0001; Figure 3b), but efficiency declined slightly from 97% to 87%, between 5 and 11 months of immersion. The lowest antifouling efficiency (82.5%) was noted in experiment #3 (BLO2001) when testing plates used in the previous year. Results from experiment #4 at Bassin Louise (BLO2003) indicated a mean efficiency of 86% in reducing zebra mussel abundance on coated plates after 5 months. In terms of biomass, the reduction efficiency was only 69.2%, with the dry biomass of zebra mussels on the coated plates (mean ± s.e. = 4.16 ± 0.86 mg) being significantly lower than that on the uncoated plates (mean ± s.e. = 13.52 ± 1.46 mg) (P = 0.0035). The coating did also significantly reduce sponge fouling by 52 to 64% during three experiments (P = 0.0015; Figure 3c). There was no significant correlation between sponge coverage and mussel abundance within each experiment (P = 0.18).

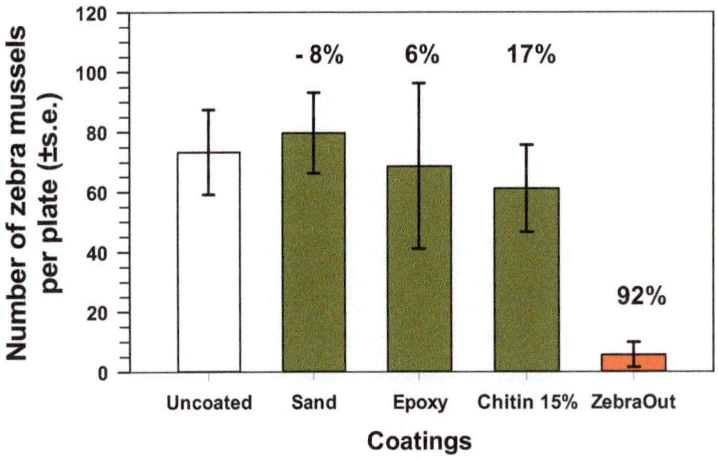

Figure 2. Abundance of zebra mussels (mean number ± s.e.) per plate underside treated with different coatings in experiment #1. Effectiveness expressed in percent change relative to uncoated polystyrene plate (control) is indicated for each coating

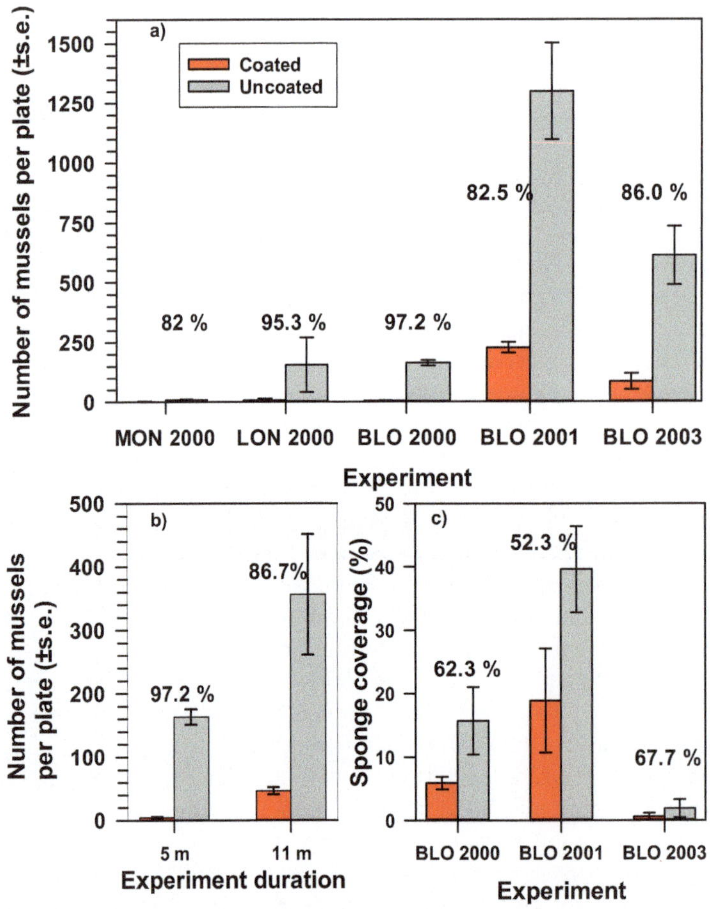

Figure 3. Differences in a) zebra mussel abundance (mean ± s.e.) on coated and uncoated plates in different experimental locations; b) zebra mussel abundance (mean ± s.e.) on coated and uncoated plates after 5-months and 11-months immersion time at Bassin Louise in experiment #2 (BLO2000); c) sponge coverage area (%, mean ± s.e.) for coated and uncoated plates in three experiments. Effectiveness expressed in percent change is indicated above every experiment

3.3.2 Zebra Mussel Size Distribution

The size distribution of zebra mussels was unimodal on the plates immersed for 11 months in experiment #2 (BLO2000) and bimodal on those submerged for 5 months in experiment #4 (BLO2003) (Figure 4). In experiment #2, the average size of mussels did not significantly ($P = 0.54$) differ between coated (mean= 5.64 ± 2.11 mm) and uncoated plates (mean= 5.99 ± 2.80 mm). The size range of mussels on coated plates was smaller ($P = 0.03$) and less variable ($P < 0.0001$) than that on uncoated plates. For experiment #4, statistical analysis performed on the two size classes (i.e. < and > 2.8 mm of shell length) separately revealed that individuals of both size classes were significantly larger on coated plates than on uncoated plates (size class 1: mean on coated plates = 2.20 ± 0.16 mm, mean on uncoated plates = 1.50 ± 0.12 mm, $P = 0.01$; size class 2: mean on coated plates = 10.00 ± 0.11 mm, mean on uncoated plates = 7.70 ± 0.11, $P< 0.0001$). Their size distribution was also different between the coated and uncoated plates (size class 1: $P = 0.008$; size class 2: $P < 0.0001$). The presence of larger mussels on the coated plates in BLO2003 contributed to the lower coating efficiency for biomass (69.2%) than for abundance (86%) estimates in this experiment.

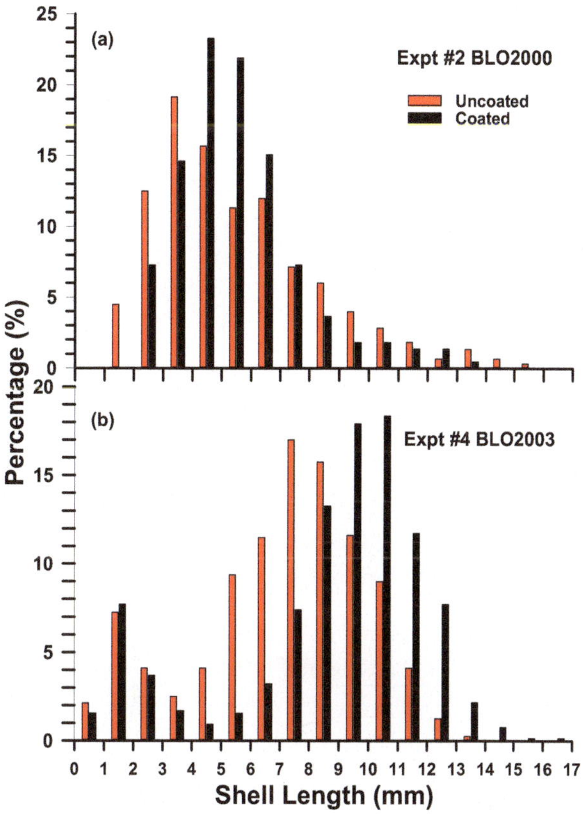

Figure 4. Zebra mussel size distribution on coated and uncoated plates immersed during a) 11 months (BLO2000) and b) 5 months (BLO2003)

3.4 Effectiveness of Coating in Seawater

Results from experiment #4 conducted at the three saltwater locations indicated clearly that the *ZebraOut*© coating was not repulsive to marine organisms. Overall, the dry biomass of organisms colonizing the coated plates was only 11% or less smaller than that measured on the uncoated plates (St-Joseph-de-la-Rive: 8.9 mg m^{-2} on coated plates and 10.2 mg m^{-2} on uncoated plates; Mont-Joli: 182.2 mg m^{-2} on coated plates and 198.7 mg m^{-2} on uncoated plates; Carleton: 140.4 mg m^{-2} on coated plates and 142.2 mg m^{-2} on uncoated plates). Differences were not statistically significant at any location ($P > 0.10$). The community composition varied between the experimental sites but was largely dominated by macroalgae at all sites, while blue mussels, sponges, sea stars, gastropods, barnacles and Caprellidae were also present on some plates. Mud was also present on some plates of all collectors immersed at Mont-Joli and Carleton.

4. Discussion

4.1 Coating Effectiveness

Results of in situ experiments demonstrated that the elastomer-based coating had a significant repulsive effect against zebra mussels. In terms of abundance, fouling by mussels on coated surface was reduced by 82 to 97% relative to control uncoated plates (Figures 2 & 3). Results on sponge coverage revealed also that the effectiveness of the coating was not restricted to zebra mussel colonization only but would apply against other freshwater organisms. The lack of toxic response in all toxicological assays done with coating leachates (Tables 2 & 3) further suggested that this repulsive effect would not be due to biocides or toxic products at the surface of the coating. The repulsive effect would presumably be due to the presence of salts within the elastomer-based coating as part of the solidification process. This was confirmed by the saltwater experiments showing that the coating was not effective against saltwater organisms and that the presence of salts within the coating was not toxic and did not stop attachment and colonization of organisms in saltwater environments. Therefore, it appeared that the effectiveness of the elastomer-based coating would be thus limited to freshwater environments and that the coating would be totally useless as an antifouling agent in marine waters.

Zebra mussels attachment depends on the substratum properties (Ackerman et al., 1996) and is modulated by various environmental factors including salinity, water turbulence and presence of biofilm (Kavouras & Maki, 2003a,b) as well as the presence of conspecifics (Kobak, 2001, 2006). Grutters et al. (2012) found that zebra and quagga mussels did not produce byssal threads in standing waters at salinities of 4 PSU and higher, implying that attachment to substrates cannot occur under these conditions. Laboratory experiments showed that the incipient lethal salinity of post-veligers and adult stages of zebra mussels is 4.5 and 2 PSU respectively (Kilgour et al., 1994), suggesting that the younger developmental stages would be slightly more tolerant to ambient salinities. Post-veliger stages were the most sensitive life stage to water salinity and may therefore limit distribution in saltwater environments. The saline condition at the immediate surface of the elastomer-based coating used in our experiment is unknown but would presumably be high enough to limit attachment of zebra mussels post-veligers and other freshwater organisms. Results from our experiment did not permit to determine if the observed repulsive effect was due to the non-attachment of post-veligers at the time of settlement or to the settlement of organisms followed by the movement and detachment of the new settlers. Presumably, the elastomer-based coating may be especially repulsive for zebra mussel larvae that detach from coated plates after the initial settlement. This first contact of mussel larvae with the substrate would be largely stochastic whereas secondary migration would reflect a choice of the juvenile or adult mussel for a preferred substrate (Folino-Rorem et al., 2006). Studies showed that zebra mussels can settle on a substratum and then search for a suitable site to attach by crawling at the surface of the substrate (Kobak & Kakareko, 2009). The proximate factors stimulating the non-attachment or detachment of zebra mussels to a substrate are unknown. It is thought that unsuitable conditions may induce mussels detachment since post-settlement migrations of byssate mussels do occur in nature (Ackerman et al., 1994). The attachment strength of mussels would be weaker in bad conditions (Kobak, 2006), which facilitates post-settlement migrations. Also, the attachment strength of blue mussels (*Mytilus edulis*) (Carrington, 2002) and zebra mussels (Kobak, 2006) was found to increase with mussel size, implying that smaller mussels may have had more difficulty remaining on coated plates versus uncoated plates.

The use of black- or white-coloured plates had no significant influence on zebra mussel abundance and sponge colonization. Previous studies using settlement plates pointed to the preference of different groups of biofoulers for dark areas (Swain et al., 2006; Kobak et al., 2009). However, the long term effect of surface colour may be reduced as biofilms develop (Swain et al., 2006). In a previous study examining the performance of a chitin-based coating (de Lafontaine et al., 2002), it was hypothesized that higher degree of substrate roughness may enhance the fixation and settlement of zebra mussels on collectors plate. The results of the present study did not confirm this hypothesis, as the abundance of zebra mussels on rough substrates (sand and 15% chitin coatings) was not significantly different than that on smooth substrates (uncoated and epoxy coatings) (Figure 2). Notwithstanding its rougher texture compared to the uncoated plates, the elastomer-based coating was highly effective to reduce zebra mussel fouling. This may also imply that coating effectiveness measured in our experiments are low estimates, as the colonization of uncoated plates with a texture similar to the elastomer-based coating could have been even greater. The lack of significant correlation between sponge coverage and mussel abundance suggested that space competition of mussels with sponges was not a confounding factor in our study. Other studies demonstrated that sponges and zebra mussels are space competitors for hard substrates, and sponges can outcompete zebra mussels as they overgrow and kill zebra mussels by interfering with their feeding and respiration activities (Molloy et al., 1997; Early & Glonek, 1999; Lauer & Spacie, 2004; Lancioni & Gaino, 2005). Our results indicating very low densities of mussels on coated plates suggested that substrate space may not be saturated, thus limiting sponge and mussel competitive interactions. It is possible that competition would have occurred given a longer exposure period or longer duration of experiment.

4.2 Colonization on Multi-Plate Collectors

Densities of zebra mussels on uncoated plates immersed during one growing season (4-5 months) were within the range of that reported in other colonization studies using plate collectors (de Lafontaine et al., 2002; Kobak, 2005), or an order of magnitude lower (Marsden & Lansky, 2000). The colonization rate and the abundance of zebra mussels on uncoated plates varied between sites and years and were extremely low at the Montreal marina site (MON2000) and absent at Ile Perrot site (IPE2000) where presumably local factors may be limiting for zebra mussel colonization. Interannual differences in zebra mussels abundance at Bassin Louise may reflect variations in zebra mussel larval production from year to year at that site. Earlier dates of collectors deployment and the longer experimental duration in the last two experiments (#3 and #4) may have also contributed to catch more spawning events and therefore increase the colonization probability yielding higher mussel densities in these experiments. Zebra mussels generally produce several spawning events per year in areas where water temperature > 12 °C (Ram et al., 1996). In the St. Lawrence River, the larval production season may extend from May to October and larval peak densities generally occur in June and July (de Lafontaine et al., 1995; Cusson & de Lafontaine, 1997).

Zebra mussel spatial distribution within collectors can be explained by mussels preference for shaded habitats (Marsden & Lansky, 2000; de Lafontaine et al., 2002; Kobak & Nowacki, 2007). Collectors were suspended within the euphotic zone at Bassin Louise where measured light level at 4 m depth corresponded to 9.4 to 13.7% of surface light (de Lafontaine et al., 2002). The much lower abundance of mussels on topside of plate A (the most light exposed surface of the collector) is likely the consequence of a high incoming light level. This plate provided shade to the topsides of plates B and C which were more colonized than plate A. As previously observed by de Lafontaine et al. (2002), the aggregation of zebra mussels at the center of the plates (excepted for the topside of plate A) may result from the movement of juvenile mussels toward the center of the plates after their settlement to reduce light exposure near the edges of the plates. However, when space for hard substrata become saturated in the most shaded areas (as it may happen on uncoated collectors), zebra mussels would then spread to the remaining space available (Marsden & Lansky, 2000), leading to a more homogeneous distribution as observed on uncoated plates in our experiments.

4.3 Zebra Mussel Size Distribution

Analysis of the size distribution of mussels between coated and uncoated plates in the different experiments did not permit to reach firm conclusion on a size-selective effect of the coating for zebra mussels. While the mean size of mussels did differ significantly between coated and control plates in experiment #2, the larger size of mussels on coated plates relative to uncoated plates for both size classes in experiment #4 (Figure 4) may likely be due to post-settlement movement. This variability between experiments could be due in part to the difference in collectors deployment. Individual collectors were separately suspended using a rope in experiment #2 while collectors were attached to a rigid frame in experiment #4, a set-up that may favor mussels translocation between the fouled frame and the collectors.

4.4 Coating Durability

When used for the first time, the antifouling efficiency of the elastomer-based coating against zebra mussels was very high ranging from 82% to 97% over one growing season (4-5 months). The efficiency was slightly reduced after 11 months (87%) compared to 5 months immersion (97%). When re-using plates for a second growing season (BLO2001), the efficiency was 82.5%, a reduction of 15% relative to that of the first year (97%). These results suggest that while the elastomer-based coating can effectively slow down zebra mussel colonization, its repulsive effect would decrease over time. Presumably, salts can be gradually leached from the coating rendering coated surfaces less repulsive to mussels post-veligers. Given that the most critical requirement for an antifouling coating is to maintain an adequate fouling protection for extended exposure periods (Sanchez & Yebra, 2009), future in situ testing should verify the efficiency of the coating over longer term period. In addition, repeated measures sampling would be needed to obtain more precise estimate of the decline in coating efficiency over time.

5. Concluding Remarks

The results of the present study clearly showed that the elastomer-based coating can be a very useful antifouling coating against zebra mussels for submerged fixed structures in confined freshwater environments such as docks and pontoons in marinas. Presumably, it could also be used inside reservoirs, pumping stations or even ships ballast tanks, with the caveat that the coating was found ineffective in marine waters. Our tests showed that leachates from the elastomer-based coating were safe with regards to drinking water guidelines and did not induce any toxic responses, making the coating absolutely innocuous for the environment. Notwithstanding the effectiveness of the coating on fixed structures in confined freshwater environments, the product should undergo future testing of its performance and resistance to abrasion when applied on mobile floating substrates such as boats and barges.

Acknowledgements

We wish to thank Michel DeBlois, from MD Technologies, for providing the *ZebraOut*© coating material, Denis Labonté for plates analysis and zebra mussels enumeration and size measurements, and François Boudreault for image analysis and sponge cover estimation. This study was funded by the aquatic invasive species research program at Centre Saint-Laurent, Environment Canada and by the St. Lawrence Action Plan.

References

Ackerman, J. D., Sim, B., Nichols, S. J., & Claudi, R. (1994). A review of the early life history of zebra mussels (*Dreissena polymorpha*): Comparisons with marine bivalves. *Canadian Journal of Zoology, 72*, 1169-1179. http://dx.doi.org/10.1139/z94-157

Ackerman, J. D., Cottrell, C. M., Ethier, C. R., Allen, D. G., & Spelt, J. K. (1996). Attachment strength of zebra mussels on natural, polymeric, and metallic materials. *Journal of Environmental Engineering, 2*, 141-148. http://dx.doi.org/10.1061/(ASCE)0733-9372(1996)122:2(141)

Aldridge, D. C., Elliott, P., & Moggridge, G. D. (2006). Microencapsulated BioBullets for the control of biofouling zebra mussels. *Environmental Science and Technology, 40*, 975-979. http://dx.doi.org/10.1021/es050614

Angarano, M. B., McMahon, R. F., Hawkins, D. L., & Schetz, J. A. (2007). Exploration of structure-antifouling relationships of capsaicin-like compounds that inhibit zebra mussel (*Dreissena polymorpha*) macrofouling. *Biofouling, 23*, 295-305. http://dx.doi.org/10.1080/08927010701371439

Angarano, M. B., McMahon, R. F., & Schetz, J. A. (2009). Cannabinoids inhibit zebra mussel (*Dreissena polymorpha*) byssal attachment: A potentially green antifouling technology. *Biofouling, 25*, 127-138. http://dx.doi.org/10.1080/08927010802592743

Bernabeu, A., Vicente, R., Peribanez, M. A., Arques, A., & Amat, A. M. (2011). Exploring the applicability of solar driven photocatalytic processes to control infestation by zebra mussel. *Chemical Engineering Journal, 171*, 490-494. http://dx.doi.org/10.1016/j.cej.2011.04.009

Bowling, J. J., Mohammed, R., Diers, J. A., & Hamann, M. T. (2010). Abundant ketone isolated from oily *Plakortis* sponge demonstrates antifouling properties. *Chemoecology, 20*, 207-213. http://dx.doi.org/10.1007/s00049-010-0049-z

Brines Miller, E., & Watzin, M. C. (2007). The effects of zebra mussels on the lower planktonic foodweb in Lake Champlain. *Journal of Great Lakes Research, 33*, 407-420. http://dx.doi.org/10.3394/0380-1330(2007)33[4 07:TEOZMO]2.0.CO;2

Bueley, C., Olender, D., & Bocking, B. (2014). In-situ trial of UV-C as an antifoulant to reduce biofouling induced measurement error. *Journal of Ocean Technology, 9*, 49-67. Retrieved from http://www.thejot.net/?page_id=837&show_article_preview=601

Callow, J. A., & Callow, M. E. (2011). Trends in the development of environmentally friendly fouling-resistant marine coatings. *Nature Communications, 2*, 244. http://dx.doi.org/10.1038/ncomms1251

Carrington, E. (2002). Seasonal variation in the attachment strength of blue mussels: Causes and consequences. *Limnololy and Oceanography, 47*, 1723-1733. http://dx.doi.org/10.4319/lo.2002.47.6.1723

Chambers, L. D., Stokes, K. R., Walsh, F. C., & Wood, R. J. K. (2006). Modern approaches to marine antifouling coatings. *Surface & Coatings Technology, 201*, 3642-3652. http://dx.doi.org/10.1016/j.surfcoat.2006.08.129

Claudi, R., & Mackie, G. L. (1994). *Practical manual for zebra mussel monitoring and control* (pp. 1-251). Boca Raton (US), CRC Press.

Cope, W. G., Bartsch, M. R., & Marking, L. L. (1997). Efficacy of candidate chemicals for preventing attachment of zebra mussels (*Dreissena polymorpha*). *Environmental Toxicology and Chemistry, 16*, 1930-1934. http://dx.doi.org/10.1002/etc.5620160923

Costa, R., Elliott, P., Saraiva, P. M., Aldridge, D., & Moggridge, G. D. (2008a). Development of sustainable solutions for zebra mussel control through chemical product engineering. *Chinese Journal of Chemical Engineering, 16*, 435-440. http://dx.doi.org/10.1016/S1004-9541(08)60101-9

Costa, R., Aldridge, D. C., & Moggridge, G. D. (2008b). Seasonal variation of zebra mussel susceptibility to molluscicidal agents. *Journal of Applied Ecology, 45*, 1712-1721. http://dx.doi.org/10.1111/j.1365-2664.2008.01555.x

Costa, R., Aldridge, D. C., & Moggridge, G. D. (2011a). Preparation and evaluation of biocide-loaded particles to control the biofouling zebra mussel, *Dreissena polymorpha*. *Chemical Engineering Research and Design, 89*, 2322-2329. http://dx.doi.org/10.1016/j.cherd.2011.02.027

Costa, R., Elliott, P., Aldridge, D. C., & Moggridge, G. D. (2011b). Enhanced mortality of the biofouling zebra mussel, *Dreissena polymorpha*, through the application of combined control agents. *Journal of Great Lakes Research, 37*, 272-278. http://dx.doi.org/10.1016/j.jglr.2011.01.005

Costa, R., Aldridge, D. C., & Moggridge, G. D. (2012). Multicomponent molluscicide mixtures for zebra mussel control. *Journal of Great Lakes Research, 38*, 317-325. http://dx.doi.org/10.1016/j.jglr.2012.03.010

Cusson, B., & de Lafontaine, Y. (1997). Présence et abondance des larves de Moules zébrées dans la rivière Richelieu et le Saint-Laurent en 1996. Environnement Canada – Région du Québec, Conservation de l'environnement, Centre Saint-Laurent, Rapport Scientifique et Technique, ST-143, 58p.

Dafforn, K. A., Lewis, J. A., & Johnston, E. L. (2011). Antifouling strategies: History and regulation, ecological impacts and mitigation. *Marine Pollution Bulletin, 62*, 453-465. http://dx.doi.org/10.1016/j.marpolbul.2011.01.012

de Lafontaine, Y., Lapierre, L., Henry, M., & Grégoire, Y. (1995). Abondance des larves de Moule zébrée (*Dreissena polymorpha*) et de Quagga (*Dreissena bugensis*) aux abords des centrales hydroélectriques de Beauharnois, Les Cèdres et Rivière-des-Prairies. Environnement Canada – Région du Québec, Conservation de l'environnement, Centre Saint-Laurent, Rapport Scientifique et Technique, ST-14, 52p.

de Lafontaine, Y., Costan, G., & Delisle, F. (2002). Testing a new anti-zebra mussel coating with a multi-plate sampler: Confounding factors and other fuzzy features. *Biofouling, 18*, 1-12. http://dx.doi.org/10.1080/08927010290017671

de Lafontaine, Y., Despatie, S. P., & Wiley, C. (2008). Effectiveness and potential toxicological impact of the PERACLEAN® Ocean ballast water treatment technology. *Ecotoxicology and Environmental Safety, 71*, 355-369. http://dx.doi.org/10.1016/j.ecoenv.2007.10.033

de Lafontaine, Y., Chambers, Y., Despatie, S. P., Gagnon, C., & Blaise, C. (2013). Effectiveness and potential environmental impact of a yeast-based deoxygenation process for treating ship ballast waters. *Water Quality Research Journal of Canada, 48*, 55-75. http://dx.doi.org/10.2166/wqrjc.2013.117

Diers, J. A., Pennaka, H. K., Peng, J., Bowling, J. J., Duke, S. O., & Hamann, M. T. (2004). Structural activity relationship studies of zebra mussel antifouling and antimicrobial agents from verongid sponges. *Journal of Natural Products, 67*, 2117-2120. http://dx.doi.org/10.1021/np040097t

Diers, J. A., Bowling, J. J., Duke, S. O., Wahyuono, S., Kelly, M., & Hamann, M. T. (2006). Zebra mussel antifouling activity of the marine natural product aaptamine and analogs. *Marine Biotechnology, 8*, 366-372. http://dx.doi.org/10.1007/s10126-005-6055-4

Early, T. A., & Glonek, T. (1999). Zebra Mussel Destruction by a Lake Michigan Sponge: Populations, in Vivo ^{31}P Nuclear Magnetic Resonance, and Phospholipid Profiling. *Environmental Science and Technology, 33*, 1957-1962. http://dx.doi.org/10.1021/es980874

Farsad, N., & Sone, E. D. (2012). Zebra mussel adhesion: Structure of the byssal adhesive apparatus in the freshwater mussel, *Dreissena polymorpha*. *Journal of Structural Biology, 177*, 613-620. http://dx.doi.org/10.1016/j.jsb.2012.01.011

Folino-Rorem, N., Stoeckel, J., Thorn, E., & Page, L. (2006). Effects of artificial filamentous substrate on zebra mussel (*Dreissena polymorpha*) settlement. *Biological Invasions, 8*, 89-96. http://dx.doi.org/10.1007/s10530-005-0330-1

Gagné, F. (2006). Acute toxicity assessment of liquid samples with primary cultures of rainbow trout hepatocytes. In C. Blaise, & J-F. Ferard (Eds.), *Small-scale freshwater toxicity investigations. Vol. 1 – Toxicity Test Methods* (pp. 453-472). Netherlands, Springer.

Greenberg, A. E., Clesceri, L. S., & Eaton, A. D. (1992). *Standard methods for the examination of water and wastewater* (18th Edition). Retrieved from http://www.mwa.co.th/download/file_upload/SMWW_1000-3000.pdf

Grutters, B. M., Verhofstad, M. J., van der Velde, G., Rajagopal, S., & Leuven, R. S. (2012). A comparative study of byssogenesis on zebra and quagga mussels: the effects of water temperature, salinity and light-dark cycle. *Biofouling, 28*, 121-129. http://dx.doi.org/10.1080/08927014.2012.654779

Hellio, C., & Yebra, D. M. (2009). *Advances in marine antifouling coatings and technologies*. Cambridge. Woodhead Publishing Ltd.

Howell, D., & Behrends, B. (2010). Consequences of antifouling coatings - The chemist's perspective. In S. Dürr & J. C. Thomason (Eds.), *Biofouling* (pp. 226-242). Oxford (UK): Wiley-Blackwell.

Karatayev, A., Burlakova, L., & Padilla, D. (2002). Impacts of zebra mussels on aquatic communities and their role as ecosystem engineers. In E. Leppäkoski, S. Gollasch, & S. Olenin (Eds.), *Invasive Aquatic Species of Europe. Distribution, Impacts and Management* (pp. 433-446). Netherlands: Springer.

Kavouras, J. H., & Maki, J. S. (2003a). The effects of natural biofilms on the reattachment of young adult zebra mussels to artificial substrates. *Biofouling, 19*, 247-256. http://dx.doi.org/10.1080/0892701032000077149

Kavouras, J. H., & Maki, J. S. (2003b). Effects of biofilm on zebra mussel postveliger attachment to artificial surfaces. *Invertebrate Biology, 122*, 138-151. http://dx.doi.org/10.1111/j.1744-7410.2003.tb00079.x

Kilgour, B. W., Mackie, G. L., Baker, M. A., & Keppel, R. (1994). Effects of salinity on the condition and survival of Zebra mussels (*Dreissena polymorpha*). *Estuaries, 17*, 385-393. http://dx.doi.org/10.2307/1352671

Kobak, J. (2001). Light, gravity and conspecifics as cues to site selection and attachment behaviour of juvenile and adult *Dreissena polymorpha* Pallas, 1771. *Journal of Molluscan Studies, 67*, 183-189. http://dx.doi.org/10.1093/mollus/67.2.183

Kobak, J. (2005). Recruitment and distribution of *Dreissena polymorpha* (Bivalvia) on substrates of different shape and orientation. *International Review of Hydrobiology, 90*, 159-170. http://dx.doi.org/10.1002/iroh.200410756

Kobak, J. (2006). Factors affecting the attachment strength of *Dreissena polymorpha* (Bivalvia). *Biofouling, 22*, 141-150. http://dx.doi.org/10.1080/08927010600691895

Kobak, J., & Nowacki, P. (2007). Light-related behaviour of the zebra mussel (*Dreissena polymorpha*, Bivalvia). *Fundamental and Applied Limnology Archiv fur Hydrobiologie, 169*, 341-352. http://dx.doi.org/10.1127/1863-9135/2007/0169-0341

Kobak, J., & Kakareko, T. (2009). Attachment strength, aggregation and movement of the zebra mussel (Dreissena polymorpha, Bivalvia) in the presence of potential predators. *Fundamental and Applied Limnology Archiv fur Hydrobiologie, 174*, 193-204. http://dx.doi.org/10.1127/1863-9135/2009/0174-0193

Kobak, J., Poznanska, M., & Kakareko, T. (2009). Effect of attachment status and aggregation on the behaviour of the zebra mussel *Dreissena polymorpha*. *Journal of Molluscan Studies, 75*, 119-126. http://dx.doi.org/10.1093/mollus/eyn046

Lancioni, T., & Gaino, E. (2005). Competition between the freshwater sponge *Ephydatia fluviatilis* and the zebra mussel *Dreissena polymorpha* in Lake Trasimeno (central Italy). *Italian Journal of Zoology, 72*, 27-32. http://dx.doi.org/10.1080/11250000509356649

Lauer, T. E., & Spacie, A. (2004). Space as a limiting resource in freshwater systems: Competition between zebra mussels (*Dreissena polymorpha*) and freshwater sponges (Porifera). *Hydrobiologia, 517*, 137-145. http://dx.doi.org/10.1023/B:HYDR.0000027342.31716.9a

Lewis, J. A. (2009). Non-silicone biocide-free antifouling solutions. In C. Hellio, & D. M. Yebra (Eds.), *Advances in marine antifouling coatings and technologies* (pp. 709-724). Cambridge, Woodhead Publishing Ltd.

Maréchal, J-P., & Hellio, C. (2009). Challenges for the development of new non-toxic antifouling solutions. *International Journal of Molecular Sciences, 10*, 4623-4637. http://dx.doi.org/10.3390/ijms10114623

Marsden, J. E., & Lansky, D. M. (2000). Substrate selection by settling zebra mussels, *Dreissena polymorpha*, relative to material, texture, orientation, and sunlight. *Canadian Journal of Zoology, 78*, 787-793. http://dx.doi.org/10.1139/z00-004

Martin, M. D., Mackie, G. L., & Baker, M. A. (1993). Control of the Biofouling Mollusc, *Dreissena polymorpha* (Bivalvia: Dreissenidae), with sodium hypochlorite and with polyquaternary ammonia and benzothiazole compounds. *Archives of Environmental Contamination and Toxicology, 24*, 381-388. http://dx.doi.org/10.1007/BF01128738

MD Technologies. (2008). *ZebraOut*. Retrieved from http://mdtech1996.blogspot.com/2008/02/zbraout.html

Molloy, D. P., Karatayev, A. Y., Burlakova, L. E., Kurandina, D. P., & Laruelle, F. (1997). Natural enemies of zebra mussels: Predators, parasites, and ecological competitors. *Reviews in Fisheries Science, 5*, 27-97. http://dx.doi.org/10.1080/10641269709388593

Molloy, D. P., Mayer, D. A., Gaylo, M. J., Morse, J. T., Presti, K. T., Sawyko, P. M., ... Griffin, B. H. (2013). *Pseudomonas fluorescens* strain CL145A - a biopesticide for the control of zebra and quagga mussels (Bivalvia: Dreissenidae). *Journal of Invertebrate Pathology, 113*, 104-114. http://dx.doi.org/10.1016/j.jip.2012.12.012

Morse, J. T. (2009). Assessing the effects of application time and temperature on the efficacy of hot-water sprays to mitigate fouling by *Dreissena polymorpha* (Pallas). *Biofouling, 25*, 605-610. http://dx.doi.org/10.1080/08927010902989245

Pimentel, D., Zuniga, R., & Morrison, D. (2005). Update on the environmental and economic costs associated with alien-invasive species in the United States. *Ecological Economics, 52*, 273-288. http://dx.doi.org/10.1016/j.ecolecon.2004.10.002

Qian, P-Y., Xu, Y., & Futesani, N. (2010). Natural products as antifouling compounds: recent progress and future perspectives. *Biofouling, 26*, 223-234. http://dx.doi.org/10.1080/08927010903470815

Ram, J. L., Fong, P. P., & Garton, D. W. (1996). Physiological aspects of zebra mussel reproduction: maturation, spawning, and fertilization. *American Zoologist, 36*, 326-338. Retrieved from http://www.jstor.org/stable/3884094

Regoli, L., Chan, H. M., de Lafontaine, Y., & Mikaelian, I. (2001). Organotins in zebra mussels (*Dreissena polymorpha*) and sediments of the Quebec City Harbour area of the St. Lawrence River. *Aquatic Toxicology, 53*, 115-126. http://dx.doi.org/10.1016/S0166-445X(00)00174-0

Sanchez, A., & Yebra, D. M. (2009). Ageing tests and long-term performance of marine antifouling coatings. In C. Hellio, & D. M. Yebra (Eds.), *Advances in marine antifouling coatings and technologies* (pp. 393-421). Cambridge, Woodhead Publishing Ltd.

Swain, G., Herpe, S., Ralston, E., & Tribou, M. (2006). Short-term testing of antifouling surfaces: the importance of colour. *Biofouling, 22*, 425-429. http://dx.doi.org/10.1080/08927010601037163

Townsin, R. L., & Anderson, C. D. (2009). Fouling control coatings using low surface energy, foul release technology. In C. Hellio, & D. M. Yebra (Eds.), *Advances in marine antifouling coatings and technologies* (pp. 693-708). Cambridge, Woodhead Publishing Ltd.

Van Appledorn, M., & Bach, C. E. (2007). Effects of zebra mussels (*Dreissena polymorpha*) on mobility of three native mollusk species. *American Midland Naturalist, 158*, 329-337.

Webster, D. C., & Chisholm, B. J. (2010). New directions in antifouling technologies. In S. Durr, & J. C. Thomason (Eds.), *Biofouling* (pp. 366-387). New-York, Wiley-Blackwell.

Wells, A. W., Meyer, A. E., Matousek, J. A., Baier, R. E., & Neuhauser, E. F. (1997). *Nontoxic foul-release coatings for zebra mussel control* (pp. 451-460). Waterpower'97, Proceedings of the International Conference on Hydropower, American Society of Civil Engineers, Atlanta (US).

PERMISSIONS

All chapters in this book were first published in ENRR, by Canadian Center of Science and Education; hereby published with permission under the Creative Commons Attribution License or equivalent. Every chapter published in this book has been scrutinized by our experts. Their significance has been extensively debated. The topics covered herein carry significant findings which will fuel the growth of the discipline. They may even be implemented as practical applications or may be referred to as a beginning point for another development.

The contributors of this book come from diverse backgrounds, making this book a truly international effort. This book will bring forth new frontiers with its revolutionizing research information and detailed analysis of the nascent developments around the world.

We would like to thank all the contributing authors for lending their expertise to make the book truly unique. They have played a crucial role in the development of this book. Without their invaluable contributions this book wouldn't have been possible. They have made vital efforts to compile up to date information on the varied aspects of this subject to make this book a valuable addition to the collection of many professionals and students.

This book was conceptualized with the vision of imparting up-to-date information and advanced data in this field. To ensure the same, a matchless editorial board was set up. Every individual on the board went through rigorous rounds of assessment to prove their worth. After which they invested a large part of their time researching and compiling the most relevant data for our readers.

The editorial board has been involved in producing this book since its inception. They have spent rigorous hours researching and exploring the diverse topics which have resulted in the successful publishing of this book. They have passed on their knowledge of decades through this book. To expedite this challenging task, the publisher supported the team at every step. A small team of assistant editors was also appointed to further simplify the editing procedure and attain best results for the readers.

Apart from the editorial board, the designing team has also invested a significant amount of their time in understanding the subject and creating the most relevant covers. They scrutinized every image to scout for the most suitable representation of the subject and create an appropriate cover for the book.

The publishing team has been an ardent support to the editorial, designing and production team. Their endless efforts to recruit the best for this project, has resulted in the accomplishment of this book. They are a veteran in the field of academics and their pool of knowledge is as vast as their experience in printing. Their expertise and guidance has proved useful at every step. Their uncompromising quality standards have made this book an exceptional effort. Their encouragement from time to time has been an inspiration for everyone.

The publisher and the editorial board hope that this book will prove to be a valuable piece of knowledge for researchers, students, practitioners and scholars across the globe.

LIST OF CONTRIBUTORS

Mohammed Saif Al-Kalbani and Martin F. Price
Centre for Mountain Studies, Perth College, University of the Highlands and Islands, Perth, UK

Mushtaque Ahme
Department of Soils, Water & Agricultural Engineering, College of Agricultural and Marine Sciences, Sultan Qaboos University, Sultanate of Oman

Asma Abahussain
Department of Natural Resources and Environment, College of Graduate Studies, Arabian Gulf University, Manama, Kingdom of Bahrain

Timothy O'Higgins
Scottish Association for Marine Sciences, University of the Highlands and Islands, Scottish Marine Institute, Argyll, UK

Shaaban A. Mwachireya
Kenya Marine & Fisheries Research Institute, PO BOX 81651, Mombasa 80100, Kenya

Tim R. McClanahan
Wildlife Conservation Society, Marine Programs, Bronx, NY. 10460, USA

Isabelle M. Cote and Brian E. Hartwick
Department of Biological Sciences, Simon Fraser University, 8888 University Drive, Burnaby, BC, V5A 1S6 Canada

Nara Somaratne
South Australian Water Corporation, 250 Victoria Square, Adelaide, SA 5000, Australia

Simon Mann
Zonge Engineering & Research Organization (Australia) Pty Ltd, 39 Raglan Avenue, Edwardstown, SA 5039, Australia

Marilia Mitidieri Fernandes de Oliveira and Nelson Francisco Favilla Ebecken
Center of Technology, Federal University of Rio de Janeiro-Civil Engineering Postgraduate Program–COPPE/UFRJ, Rio de Janeiro, Brazil

Jorge Luiz Fernandes de Oliveira and Marina Aires
Geography Postgraduate Program, Geoscience Institute of Fluminense Federal University-UFF, Niterói, Brazil

José Maria de Castro Jr.
Defense and Civil Security Postgraduate Program of Fluminense Federal University-UFF, Niterói, Brazil

Immaculate Mogotsi
Gender Research Programme, Multidisciplinary Research Centre, University of Namibia, Namibia

Selma Lendelvo
Life Science Division, Multidisciplinary Research Centre, University of Namibia, Namibia

Margaret Angula
Department of Geography, History and Environmental Studies, Faculty of Humanities and Social Sciences, University of Namibia, Namibia

Jesaya Nakanyala
Department of Wildlife Management, Faculty of Agriculture and Natural Resources, University of Namibia, Namibia

Mohammad R. Masnavi
Faculty of Environment, Department of Environmental Design, University of Tehran, Tehran, Iran

Neda Amani
Faculty of Environment, Department of Environmental Design, University of Tehran, Tehran, Iran

Ali Ahmadzadeh
Pars Special Economic Energy Zone Org, Asalouyeh, Iran

Henry Ojobo
Department of Architecture, Faculty of Environmental Sciences, Kaduna State University, Nigeria

Bukar Usman Wakawa and Aminu Umar
Department of Architecture, School of Environmental Technology, Abubakar Tafawa Balewa University, Nigeria

Lucia Chehade, Audrey Caron and Céline Aguer
Institut de recherche de l'Hôpital Montfort, Ottawa, ON K1K 0T2, Canada
Faculty of Medicine, Biochemistry, Microbiology and Immunology Department, University of Ottawa, Ottawa,ON K1H 8L1, Canada

Hongyan Liao
Finance Department of International Business School, Jinan University, Zhuhai, China

Yuabiao Zhang, Zhifeng Chen and Zexin Meng
Packaging Engineering Institute, Jinan University, Zhuhai, China

Sevinc A. Sirdas and Ali Behzat Diriker
Department of Meteorological Engineering, Faculty of Aeronautics and Astronautics, Istanbul Technical University, Turkey

Vinay Kumar
Department of Earth, Ocean & Atmospheric Sciences, Florida State University, Tallahassee, FL, 32306, USA

Rose Jane Peras
Department of Social Forestry and Forest Governance, College of Forestry and Natural Resources, University of the Philippines Los Banos, Philippines
Global Forest Environmental Studies, Graduate Schol of Agricultural and Life Sciences, The University of Tokyo, Japan

Juan Pulhin
Department of Social Forestry and Forest Governance, College of Forestry and Natural Resources, University of the Philippines Los Banos, Philippines

Makoto Inoue and Abrar Jurar Mohammed
Global Forest Environmental Studies, Graduate Schol of Agricultural and Life Sciences, The University of Tokyo, Japan

Kazuhiro Harada
Graduate School of Bioagricultural Sciences, Division of Regional Resources Management, Lab. of Forest Resources Utilization, Nagoya University, Japan

Masatoshi Sasaoka
Graduate School of Letters, Hokkaido University, Japan

Kahsay N. Zeraebruk, Alfred O. Mayabi and John M. Gathenya
College of Engineering and Technology, Jomo Kenyatta University of Agriculture and Technology, Kenya

Asamere Wolde, Tadesse Amsalu and Molla Mekonnen Alemu
35 Frazer Street Wilkinson Road, Freetown, Sierra Leone

John Warui Kiringe
Habitat Planners, P.O. Box 10982-00100 Nairobi, Kenya

Francis Mwaura
Department of Geography & Environmental Studies, University of Nairobi, Box 30197-00100, Nairobi, Kenya

Fiesta Warinwa
African Wildlife Foundation (AWF), P.O. Box 310-00502, Nairobi, Kenya

Yves de Lafontaine
Department of Fisheries and Oceans Canada, Institut Maurice-Lamontagne, Canada

Julie Veillette
Environment and Climate Change Canada, Centre Saint-Laurent, Canada

Index

A

Agricultural Production, 1-2, 152, 156, 189
Air Pollution, 53-55, 61-64
Air Quality, 53-56, 61-64
Animal Husbandry, 2
Arima Model, 167, 169-171, 173, 178
Atmospheric Co2, 17, 31
Atmospheric Pollutants, 53, 86

B

Bicarbonate, 3, 5, 7, 10-11, 14, 34
Bio Region Theory, 79
Biodiversity, 27, 31, 67, 77-78, 80, 82, 84, 86, 88, 152-153, 155, 159-160, 162, 181, 198, 207-209
Bp Neural Network, 167, 169-170, 178-179

C

Cbnrm Activities, 66
Clean Development Mechanism, 181, 190, 193
Climate Change, 12, 17-18, 28, 31, 33, 35, 37, 54, 63-64, 67, 75, 121, 134-135, 139, 150-153, 158, 161-162, 181, 193-194, 206, 208-209
Community Forestry, 65-67, 69-70, 74-75, 153, 193-194
Conductivity, 1, 4, 6, 9, 14-15, 49, 214-215
Conduit Flow Path, 48
Conservation, 11-12, 14, 17-18, 36, 66-69, 74-75, 84, 86-89, 152-160, 162-165, 178, 183, 188, 193, 195-196, 205-208, 221
Coral Reef, 17-19, 31, 33-37, 82
Coral Skeletal, 18, 20, 25, 28-29, 34
Crop Toxicity, 2
Cyanobacteria Blooms, 7

D

Degradation, 17, 28, 31, 37, 61, 65, 67, 82-84, 101-102, 104, 152-153, 155, 160, 163-165, 167, 183-184, 188, 196, 205, 207
Dissolved Oxygen, 4, 6-7, 214
Distilled Water, 20
Drainage System, 197, 199, 205

E

Eco-tourism, 77, 191
Ecological, 17-18, 31, 35, 66-67, 71, 74, 77-89, 107, 165, 167-168, 172, 179, 181, 185, 188, 196-197, 206, 221, 223
Ecological Design, 80-81, 85-86
Ecological Disturbance, 77
Ecological Landscape Design, 77, 89
Electrical Conductivity, 1, 4, 6, 9, 14, 49
Environmental Degradation, 28, 61, 167, 188, 205
Environmental Resources, 54, 65
Extreme Weather Events, 135-136, 139, 150

F

Falkenmark Indicator, 167-168, 171
Fisheries, 17, 31, 35, 76, 89, 196, 209, 223
Forest Resource Management, 65, 67-68, 71
Forest Resources, 65-75, 152, 158, 180, 189
Freshwater Lens, 38-40, 48-50

G

Gas Emissions, 53-54, 62
Global Climate Change, 17-18, 28, 31, 135
Greenhouse Gas, 54, 62, 153
Groundwater, 1, 12-14, 19, 38-41, 45-46, 48-52, 83, 167, 172-173, 196

H

Harvested Resources, 73, 156
Harvesting Methods, 71
Hazards, 1, 3, 5, 8-9, 11, 181
Heavy Rain, 135, 146-147
Humidity, 57, 91-92, 94-98, 100
Hydro-chemical, 3, 7
Hydrochemical Water Quality, 5
Hydrodynamic Energy, 17-18, 29-30

I

Instrumental Techniques, 1, 5
Invasive Species, 86, 209, 220, 223
Irrigation Water, 1-2, 5-10, 13-16, 206

K

Kirisia Forest Watershed, 195-196, 198-205

L

Local Livelihoods, 155, 195-197, 208
Lto Cycle, 54, 57

M

Mangrove Communities, 77-78, 81-87, 89
Marine Environments, 209
Meteorological Models, 53
Mitigation, 62-63, 92-93, 153, 155, 158, 209, 221

N

Natural Ecosystems, 2, 77, 83, 85, 206

Natural Regeneration, 18, 86, 152, 156-157, 159, 164, 180-182, 187, 191-193

Nutrients, 17-19, 26, 29-30, 33, 36, 82, 84, 104, 114

O

Ocean Acidification, 17, 28, 30-31, 33-34, 36

P

Persistent Organic Pollutants, 101, 113-116, 119

Polluted Water, 81

R

Recrystallized Limestone, 38, 46

Reforestation, 152-153, 156-165, 180-183, 185-188, 192

S

Salinity, 1-3, 5, 9, 11, 14-15, 20-21, 39, 81, 84, 204, 218, 222

Scarcity, 1, 121, 167-168, 172, 178, 196, 198

Sea Surface Temperature, 17, 139

Sediment, 17-18, 20-21, 26-27, 29-30, 32, 34-35, 37, 47, 84, 87, 123

Sodium Adsorption Ratio, 3, 5, 9, 14-15

Soil Productivity, 2

Sustainable Forest, 65-67, 70, 72, 76, 154, 158, 193, 207

Sustainable Forest Management, 65-66, 70, 72, 76, 154, 158, 193, 207

Sustainable Livelihoods, 152-155, 159-161, 164-165

Sustainable Management, 2, 12, 38, 87, 131, 153-154

T

Temperature, 12, 17, 20-21, 28, 32, 34, 36-37, 48, 54-55, 57, 81, 83, 91-92, 94-100, 124, 135-136, 139, 143-144, 146-147, 150, 182, 205, 211, 219, 222-223

Troposphere, 56, 135, 149

U

Urbanization, 2, 91, 133, 205

W

Wastewater Reuse, 1, 5-7, 11

Water Quality Index, 1, 3, 5, 10-15

Water Resources, 1-2, 5, 7, 10-13, 51, 83, 120-121, 123, 126, 128, 131, 133-134, 157, 167-169, 171-173, 176-179, 196, 198, 200-201, 204-208

Water Supply, 40-43, 120-125, 127-129, 131-134, 167, 171-174, 177-179, 183, 188, 200, 202, 204